NEURAL CELL SPECIFICATION

SPECIFICATION

Molecular Mechanisms and Neurotherapeutic
Implications

ALTSCHUL SYMPOSIA SERIES

Series Editors: Sergey Fedoroff and Gary D. Burkholder

NEURAL CELL SPECIFICATION

Molecular Mechanisms and Neurotherapeutic Implications

Edited by

Bernhard H. J. Juurlink
Patrick H. Krone
William M. Kulyk
Valerie M. K. Verge

and

J. Ronald Doucette

University of Saskatchewan
Saskatoon, Saskatchewan, Canada

SPRINGER SCIENCE+BUSINESS MEDIA, LLC

Library of Congress Cataloging-in-Publication Data

On file

Proceedings of the Third Altschul Symposium on Neural Cell Specification: Molecular Mechanisms and Neurotherapeutic Implications, held May 12–14, 1994, in Saskatoon, Saskatchewan, Canada

ISBN 978-0-306-45185-0   ISBN 978-1-4615-1929-4 (eBook)
DOI 10.1007/978-1-4615-1929-4
© 1995 Springer Science+Business Media New York
Originally published by Plenum Press, New York in 1995

FOREWORD

The last decades have witnessed a radical change in our views on central nervous system damage and repair. This change is not only due to the emergence of new powerful tools for the analysis of the brain and its reactions to insults, but it also reflects a conceptual change in the way we approach these problems. As an illustration to this development, it is instructive to go back to the proceedings of a meeting at the NIH in 1955 edited by William F. Windle, which summarizes the disillusioned and pessimistic view on CNS regeneration prevailing at the time. While this generation of researchers were well aware of the issues at stake, they felt they had reached the end of the road; the approaches they had pursued had got stuck and the tools available could not take them any further. I can very well imagine that the participants, most of them leaders in the field, left that conference feeling they had heard their field being sentenced to death.

In this perspective, the developments we have witnessed during the last few decades signal no less than rebirth. Research on neural development, damage, and repair is the aspect of neuroscience that has benefited most from the application of new tools and concepts generated by modern cell biology and molecular genetics. The recent revival can be traced back to the first descriptions of the modifiability of synaptic connectivity and of lesion-induced neuronal growth phenomena in the adult mammalian CNS, which emerged in the late sixties and early seventies (Raisman, 1969; Wall and Egger, 1971; Lynch et al., 1972; Tsukahara et al., 1974). This was followed by the demonstration that regeneration, growth, and functional recovery in the damaged CNS can be promoted by transplantation of neural tissue or cells derived from the central or peripheral nervous system (Björklund et al., 1976, 1979; Lund and Hauschka, 1976; Perlow et al., 1979; Richardson et al., 1980; Krieger et al., 1982), and by the discovery of new classes of molecules with potent neurotrophic and neurotropic actions, or growth inhibitory properties in the developing and adult nervous system (Edelman, 1986; Caroni and Schwab, 1986; Barde, 1990; Stöckli et al., 1989). These developments form the basis of some of the most active research strategies for CNS neuroprotection and repair pursued today.

Some of the most exciting future developments in this field will undoubtedly come from the interface between molecular genetics, developmental neurobiology, and neuroregeneration research. The editors of the present volume have been highly successful in capturing some of the central themes in this emerging new interdisciplinary research enterprise. Research on the detailed molecular mechanisms involved in early neural development, covered in some of the most interesting chapters in the book, has opened up exciting new possibilities. Much of the initial work on pattern formation and neuronal specification during early embryogenesis was carried out in Drosophila and other invertebrates. Most remarkably, however, work carried out during the last few years has shown that these families of regulatory genes, many of which code for proteins that act as transcription factors, are highly conserved throughout the animal kingdom. It is likely therefore, that essentially the same molecular mechanisms operate during early development of the nervous system in mammals, birds, fish, and flies. As a consequence, new families of molecules that are critically involved in the regulation of neuronal diversity and phenotypic differentiation in mammals are currently being unraveled at an amazing pace. In the near future we may thus get at least partial answers to some essential questions, such as the mechanisms by which ectodermal cells decide to become neural precursors; the mechanisms by which early neural progenitors choose to become either neurons or glia; and the mechanisms of differentiation of neuronal progenitors into specific neuronal subtypes.

A second set of equally fundamental questions that are discussed in the book relate to the regulation of cell proliferation, cell death, migration, axonal growth, guidance and specification of neuronal connectivity. These are basic neurobiological phenomena that are critical for our understanding not only of how the nervous system is formed during development, but also for the understanding of degeneration, regeneration, and repair in the adult CNS.

The neurotherapeutic implications of all this (which is my own particular interest) are far-reaching. Indeed, the first applications of these emerging new technologies in clinical trials may not be so distant in the future. Intracerebral transplantation of fetal CNS tissue has already reached the clinical stage. Embryonic neurons and neuroblasts are powerful sources of cells for intracerebral transplantation. They have a remarkable capacity to grow, integrate, mature, and function after implantation into developing or adult injured brains. In particular, transplants of fetal dopamine neurons obtained from the developing midbrain have been shown to reverse many of the neurological deficits seen in animal models of Parkinson's disease, and this approach has been shown to be feasible also in patients suffering from this disease.

However, for the further development of the cell transplantation technique it is essential to find alternative sources of cells, and to find ways to increase the yield and growth capacity of progenitors or neuroblasts obtained from limited amounts of embryonic or fetal starting material. Deeper insights into the molecular mechanisms underlying cell specification, growth, and differentiation, which are leading themes in several chapters in this book, will greatly promote the development of rational cell transplantation and genetic engineering strategies that may ultimately form the basis for efficient new therapies, not only in neurodegenerative diseases but also in conditions of ischemic and traumatic insults to the CNS. According to the figures published by the U.S. Office of Technological Assessment in 1990, some 10-12 million Americans suffer from neurodegenerative diseases, stroke, or traumatic injury of the brain or spinal cord, with an estimated cost for the U.S. society of more than 100 billion dollars in medical expenses and lost income. Since current medicine has little to offer in the way of effective therapy, the search for new approaches to the treatment of CNS damage and repair is clearly of great importance.

The present volume is an excellent introduction to some of the most interesting current developments and the remarkable progress that is being made in this exciting field. For those readers who are about to explore the many interesting chapters in the book I can promise most stimulating reading. When you are finished, I may suggest you acquire a spare copy of Windle's 1955 book to be placed next to this one. Together they will tell you that 40 years of research can make a difference.

Anders Björklund

References

Barde YA (1990): The nerve growth factor family. Progr Growth Factor Res 2: 237-248.

Björklund A and Stenevi U (1979): Reconstruction of nigrostriatal dopamine pathway by intracerebral nigral transplants. Brain Res. 177: 555-560.

Björklund A, Stenevi U and Svendgaard N (1976): Growth of transplanted monoaminergic neurones in the adult hippocampus along the perforant path. Nature 262: 787-90.

Caroni P and Schwab ME (1988): Antibody against myelin-associated inhibitor of neurite growth neutralizes non-permissive substrate properties of CNS white matter. Neuron 1: 85-96.

Edelman GM (1986): Cell adhesion molecules in the regulation of animal form and tissue pattern. Ann Rev Cell Biol 2: 81-116.

Krieger DT, Perlow MJ, Gibson MJ, Dames TF, Zimmerman EA, Ferin M and Charlton HM (1982): brain grafts reverse hypogonadism of gonadotropin releasing hormone deficiency. Nature 298: 468-471.

Lund RD and Hauschka SD (1976): Transplanted neural tissue develops connections with host brain. Science 193: 582-84.

Lynch G, Matthews DA, Mosko S, Parks T and Cotman C (1972): Induced acetylcholinesterase-rich layer in rat dentate gyrus following entorhinal lesions. Brain Res 42: 311-318.

Perlow MJ, Freed WJ, Hoffer BJ, Seiger A, Olson L and Wyatt RJ (1979): Brain grafts reduce motor abnormalities produced by destruction of nigrostriatal dopamine system. Science 204: 643-647.

Raisman G (1969): Neuronal plasticity in the septal nuclei of the adult rat. Brain Res 14:25-48.

Richardson PM, McGuinness UM and Aguayo AJ (1980): Axons from CNS neurones regenerate into PNS grafts. Nature 284: 264-265.

Stöckli KA, Lottspeich F, Sendtner M, Masjakowsky P, Carroll P, Götz RL, Lindholm D and Thoenen H (1989): Molecular cloning, expression and regional distribution of ciliary neurotrophic factor. Nature 342: 920-923.

Tsukahara N, Hultborn H and Murakami F (1974): Sprouting of cortico-rubral synapses in the red nucleus neurones after destruction of the nucleus interpositus of the cerebellum. Experientia (Basel) 30: 57-58.

Wall PD and Egger MD (1971): Formation of new connexions in adult rat brains after partial deafferentation. Nature 232: 542-545.

Windle WF [Editor] (1955): "Regeneration in the Central Nervous System," Charles C. Thomas, Publ., Springfield, IL, U.S.A.

CONTENTS

PATTERN FORMATION IN THE VERTEBRATE CNS

GENETIC DETERMINANTS OF NEURAL CELL FATE

NEURAL CELL DIFFERENTIATION

STRATEGIES FOR NEURAL CELL REPLACEMENT IN
NEURODEGENERATIVE DISORDERS

NEURAL CELL SPECIFICATION

Molecular Mechanisms and Neurotherapeutic Implications

PATTERN FORMATION IN THE VERTEBRATE CNS

Hox GENE FUNCTION AND THE DEVELOPMENT OF THE HEAD

M. Mark, F.M. Rijli, T. Lufkin, P. Dollé, P. Gorry, and P. Chambon

Institute de Génétique et de Biologie Moléculaire et
Cellulaire, CNRS/INSERM/ULP/Collège de France,
BP 163 - 67404 ILLKIRCH-CEDEX, C.U. de Strasbourg,
FRANCE

INTRODUCTION

HOM/*Hox* genes are a family of genes (the homeogene family) that show similarities in structure, organisation into complexes and expression patterns. They code for transcriptional regulators which are thought to function at the top of a genetic hierarchy that controls the relative position of the cells along the embryonic axes. First discovered in *Drosophila* and then found in vertebrates, they were recently identified in the genome of animals as diverse as nematodes, leeches, amphioxus and hydra, suggesting that a HOM/*Hox* gene cluster already existed in the common ancestor of all animals. The presence of HOM/*Hox* cluster(s) has been proposed as one of the characters defining the Kingdom Animalia (i.e. the Zootype: reviewed in Slack, 1993; Holland, 1992; Thorogood, 1993; Kappen et al., 1993). Interphyletic comparisons between insect HOM genes and vertebrate *Hox* genes are well documented (reviewed in McGinnis and Krumlauf, 1992; Botas, 1993). In vertebrates there are four paralogous clusters of Hox genes, referred to as Hox A, B, C, and D, each showing clear structural homology to the prototypic homeotic complex (HOM-C) of *Drosophila*. During ontogenesis, *Hox* genes are expressed in the neurectoderm and paraxial mesoderm in specific but overlapping domains that extend from the caudal end of the embryo to a sharp rostral limit. There is a correlation (termed spatial colinearity) between the position of a *Hox* gene within its cluster and the location of its rostral limit of expression. Significantly, this spatial colinearity also exists in *Drosophila*. Both vertebrate *Hox* and insect HOM gene expression domains respect metameric boundaries. Loss-of-function mutations (i.e. generated by gene disruption) and gain-of-function mutations (i.e. generated by ectopic gene expression) of both mouse *Hox* genes and *Drosophila* HOM genes often cause a transformation of specific metameres [serially homologous anatomical units, repeated along the rostrocaudal axis such as somites (mouse) or

Neural Cell Specification: Molecular Mechanisms and Neurotherapeutic Implications
Edited by Juurlink *et al.*, Plenum Press, New York, 1995

3

parasegments (*Drosophila*)] into the likeness of their neighbours, indicating that these genes are involved in the specification of the phenotype of a given segment according to its position along the rostrocaudal axis. Since the segmented body plans of mouse and *Drosophila* have arisen totally independently during evolution, it appears that at least part of HOM/*Hox* genes network has been co-opted to impart morphogenetic segment identity in animal groups employing different segmentation strategies (Holland, 1990).

It has been proposed that the vertebrate head (comprising the brain and skull case, the cephalic sense organs and their associated capsules, the derivatives of the frontonasal processes and of the first 3 pharyngeal arches) is, in evolutionary terms, a recent structure, that has arisen largely de novo rather than by the modification of a preexisting structure, since elements relevant to its formation (i.e. the neural crest and the ectodermal placodes) are absent in the immediate vertebrate ancestors (protochordates; the prototype of modern protochordates is Amphioxus). Thus, in the transition from invertebrates to vertebrates, three important events have occurred: i) the appearance of ectodermal placodes that make major contribution to the sense organs and to the sensory ganglia of the cranial nerves, ii) the appearance of neural crest cells (NCC) from which most of the skeleton of the head is derived, iii) the segmentation of the brain which is particularly well documented at the level of the rhombencephalon, where it appears to correlate with that of its derived neural crest cells (reviewed in: Gans and Northcutt, 1983; Langille and Hall, 1989; Keynes and Lumsden, 1990; Figdor and Stern, 1993; Puelles and Rubenstein, 1993).

The mesenchymal neural crest (also referred to as mesectoderm or ectomesenchyme) is the only source of skeletal elements and connective tissues (including cartilage, bone, tendon, smooth muscles of arterial walls, tooth's mesenchyme and connective tissue of striated muscles) in the frontonasal, periocular and branchial regions of the head. In addition, most of the skull bones are of neural crest origin (reviewed in: Noden, 1988; Le Douarin et al., 1993; Lumsden, 1988). Interestingly, some populations of cranial mesenchymal NCC are prepatterned, i.e. imprinted with morphogenetic information, before they leave the neural tube. Noden (1983) showed that grafting premigratory midbrain/rostral hindbrain NCC (which are normally destined to populate the first pharyngeal arch) to the second and third arch in the chick embryo, resulted in duplication of first arch derived membrane bones and cartilages. The main conclusion from Noden's grafting experiment was that the morphogenetic fate of 1st pharyngeal arch NCC, which is expressed in the shape and number of the skeletal elements that they will form, has been already determined within the neural plate, thus at a stage prior to migration. Cranial NCC transplantation experiments have also revealed that the mesectodermal cells, per se, exert a dominant effect in craniofacial patterning, since they guide the assembly of striated muscle cells which are derived from paraxial mesoderm (reviewed in Noden, 1988). The NCC, in conjunction with the cells of the ectodermal placodes, also contribute to the sensory ganglia of the cranial nerves. It is unknown whether the fate of these neurogenic NCC is imprinted before migration.

The hindbrain of vertebrate embryos is transiently divided along the rostrocaudal axis into a series of bulges termed rhombomeres (reviewed in: Lumsden, 1990; Keynes and Lumsden, 1990). For instance, in the mouse there are seven such neuroepithelial swellings which are present only between 9.0 and 10.5 days post-coitum (dpc). From the time when rhombomeres 2 to 6 (R2 to R6) are first morphologically recognisable, they represent units of cell lineage restriction. All the descendants of a single dividing neuroepithelial cell are confined to the rhombomere in which they are born (Fraser et al., 1990). Differences in cell adhesion

ERRATUM

PREFACE

This volume is comprised of papers presented at the Third International Altschul Symposium: *Neural Cell Specification: Molecular Mechanisms and Neurotherapeutic Implications*. The symposium was held in Saskatoon, Saskatchewan, Canada, in May 1994, in memory of Rudolph Altschul, a graduate of the University of Prague and a pioneer in the fields of the vascular and nervous systems. Dr. Altschul was Professor and Head of the Department of Anatomy at the University of Saskatchewan from 1955 to 1963. The biennial Altschul Symposia are made possible by an endowment left by Anni Altschul, supplemented by other contributions.

The goal of the Third Altschul Symposium was to promote a critical discussion of the molecular mechanisms that regulate the specification of neural progenitor cells and also of the therapeutic strategies currently being developed for neural cell replacement. The symposium interfaced molecular genetics, developmental neurobiology, and neuroregeneration research. It provided a forum for the interchange of ideas between basic and clinical scientists working in these diverse fields of neurobiology, with the anticipation that this interchange would serve as a catalyst for the evaluation and development of more effective neurotherapeutic strategies. Readers of this volume may well agree with Anders Björklund (Foreword) that we have achieved success with the objectives set forth for the symposium.

We are grateful to the following sponsors of the symposium: the College of Medicine, University of Saskatchewan; the Office of the President, University of Sakatchewan; The Medical Research Council of Canada; The March of Dimes Birth Defects Foundation; The Spinal Cord Research Foundation (Paralyzed Veterans of America). We are also indebted to the following companies for their generous support: Amersham Canada, Inc.; Blackwell Scientific Ltd.; Canadian Life Technologies, Inc.; Ciba Geigy Canada; DiaMed Labs, Inc.; Dupont Canada, Inc.; Eli Lilly Canada, Inc.; Leica Canada, Inc.; Medical Arts Laboratory–MDS Health Group Company; Professional Diagnostic, Inc.; Promega Corporation; Research Biochemicals International; Sero-Tec Ltd.; Sin-Can International Biological Sciences, Inc.; Summit Biotechnology; Upjohn Company of Canada; VWR Scientific of Canada Ltd. We thank the members of our Scientific Advisory Committee: P. Grüss, R. Krumlauf, R. McKay, L. Olson, J. Sladek, Jr., and S. Weiss. Finally we thank the many members of the Department of Anatomy who assisted with the Symposium.

Bernhard H.J. Juurlink
William M. Kulyk
Patrick H. Krone
Valerie M.K. Verge
Ronald Doucette

Neural Cell Specification: Molecular Mechanisms and Neurotherapeutic Implications
Bernhard H.J. Juurlink *et al.*, Plenum Press, New York, 1995

properties underlie the restriction of cell movements across rhombomeric boundaries, as demonstrated by the regeneration of boundaries and inhibition of cell mixing when even and odd numbered rhombomeres are juxtaposed (Guthrie and Lumsden, 1991; Guthrie et al., 1993). Rhombomeric segmentation correlates with the spatial arrangement of neuronal structures (Lumsden and Keynes, 1989; Simon and Lumsden, 1993; Clarke and Lumsden, 1993). For example, the roots of the branchial motor nerves project from alternate rhombomeres and their neurons differentiate from specific contiguous rhombomeres, i.e. R1, R2, and R3 for trigeminal neurons, R4 and R5 for facial neurons, and R6 and R7 for glossopharyngeal neurons (in the mouse), innervating the first, second and third pharyngeal arches respectively (Lumsden and Keynes, 1989; Carpenter et al., 1993). These observations and the results from rhombomere transplantation experiments (Guthrie et al., 1992; Guthrie and Lumsden, 1992; Kuratani and Eichele, 1993) suggest that rhombomeres contain intrinsic positional information that is important for patterning of their neuronal derivatives. The segmental organisation of the hindbrain contributes to the patterning of the NCC. Mesenchymal NCC emerging from the hindbrain and caudal midbrain migrate in a segmental pattern that correspond to the location of the pharyngeal arches (Lumsden et al., 1991; Serbedzija et al., 1992; Sechrist et al., 1993). This pattern consists of three distinct streams of cells which originate mainly i) from the caudal midbrain and R1 and R2, ii) from R4, and iii) from R6, and populate the first (maxillomandibular), second (hyoid) and third pharyngeal arches respectively. These three streams are separated by two NCC deficient regions located opposite R3 and R5. Since even numbered rhombomeres control the apoptotic elimination of NCC formed from R3 and R5 (Graham et al., 1993), the separate streams of NCC are produced by mechanisms which are, at least in part, intrinsic to the neural epithelium. NCC populating the first 3 pharyngeal arches (or branchial region of the head) make an important contribution to the head skeleton (reviewed: in Noden, 1988; Le Douarin et al., 1993) since, for example in mammals, they form part of the skull case (i.e. the alisphenoid and squamosal bones are derived from the 1st pharyngeal arch mesectoderm), the jaw bones, the 3 middle ear ossicles and the hyoid bone. NCC migration from specific rhombomeres also contribute to the sensory ganglia of the cranial nerves: the trigeminal, acoustico-facial and superior glossopharyngeal/vagus ganglia are contributed by NCC cells from R2, R4 and R6 respectively (Lumsden et al., 1991).

In situ hybridization analysis of *Hox* gene expression domains in rhombomeres, rhombencephalic NCC and pharyngeal arch mesenchyme has suggested that combinations of *Hox* genes (i.e. a *Hox* combinatorial code) endow each rhombomere with a positional identity which is then transferred to the pharyngeal arches by the neural crest to specify their patterning program (reviewed in: Krumlauf, 1993; Wilkinson, 1993; Hunt and Krumlauf, 1992). In general, part of the *Hox* code expressed in any specific rhombomere is also expressed in the NCC migrating from that rhombomere. However, although the *Hoxa-2* gene is expressed in R2 (Krumlauf, 1993 and references therein; Prince and Lumsden, 1994), no *Hox* genes are expressed in the NCC populating the first pharyngeal arch. Interestingly, *Hoxa-2* expression appears to be independently regulated in the neural tube and prospective R2-derived NCC (Prince and Lumsden, 1994). The distribution of *Hox* gene transcripts in the rhombencephalon and the pharyngeal arches is illustrated in Fig. 1. Two features must be stressed: i) *Hox* genes have restricted domains of expression exhibiting sharp boundaries that map to distinct rhombomere junctions or specific pharyngeal arch interfaces; ii) the *Hox* paralogs (i.e. the genes occupying the same relative positions in different clusters) have similar expression domains in the hindbrain as well as in the pharyngeal arches with the exceptions of the *Hoxa-1* and *Hoxb-1*, *Hoxa-2* and *Hoxb-2* genes.

Disruption of *Hox* genes in the mouse via homologous recombination in murine embryonic stem (ES) cells (Chisaka and Capecchi, 1991) provides a powerful tool for gaining insights into the developmental role of these genes. We review here the phenotypic consequences of the disruption of *Hox* genes normally expressed in the rhombencephalon and the branchial region of the head.

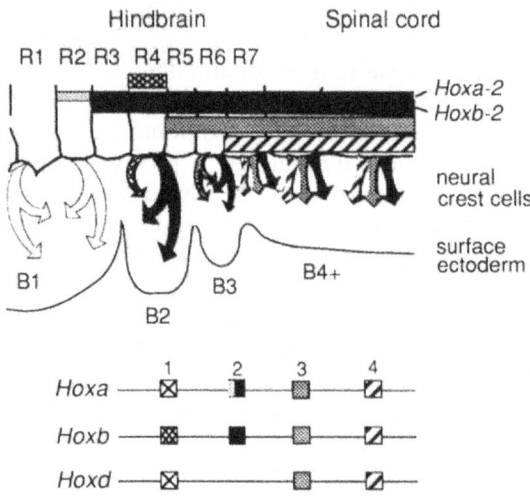

Figure 1. (a) *Hox* gene expression domains in the hindbrain and pharyngeal arches of a 9.5 dpc mouse embryo. The expression patterns of paralogue genes in rhombomeres (R1 to R7), neural crest cells and pharyngeal (branchial) arches (B1 to B4) are represented by various codes of grey, as indicated on the scheme below. The filled arrows represent *Hox* gene expressed in gangliogenic and mesenchymal NCC migrating from specific rhombomeres into the pharyngeal arches (B2, B3 and B4). The *Hoxa-1* gene is already down regulated at this stage, but at earlier stages shows the same rostral expression boundary (the future R3/R4 boundary) as its paralogue *Hoxb-1*. Note that the expression domain of *Hoxa-2* extends to the R1/R2 junction, but that there is no *Hox* gene expressed in neural crest cells migrating from R1 and R2 into the first pharyngeal arch (white arrows). Scheme adapted from Krumlauf (1993).

Hoxa-3 IS REQUIRED FOR THE DEVELOPMENT OF THIRD PHARYNGEAL ARCH MESECTODERMAL DERIVATIVES

In wild-type mice, the expression domain of *Hoxa-3* extends from the R4/R5 boundary to the caudal end of the embryo. *Hoxa-3* is also expressed in the NCC originating from R6 and R7 and in the mesenchyme of the third and the fourth pharyngeal arches (Krumlauf, 1993 and references therein; Fig.1). *Hoxa-3* null mutant mice exhibit defects in structures derived from mesenchymal NCC originating from R6 and R7, including agenesis of the thymus and parathyroid glands and absence of the carotid arteries (Chisaka and Capecchi, 1991; Condie and Capecchi, 1993). The origin of the abnormalities found at more rostral levels of the *Hoxa-3* mutant fetuses, such as the shortening of the mandible and maxilla is unclear. It is conceivable that these rostral defects might be secondary to the absence of the carotid arteries, since these vessels normally make a major contribution to the vascularisation of facial structures.

RHOMBOMERE SEGMENTATION IS LOCALLY ALTERED IN *Hoxa-1* NULL MUTANT MICE

In normal mice, the expression domain of the *Hoxa-1* gene at 7.5 and 8.0 dpc extends from the anterior boundary of the prospective R4 to the caudal end of the embryo. *Hoxa-1* is expressed only for a short period in the rhombencephalon, since it is no longer detected at 8.5 dpc, i.e. 12 hours before the formation of the R4/R5, R5/R6 and R6/R7 rhombomeric boundaries (reviewed: in Krumlauf, 1993; Hunt and Krumlauf, 1992).

Instead of the normal seven rhombomeres (R1 to R7), *Hoxa-1* null embryos have only five rhombomere-like structures (RI to RV) (Lufkin et al., 1991 ; Mark et al., 1993). The first 3 rhombomeres (RI to RIII) appear to be normal in terms of expression patterns of cellular retinoic acid binding protein I (CRABPI) and of other positional molecular markers (i.e. *Hoxa-2, Hoxb-2* and *Krox-20*; Dollé et al., 1993; Fig. 2a, b). In contrast, the *Hoxa-1* mutant rhombencephalic territories homologous to R4 and R5 are reduced, resulting in a shortening of the rhombencephalon, and remnants of these rhombomeres have fused caudally with R6 to form a single rhombomeric structure, RIV (Dollé et al., 1993; Mark et al., 1993). Indeed, in the rhombencephalon of *Hoxa-1* mutant embryos, the expression domain of *Hoxb-1* which, at 9.0 dpc can be regarded as a R4 molecular marker, is abnormally short. Furthermore, the caudal domain of *Krox-20* expression, which corresponds to R5, is practically absent (Fig. 2a, b) and, accordingly, the NCC deficient region that is normally present between the future acousticofacial ganglion (C7/8 in Fig. 2b) and the superior glossopharyngeal/vagal ganglion (C9/10 in Fig. 2b) is lacking. The mutant fourth rhombomere (RIV, Fig. 2b) coexpresses R4-specific positional markers (*Hoxb-1*, Fig. 2a, b) and a R6-specific combination of markers (*Hoxa-3* and *Hoxb-3* without *Krox-20* expression; Fig. 2a, b) in a short rostral part and a larger caudal part of this segment respectively (Dollé et al., 1993). The fusion of newly juxtaposed rhombomeres 4 and 6 into a single RIV is in agreement with the finding that experimental juxtaposition of R4 and R6 does not lead to the formation of a rhombomeric boundary (Guthrie and Lumsden, 1991).

Other structures whose development is altered in *Hoxa-1* mutants include the motor nuclei of the facial and abducens nerves, the inner ear and the sensory portion the glossopharyngeal and vagus nerves (Lufkin et al., 1991; Mark et al., 1993; Chisaka et al., 1992; Carpenter et al., 1993). All these abnormalities might be secondary to the shortening of the *Hoxa-1* mutant rhombencephalon.

In agreement with the small size of R4 and near absence of R5 in *Hoxa-1* embryos, the motor neurons of the facial nerve (derived from R4 and R5, Carpenter et al., 1993) appear scarce and those of the abducens nerve (derived from R5, Baker and Noden, unpublished results) are absent. That the absence of R5 is directly responsible for the loss of abducens nerve motor neurons is supported by the phenotypic analysis of mice homozygous for the *kreisler* (*kr*) gene, as well as for a null mutation in the *Krox-20* gene: in these mutants, similarly to the *Hoxa-1* mutants, the lack or severe reduction of R5 correlates with the agenesis of the abducens nerve motor nucleus (Frohman et al., 1993; Scheider Maunoury et al., 1993; Lumsden, personal communication). On the other hand, the size of the motor nuclei of the trigeminal and glossopharyngeal nerves, which are derived form R1 to R3 and R6 and R7 respectively, are normal in the *Hoxa-1* mutant mice.

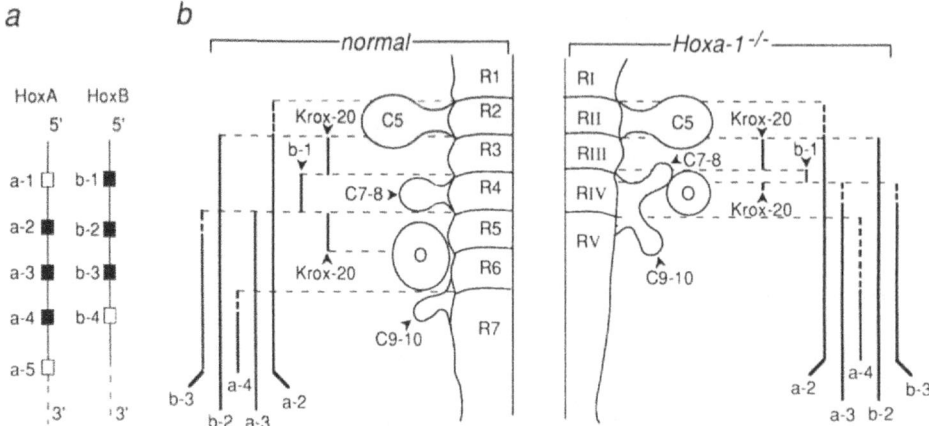

Figure 2. (a) Diagram representating the 3' regions of the HoxA and HoxB complexes. Paralogs genes face each other horizontally. Black squares indicate the genes whose expression was analyzed by in situ hybridization in *Hoxa-1* null embryos. (b) Diagrams of the normal and *Hoxa-1* null mutant rhombencephalon. C5, C7-8 and C9-10, cells of the gangliogenic NCC of the trigeminal nerve (5th pair of cranial nerves), facial and acoustic nerves (7th and 8th pairs), and glossopharyngeal and vagus nerves (9th and 10th pairs). R1 to R7 and RI to RV, rhombomeres (see text); O, otocyst. The extent of the expression domains of the different genes is denoted by vertical lines. Interrupted vertical lines indicate a faint expression level and/or an ill-defined expression boundary. For further details, see Dollé et al. (1993).

In *Hoxa-1* mutant embryos, the otocyst, which represents the common anlage of the membranous labyrinth (i.e. the epithelial inner ear) and of the vestibuloacoustic (8th cranial nerve) sensory neurons, is much smaller than normal and bears no endolymphatic duct. These early defects are likely to directly account for the aplasia of the membranous labyrinth, including the complete agenesis of the cochlea, for the absence of the spiral (acoustic) ganglion, and for the smaller size of the vestibular ganglion ; and, indirectly, for the disruption of the cartilaginous otic capsule whose formation is induced by the otocyst (references in Mark et al., 1993). The otocyst forms from an ectodermal placode located opposite to (and presumably induced by) R5 and R6 (references in Mark et al., 1993). That the defects of the inner ear are probably secondary to the shortening of the *Hoxa-1* mutant rhombencephalon is supported by the several observations. Firstly, *Hoxa-1* is never expressed in the otocyst nor in the placodal ectoderm from which this structure arises. Secondly, ear defects similar to those observed in *Hoxa-1* mutant mice are caused by ectopic otic vesicle transplantations in birds and amphibians. Thirdly, another mouse mutant, *kreisler,* displays rhombencephalic abnormalities which preceed similar inner ear defects (references in Mark et al. 1993). The *int-2* gene is expressed in R5 and R6 and codes for an ontogenic factor, FGF3. In *Hoxa-1* mutant embryos, the level of *int-2* expression in the rhombomere located opposite the otocyst (i.e. RIV) might be decreased (Mark et al., 1993). However, the size of the otocyst and the differentiation of the cochlea are not affected in *int-2* null mice (Mansour et al., 1993). This indicates that, in *Hoxa-1* null mice, possible alterations of the *int-2* dependent developmental pathway cannot account for all the inner ear defects.

In *Hoxa-1* mutant embryos, the glossopharyngeal and vagus nerves often fail to connect with the rhombencephalon and with the target organs represented by the third (for the glossopharyngeal nerve) and fourth (for the vagus nerve) pharyngeal arches. The rostral shift of the glossopharyngeal and vagus NCC-derived bipolar neuroblasts, which is secondary to the shortening of the rhombencephalon, might separate them from putative neurite growth signals produced dorsally by the neurectoderm and ventrally by the mesenchyme of the pharyngeal arches. It is also possible that the gangliogenic glossopharyngeal and vagus NCC which are normally derived from R6 and R7 might be incorrectly specified. However, since *Hoxa-1* expression has not been detected into migrating NCC, their specification by *Hoxa-1* would then have to occur at a stage preceeding migration from the neural plate (see Mark et al., 1993 and references therein)

It is noteworthy that the mesectodermal derivatives originating from R4, (stapes, styloid bone and lesser horn of the hyoid bone) are usually all identifiable in *Hoxa-1* mutant mice. The apparent lack of the stapes in some mutants (our unpublished data) is likely to reflect a secondary mechanical problem caused by the proximity of this ossicle to the highly distorted otic capsule. These data and the apparent lack of defects in regions of the *Hoxa-1* mutant mice caudal to R6 indicate that the functional domain of *Hoxa-1* (i.e. R4, R5 and perhaps part of R6) is restricted to a region of the head corresponding to its more rostral expression domain and does not directly affect the mesectodermal derivatives originating from this region.

DISRUPTION OF *Hoxa-2* RESULTS IN A HOMEOTIC TRANS-FORMATION OF SECOND TO FIRST PHARYNGEAL ARCH IDENTITY

Hoxa-2 is the only *Hox* gene to be expressed at the level of R2, and this expression is seen in the neurectoderm, but not in migrating NCC (Prince and Lumsden, 1994; Krumlauf, 1993 and references therein; Fig.1). This *Hoxa-2* pattern of expression in R2 provides a unique opportunity to study the role of a *Hox* gene in rhombomeric segmentation and specification processes, in the absence of any other *Hox* gene expression. Moreover, *Hoxa-2* and its paralogue *Hoxb-2* are the only *Hox* genes to be expressed in the mesenchyme of the second pharyngeal arch (Krumlauf, 1993 and references therein; Fig.1). Thus, disrupting the *Hoxa-2* gene (Rijli et al., 1993; Gendron-Maguire et al., 1993) also offers a unique opportunity to examine the role of *Hox* genes in branchial arch patterning.

The *Hoxa-2* null fetuses exhibited bilateral skeletal abnormalities (Fig.3) which were restricted to the head region (Rijli et al., 1993). The NCC-derived skeleton of the 2nd pharyngeal arch which include the stapes (i.e. the medial middle ear ossicle, S), the styloid bone (SY) and the lesser horn of the hyoid bone (LH) (Noden, 1988 and references therein) was selectively lacking. In contrast, an ectopic caudal set of first arch skeletal elements was present, mostly as a mirror image of its orthotopic counterpart. This ectopic set comprised : (i) within the middle ear region, a supernumerary incus (I2), malleus (M2), truncated Meckel's cartilage (MC2) and tympanic bone (T2), (ii) outside the middle ear region, a small supernumerary squamosal bone (SQ2) and an additional rod of cartilage. The latter had no direct wild-type counterpart. However, its shape and anatomical relationships indicated that it might be homologous to the reptilian pterygoquadrate (upper jaw) cartilage from which the mammalian incus (I), pterygoid (P) and part of alisphenoid (AS) bones are thought to have evolved (De Beer, 1985). Thus the complex formed by the pterygoquadrate element and the

9

caudal incus in *Hoxa-2* mutants (P2, Q and I2) might correspond to a mirror image of an atavistic skeletal formation in which the incus had already appeared while part of the quadrate bone was still present. Such a skeletal arrangement indeed existed in therapsids, the reptilian ancestors of modern mammals (Olson, 1959; Allin, 1975; Presley, 1989).

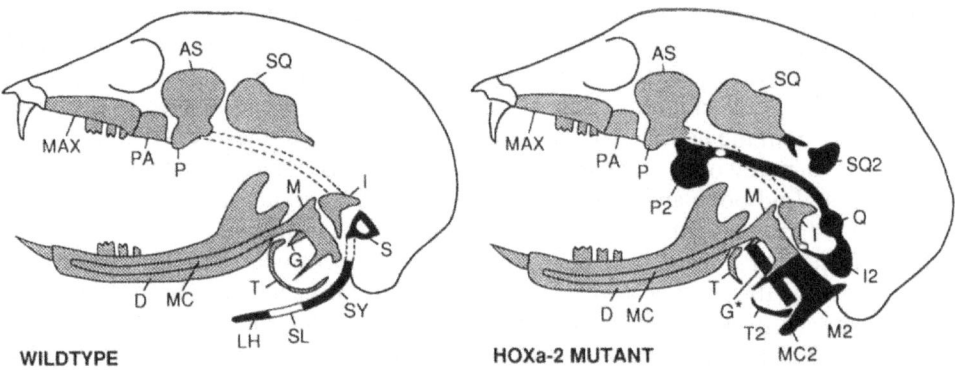

Figure 3. Schematic representation of the relationships between some skeletal elements derived from the first pharyngeal arch (in grey) and from the second pharyngeal arch (in black) in wild-type and *Hoxa-2* null mutant 18.5 dpc fetuses, seen from the lateral aspect. Note that in wild-type fetuses the stapes (S), styloid bone (SY), stylohyoid ligament (SL) and lesser horn of the hyoid bone (LH), all derive from the cartilage of the second arch (Reichert's cartilage). AS, alisphenoid bone; D, dentary bone; G and G*, wild-type and mutant gonial bones, respectively. I and I2, orthotopic and ectopic incus respectively; LH, lesser horn of the hyoid bone ; M and M2, orthotopic and ectopic malleus respectively; MAX, maxillary bone; MC and MC2 orthotopic and ectopic Meckel's cartilage respectively; P and P2, orthotopic and ectopic pterygoid bones, respectively; PA, palatine bone; Q, pterygoquadrate cartilage; S, stapes; SL, stylohyoid ligament; SQ and SQ2 orthotopic and ectopic squamosal bones respectively; SY, styloid bone; T and T2, orthotopic and ectopic tympanic bones. For further details, see Rijli et al. (1993).

Based solely on these data from skeletal analyses, two mutually exclusive interpretations could be put forward to account for the *Hoxa-2* mutant phenotype. The first interpretation is that second arch NCC, which normally arise from R4, either never existed or never migrated and underwent cell death. The *Hoxa-2* ectopic skeletal elements would then derive from NCC carrying a first arch identity which have migrated to the second arch from a more rostral axial level corresponding to the mesencephalon and/or metencephalon (i.e. the 3 rostral rhombomeres). Alternatively, the phenotypic specification of R4-derived mesectodermal cells might have been altered in such a way that these cells now express a first pharyngeal arch skeletogenic program. The following observations strongly support this latter possibility: i) the segmented anatomy of the rhombencephalon and pharyngeal arches is normal in *Hoxa-2* mutant embryos, ii) the molecular identity of rhombomeres and pharyngeal arches is also normal, as assessed by in situ hybridization using antisense probes for different *Hox* genes and for the *Krox-20* gene. Altogether, our data indicate that disruption of the *Hoxa-2* gene results in a homeotic transformation of second to first pharyngeal arch identity (Rijli et al., 1993).

The *Hoxa-2* mutation had no apparent effect on gangliogenic NCC derivatives. In addition it had little effect on the neurectoderm, since the R1 to R3-derived motor nucleus of the trigeminal nerve was apparently normal, and the R4 and R5-derived motor nucleus of the facial nerve was only occasionally affected which might reflect a secondary defect. Thus the functional domain of the *Hoxa-2* gene appears to be restricted to the mesectoderm of the 2nd pharyngeal arch, in the region corresponding to its most anterior domain of expression.

Hox GENES AND THE SEGMENTATION OF THE HINDBRAIN

Hoxa-1 expression in the prospective R4 to R6 region is briefly transient and occurs early, i.e. prior to the restriction of *Hoxb-1* expression to the presumptive R4 and of *Krox-20* expression to the presumptive R5, and to the subsequent morphological segmentation of R4, R5 and R6. There are two explanations for the abnormal segmentation in the R4 to R6 region of the *Hoxa-1* mutant rhombencephalon (Mark et al., 1993; Carpenter et al., 1993; Wright, 1993). In the first scenario, segmentation would not be directly affected in the absence of *Hoxa-1*. Instead, some of the presumptive R4 cells and most of the presumptive R5 cells may not be properly specified and may adopt the fate of those of neighbouring rhombomeres, leading to the establishment of abnormal rhombomeric boundaries. R4 and R5 may also fail to fully develop, because the normal function of *Hoxa-1* would be to control the activity of mitogenic and/or of survival factor(s) required to maintain an earlier specified segmental pattern. In the absence of these factors some R4 and most of R5 cells might fail to divide and/or undergo apoptosis. Alternatively, the neuroepithelial cells corresponding to the pre-R4 and pre-R5 might have lost some specific adhesive properties and instead of remaining clustered they would be dispersed along the rostrocaudal axis of the rhombencephalon, leading to the partial disappearance of R4 and R5. Extensive cell mixing of committed neuronal precursors might explain the facial motor neuron ectopias in R3, and R6 and R7 reported by Carpenter et al. (1993). A second scenario supposes the direct involvement of *Hoxa-1* in the early segmentation of the central hindbrain. Indeed, hindbrain segmentation begins with the formation of the two constrictions which delineate the region from which R4 and R5 will originate (Lumsden, 1990 and references therein) and this region corresponds to the *Hoxa-1* functional domain. In *Drosophila*, the division of the embryo into parasegments is controlled by segmentation genes, some of which act upstream of homeotic genes (Gaul and Jäckle, 1987; Pankratz and Jäckle, 1990). In contrast, homeotic gene products are not involved in the segmentation process: mutations of *Drosophila* homeotic genes affect the phenotype but not the number of segments (Lawrence, 1992). The rhombencephalon is a region of the vertebrate embryo whose segmentation is similar (in ontogenic terms) to that of insects (Fraser et al., 1990). Aside from the *krüppel* related gene *Krox-20* (Swiatek and Gridley, 1993; Schneider Maunoury et al., 1993), homologues of the *Drosophila* segmentation genes have not been found with roles in vertebrate segmentation. Since *Hoxa-1* mutant mice have apparently lost two rhombomeres, that is two morphological segments, *Hoxa-1* might function more like a *Drosophila* segmentation gene than as a segment identity (homeotic selector) gene. Note, however, that no *Hox* genes are ever expressed in R1 and that the results of the *Hoxa-2* knockout clearly show that the formation of R2 does not require the expression of *Hox* genes. Whether *Hoxa-1* expression is implicated in the segmentation process of the R4-R6 region remains an open question.

Hox GENES AND THE PATTERNING OF THE CRANIAL MESECTODERM

The *Hoxa-2* null mutation selectively changes the fate of second arch mesectodermal cells. Interestingly, disrupting its neighbour, *Hoxa-3*, selectively affects third arch mesectodermal cells. These data support the view that in the branchial region of the head different populations of mesectodermal cells are specified by distinct *Hox* genes. In contrast, disrupting *Hoxa-1* has apparently no direct effect on any cranial mesectodermal derivative.

The results of the *Hoxa-2* knockout support Noden's conclusion that some first pharyngeal arch mesectoderm cells are embodied with cell autonomous patterning information (Noden, 1983). The symmetrical arrangement of the duplicated skeletal elements in the middle ear region of the *Hoxa-2* mutants is intriguing. Examination of embryos in which short segments of neural crest were rotated along the rostrocaudal or dorsoventral axis has revealed that pharyngeal arch skeletal tissues developed with a normal polarity, thus suggesting that the mesectoderm receives polarizing cues from the environment into which it migrates, possibly from the pharyngeal arch ectoderm (Noden, 1983 and references therein). This raises the possibility that a zone of polarizing activity might exist at the first/second arch interface (the region of the first pharyngeal cleft) which might spread an inductive signal both rostrally (to the first arch mesenchyme) and caudally (to the second arch mesenchyme), in a symmetrical fashion. This hypothetical signal might be normally used by first and second arch NCC surrounding the first pharyngeal cleft to fine tune the assembly of their skeletal derivatives in the middle ear region. Alternatively, only the first arch derived mesenchyme might be able to respond to the signal, due to the expression of first arch specific receptors or responsive genes. In *Hoxa-2* mutants, similarly programmed first and second pharyngeal arch NCC would then similarly interpret the symmetrical polarizing signal, resulting in mirror image duplications.

Another interesting outcome of our study is that only a subset of first arch derivatives are duplicated in the *Hoxa-2* mutants. The lack of duplications of the dentary, maxillary and palatine bones and of teeth, which are all first arch derivatives, might either reflect the existence of first arch-specific osteogenic and odontogenic ectodermal signals (see Hall, 1987; Lumsden, 1988) or the selective duplication of R1 and R2 mesenchymal NCC-derived skeletal elements in the *Hoxa-2* mutants. This latter possibility is strongly supported by the observation that rostral hindbrain (R1 and R2) NCC contributes mainly to the proximal part of the skeleton derived from the first arch in the chick (Noden, 1978), i.e. to the elements homologous to those which are duplicated in the *Hoxa-2* mutants. Thus, in the absence of a functional *Hoxa-2* gene, the R4 derived mesenchymal NCC might be reprogrammed to a R1 and R2 phenotype, but not to a more rostral mesencephalic one. Alternatively, the duplication of the elements derived from NCC of more rostral origin might be selectively suppressed by *Hoxb-2* expression in the second arch, which would exert a posteriorizing effect (see Rijli et al., 1993).

Hox GENES AND THE EVOLUTION OF THE HEAD

The homeotic transformation of second to first pharyngeal arch identity in *Hoxa-2* null mutants reveals that the morphogenetic program of the R1 and R2-derived NCC corresponds to a Ground (or default) skeletogenic Patterning Program (GPP) which is common to mesenchymal NCC of at least the first and second pharyngeal arches and does not require *Hox* gene expression. Note that this GPP may correspond to only a fraction of the normal first arch skeletogenic program (see above and Rijli et al., 1993). Interestingly, grafts of presumptive frontonasal quail embryo NCC in place of presumptive second arch NCC of host chick embryos give rise to ectopic first arch skeletal elements. Thus the GPP might also be present in the frontonasal mesectodermal cells, and emerge only when these cells are removed from the patterning influence of the forebrain neurectoderm (Noden, 1983).

In wild-type mice the GPP is respecified by *Hoxa-2* which, like *Drosophila* homeotic genes, acts as a selector gene to yield the NCC second arch specific morphogenetic program. In the absence of a functional *Hoxa-2* gene, the second arch mesenchymal NCC apparently execute a GPP which corresponds to an ancestral one, since it results in the appearance of a reptilian pterygoquadrate element. Moreover, the presence of both a quadrate remnant and an incus among the duplicated elements suggests that the mouse GPP corresponds to that of the therapsid phase of mammalian evolution (see above). Subsequently, the present day mammalian first arch skeletal pattern has been generated from the therapsid pattern by a process involving a *Hox* gene independent genetic system, resulting in the disappearance of the quadrate. We now have evidence that this process involves the expression of retinoic acid responsive genes (Lohnes et al, 1994).

That the first two mammalian arches are embodied with a GPP is not completely unexpected, since they are derived from the branchial basket of agnathan ancestors [lamprey-like, jawless vertebrates lying in the phylogenetic hierarchy between the "headless" protochordates and the gnathostome (jawed) vertebrates] that was formed by a series of identical cartilages supporting the gills. Agnathans arches may in fact correspond to the primordial NCC skeletal ground pattern, which was subsequently modified during gnasthostome radiation to yield the present day mouse skeletal ground pattern (Langille and Hall, 1989 and references therein). It is tempting to speculate that the *Hoxa-2* gene might have been also expressed in the agnathans rostral most gill bearing arch (the homologue of the first arch of gnathostomes; de Beer, 1985; Langille and Hall, 1989 and references therein), and that the retreat of *Hoxa-2* expression from NCC emigrating into the first arch (leaving behind a vestigial expression domain in R2) might have been a prerequisite for subsequent modification of the primordial GPP towards a jaw morphogenetic program. In the same context we note that ectopic expression of a single *Hox* gene (*Hoxd-4*) (Lufkin et al., 1992) in the occipital somites is sufficient to transform the occipital bones into vertebrae. Since the agnathans had occipital vertebrae instead of occipital bones, a primordial vertebral GPP might have been modified during the transition from agnathans to gnathostomes to yield the occipital bones.

ACKNOWLEDGEMENTS

We would like to thank Drs. A. Dierich and M. LeMeur for their collaboration; Dr. S. Ward for proof reading the manuscript; B. Weber, C. Fisher and V. Giroult for technical help; C. Werlé for help with illustrations and the secretarial staff for typing the manuscript. This work was supported by the Institut National de la Santé et de la Recherche Médicale, the Centre National de la Recherche Scientifique, the Centre Hospitalier Universitaire Régional, the Assocation pour la Recherche sur le Cancer and the Fondation pour la Recherche Médicale.

REFERENCES

Allin EF (1975): Evolution of the mammalian middle ear. J Morph 147:403-438.

Botas J (1993): Control of morphogenesis and differentiation by HOM/*Hox* genes. Curr. Opinion Cell Biol 5: 1015-1022.

Carpenter EM, Goddard JM, Chisaka O, Manley and Cappecchi R (1993): Loss of *Hox-A1* (*Hox-1.6*) fuction results in the reorganization of the murine hindbrain. Development 118: 1063-1075.

Chisaka O and Capecchi MR (1991): Regionally restricted developmental defects resulting from targeted disruption of the mouse homeobox gene *Hox-1.5*. Nature 350: 473-479.

Chisaka O, Musci TS and Capecchi MR (1992): Developmental defects of the ear, cranial nerves and hindbrain resulting from targeted disruption of the mouse homeobox gene *Hox-1.6*. Nature 355: 516-520.

Clarke JDW and Lumsden A (1993): Segmental repetition of neuronal phenotype sets in the chick embryo hindbrain. Development 118: 151-162.

Condie BG and Capecchi MR (1993): Mice homozygous for a targeted disruption of *Hoxd-3* (*Hox-4.1*) exhibit anterior transformations of the first and second cervical vertebrae, the atlas and the axis. Development 119: 579-595.

de Beer G (1985): "The Development of the Vertebrate Skull". The University of Chicago Press, Chicago.

Dollé P, Lufkin T, Krumlauf R, Mark M, Duboule D and Chambon P (1993): Local alterations of *Krox-20* and *Hox* gene expression in the hindbrain suggest lack of rhombomeres 4 and 5 in homozygote null *Hoxa-1* (*Hox1-6*) mutant embryos. Proc Natl Acad Sci USA 60: 7666-7670.

Figdor MC and Stern CD (1993): Segmental organization of embryonic diencephalon. Nature 363: 630-634.

Fraser S, Keynes R and Lumsden A (1990): Segmentation in the chick embryo hindbrain is defined by cell lineage restrictions. Nature 344: 431- 435.

Frohman MA, Martin GR, Cordes SP, Hamalek LP and Barsh,GS (1993): Altered rhombomere specific gene expression and hyoid bone differentiation in the mouse segmentation mutant, *kreisler (kr)*. Development 117: 925-936.

Gans C and Northcutt RG (1983): Neural crest and the origin of vertebrates: a new head. Science 220: 268-274.

Gaul U and Jäckle H (1987): How to fill a gap in the *Drosophila* embryo. Trends Genet 3: 127-131.

Gendron-Maguire M, Mallo M, Zhang M and Gridley T (1993): *Hoxa-2* mutant mice exhibit homeotic transformation of skeletal elements derived from cranial neural crest. Cell 75: 1317-1331.

Graham A, Heyman I and Lumsden A (1993): Even numbered rhombomeres control the apoptotic elimination of neural crest cells from odd numbered rhombomeres in the chick hindbrain. Development 119: 233-245.

Guthrie S and Lumsden A (1991): Formation and regeneration of rhombomere boundaries in the developing chick hindbrain. Development 112: 221-230.

Guthrie S and Lumsden A (1992): Motor neuron pathfinding following rhombomere reversals in the chick embryo hindbrain. Development 114: 663-673.

Guthrie S, Muchamore I, Kuroiwa A, Marshall H, Krumlauf R and Lumsden A (1992): Neuroectodermal autonomy of *Hox-2.9* expression revealed by rhombomere transpositions. Nature 356: 157-159

Guthrie S, Prince V and Lumsden A (1993): Selective dispersal of avian rhombomere cells in orthotopic and heterotopic grafts. Development 118: 527-538.

Hall BK (1987): Earliest evidence of cartilage and bone development in embryonic life. Clin Orthop Rel Res 225: 255-272.

Holland P (1990): Homeobox genes and segmentation: co-option, co-evolution, and convergence. Seminars in Dev Biol: 135-145.

Holland P (1992): Homeobox genes in vertebrate evolution. BioEssays 14: 267-273.

Hunt P and Krumlauf R (1992): *Hox* codes and positional specification in vertebrate embryonic axes. Annu Rev Cell Biol 8: 227-256.

Kappen C, Schughart K and Ruddle FH (1993): Early evolutionary origin of major homeodomain sequences classes. Genomics 18: 54-70.

Keynes R and Lumsden A (1990): Segmentation and the origin of regional diversity in the vertebrate central nervous system. Neuron 2: 1-9.

Krumlauf R (1993): *Hox* genes and pattern formation in the branchial region of the vertebrate head. Trends Genet 9: 106-112.

Kuratani SC and Eichele G (1993): Rhombomere transplantation repatterns the segmental organization of cranial nerves and reveals cell autonomous expression of a homeodomain protein. Development 117: 105-117.

Langille RM and Hall BK (1989): Developmental processes, developmental sequences and early vertebrate phylogeny. Biol Rev 64: 73-91.

Lawrence PA (1992): "The Making Of A Fly. The Genetics Of Animal Design," Blackwell Scientific Publications, Oxford.

Le Douarin N, Ziller C and Couly G (1993): Patterns of neural crest derivatives in the avian embryos: In vivo and in vitro studies. Dev Biol 159: 24-49.

Lohnes D, Mark M, Mendelsohn C, Dollé P, Dierich A, Gorry P, Gansmuller A and Chambon P (1994): Function of the retinoic acid receptors (RARs) during development. I. Craniofacial and skeletal abnormalities in RAR double mutants. Development 120: 2723-2748.

Lufkin T, Dierich A, LeMeur M, Mark M and Chambon P (1991): Disruption of the *Hox-1.6* homeobox gene results in defects in a region corresponding to its rostral domain of expression. Cell 66: 1105-1119.

Lufkin T, Mark M, Hart CP, Dollé P, LeMeur M and Chambon P (1992): Homeotic transformation of the occipital bones of the skull by ectopic expression of a homeobox gene. Nature 359: 835-841.

Lumsden AGS (1988): Spatial organization of the epithelium and the role of neural crest cells in the initiation of the mammalian tooth germ. Development 103: 155-169.

Lumsden A (1990): The cellular basis of segmentation in the developing hindbrain. Trends Neurosci 13: 329-335.

Lumsden A and Keynes R (1989): Segmental patterns of neuronal development in the chick brain. Nature 337: 424-428.

Lumsden A, Sprawson N and Graham A (1991): Segmental origin and migration of neural crest cells in the hindbrain region of the chick embryo. Development 113: 1281-1291.

Mansour SL, Goddard JM and Capecchi MR (1993): Mice homozygous for a targeted disruption of the proto oncogene *int-2* have developmental defects in the tail and inner ear. Development 117: 13-28.

Mark M, Lufkin T, Vonesch JL, Ruberte E, Olivo JC, Dollé P, Gorry P, Lumsden A and Chambon P (1993): Two rhombomeres are altered in *Hoxa-1* mutant mice. Development 119: 319-338.

McGinnis W and Krumlauf R (1992): Homeobox genes and axial patterning. Cell 68: 283-302.

Noden DM (1978): The control of avian cephalic neural crest cytodifferentiation.I. Skeletal and connective tissues. Dev Biol 67: 296-312.

Noden DM (1983): The role of the neural crest in patterning of avian cranial skeletal, connective, and muscle tissues. Dev Biol 96: 144-165.

Noden DM (1988): Interactions and fates of avian craniofacial mesenchyme. Development 103(Suppl.): 121-140.

Olson EC (1959): The evolution of mammalian characters. Evolution 13: 344-353.

Pankrats MJ and Jäckle H (1990): Making stripes in the *Drosophila* embryo. Trends Genet 6: 287-292.

Presley R (1989): Ontogeny and the evolution of the mammalian jaw complex. In Wake DB and Roth G (eds): "Complex Organismal Functions: Integration and Evolution in Vertebrates," John Wiley and Sons, pp 53-61.

Prince V and Lumsden A (1994): *Hoxa-2* expression in normal and transposed rhombomeres:independent regulation in the neural tube and neural crest. Development 120: 911-923.

Puelles L and Rubenstein JLR (1993): Expression patterns of homeobox and other putative regulatory genes in the embryonic mouse forebrain suggest a neuromeric organization. Trends Neurosci 16: 472-478.

Rijli PM, Mark M, Lakkaraju S, Dierich A, Dollé P and Chambon P (1993): A homeotic transformation is generated in the rostral branchial region of the head by disruption of *Hoxa-2*, which acts as a selector gene. Cell 75: 1333-1349.

Schneider Maunoury S, Topilko P, Seitanidou T, Levi G, Cohen Tannoudji M, Pournin S, Babinet C and Charnay P (1993): Disruption of *Krox-20* results in alteration of rhombomeres 3 and 5 in the developing hindbrain. Cell 75: 1199-1214.

Sechrist J, Serbedzija GN, Scherson T, Fraser SE and Bronner-Fraser M (1993): Segmental migration of the hindbrain neural crest does not arise from its segmental generation. Development 118: 691-703.

Serbedzija GN, Bronner Fraser M and Fraser SE (1992): Vital dye analysis of cranial neural crest cell migration in the mouse embryo. Development 116: 297-307.

Simon H and Lumsden A (1993): Rhombomere specific origin of contralateral vestibulo acoustic efferent neurons and their migration across the embryonic midline. Neuron 11: 209-220.

Slack JMW, Holland PWH, and Graham CF (1993): The zootype and the phylotypic stage. Nature 361: 490-492.

Swiatek PJ and Gridley T (1993): Perinatal lethality and defects in hindbrain development in mice homozygous for a targeted mutation of the zinc finger gene *Krox-20*. Genes Dev 7: 2071-2084.

Thorogood P (1993): The problems of building a head. Current Biol 3: 705-707.

Wilkinson DG (1993): Molecular mechanisms of segmental patterning in the vertebrate hindbrain and neural crest. BioEssays 15: 499-505.

Wright CVE (1993): *Hox* gene in the hindbrain. Current Biol 3: 618-621.

GENETIC MECHANISMS RESPONSIBLE FOR PATTERN FORMATION IN THE VERTEBRATE HINDBRAIN: REGULATION OF Hoxb-1

Michèle Studer[1], Heather Marshall[1], Heike Pöpperl[1], Atsushi Kuroiwa[2] and Robb Krumlauf[1]

[1]Laboratory of Developmental Neurobiology, National Institute for Medical Research, The Ridgeway, Mill Hill, London NW7 1AA, U.K. [2]Department of Molecular Biology, School of Science, Nagoya University, Chihusa-Ku, Nagoya, JAPAN.

INTRODUCTION

During development of the vertebrate nervous system a process of segmentation, that will give rise to the generation of morphologically repeated units called rhombomeres (r), occurs in the hindbrain (reviewed in Lumsden, 1990; Wilkinson and Krumlauf, 1990). The formation of rhombomeres is correlated with the process of neurogenesis involving the reticular formation and the branchial motor system. Each branchial motor nucleus occupies a distinct position in the hindbrain and is derived from neurons in two adjacent rhombomeres. These neurons lie in register with the appropriate branchial arch, in a two-segment repetition pattern (Lumsden and Keynes, 1989). Boundaries between even and odd numbered rhombomeres are formed progressively in an order that does not follow a strict anterior to posterior progression (Vaage, 1969; Lumsden, 1990). To understand more about the establishment and formation of rhombomere boundaries, cell lineage studies have been performed in the chick. Single cell labelling experiments have shown that cell mixing only occurs between neighbouring segments before the boundaries between future odd and even numbered rhombomeres are formed (Wilkinson et al., 1989a; Guthrie and Lumsden, 1991; Guthrie et al., 1993). After boundary formation, rhombomeres become lineage-restricted cellular compartments, where cells are committed to a specific segment, hence each segment can maintain a distinct regional identity (Fraser et al., 1990; Birgbauer and Fraser, 1994).

Evidence for a class of genes involved in segmentation and specification of segment identity during embryogenesis comes from the work in *Drosophila*, where mutations in segmentation and homeotic genes (*HOM-C*) were shown to affect the

Neural Cell Specification: Molecular Mechanisms and Neurotherapeutic Implications
Edited by Juurlink *et al.*, Plenum Press, New York, 1995

17

fate of segments (Lewis, 1978; Akam, 1987; Ingham, 1988). A common motif between *Drosophila* and vertebrate, the homeobox, has been identified and isolated from the genome of many species, including that of the human (reviewed in Scott et al., 1989; McGinnis and Krumlauf, 1992). It is now generally accepted that the *Hox* homeobox-containing genes represent a network involved in regulating pattern formation during development. In vertebrates there are four different gene clusters, *HoxA* to *D*, located on different chromosomes (Scott, 1992, for nomenclature). These vertebrate complexes are highly conserved and have arisen by duplication and divergence from a common ancestor, so that genes in each of the four clusters are related to specific members in the other clusters, referred to 'paralogous genes' (Duboule and Dollé, 1989; Graham et al. 1989; Boncinelli et al. 1991; Krumlauf, 1992). Extensive homology between the different clusters and the different species have suggested that location of important regions involved in the regulation of the genes might be conserved during evolution. A conservation on the basis of regulation is also suggested by a common property of *Hox/HOM-C* genes, 'colinearity', which is a spatial and temporal correlation between the order of the genes in the cluster and their expression patterns along the anterior-posterior (A-P) axis of the embryo (Duboule and Dollé, 1989; Graham et al., 1989). The most 3' genes in a cluster have more anterior limits of expression, are expressed earlier (Izpisua-Belmonte et al., 1991) and respond to lower doses of retinoic acid (RA) than their 5' neighbours (Simeone et al., 1990, 1991; Papalopulu et al., 1991; Dekker et al., 1993). A recent example arguing for conservation of regulatory elements is shown by the *Drosophila* gene *Deformed* and its mouse homologue (Awgulewitsch and Jacobs, 1992; Malicki et al., 1992).

In the hindbrain, in situ analysis has shown that *Hox* gene expression is established at early neural plate stages prior to the formation of rhombomere boundaries. At 9.5 days post coitum (dpc), at a stage when rhombomeres are well formed, the anterior expression border of the most 3' *Hox* genes maps to specific rhombomere boundaries (Wilkinson et al., 1989b; Hunt et al., 1991). The expression data, together with mutational analysis of *Hoxa-1* (Lufkin et al., 1991; Chisaka et al., 1992; Carpenter et al., 1993; Dollé et al., 1993; Mark et al., 1993) which has shown a disruption in the hindbrain organization, suggest a segmental patterning role of *Hox* genes in the hindbrain. Moreover grafting experiments in chick embryos have linked cell autonomous *Hox* expression to rhombomere formation (Guthrie et al., 1992; Kuratani and Eichele, 1993).

It has been found that two members of the labial group, *Hoxa-1* and *Hoxb-1*, already have a relatively sharp boundary in the neural plate, in the region which will correspond to the future r3/4 boundary (Sundin and Eichele, 1990; 1992, Muchamore unpublished) at a stage where cell lineage tracing experiments show a high degree of cell mixing. This could suggest that sharp boundaries of *Hox* gene expression may be independent of cell sorting and instead maintained by other processes like cell death or cell respecification. Cell-cell interactions by a signalling pathway and/or downregulation of specific genes could be part of the mechanisms involved in change of cell fate. There is little information which addresses the cellular and molecular mechanisms involved in this process.

Hoxb-1 EXPRESSION IS PROGRESSIVELY RESTRICTED TO R4

Hoxb-1 is the first member of the *Hox-B* complex to be expressed in the embryo (Murphy et al., 1989; Wilkinson et al., 1989b; Frohman et al., 1990; Murphy and Hill, 1991) and displays the strongest response to RA in cell culture Simeone et al., 1990; Papalopulu et al., 1991; Dekker et al., 1993) and embryos (Morriss-Kay et al., 1991; Conlon and Rossant, 1992; Marshall et al., 1992). The early phase of *Hoxb-1* expression is first observed at 7.0-7.25 dpc in mesoderm, then in the overlying ectoderm with a posterior to anterior progression (Murphy et al., 1989; Wilkinson et al., 1989b, Frohman et al., 1990). From 7.5 dpc to 8.0 dpc expression extends from the most posterior regions of the primitive streak to a sharp boundary at the presumptive r3/4 boundary. This early phase is similar to the one observed in the paralogous gene, *Hoxa-1* (Hunt et al., 1991; Murphy and Hill, 1991). In both cases this early expression regresses posteriorly until it gradually disappears. However, in the case of *Hoxb-1* the most anterior domain of expression persists and is restricted to r4 in the hindbrain by 9.5 dpc.

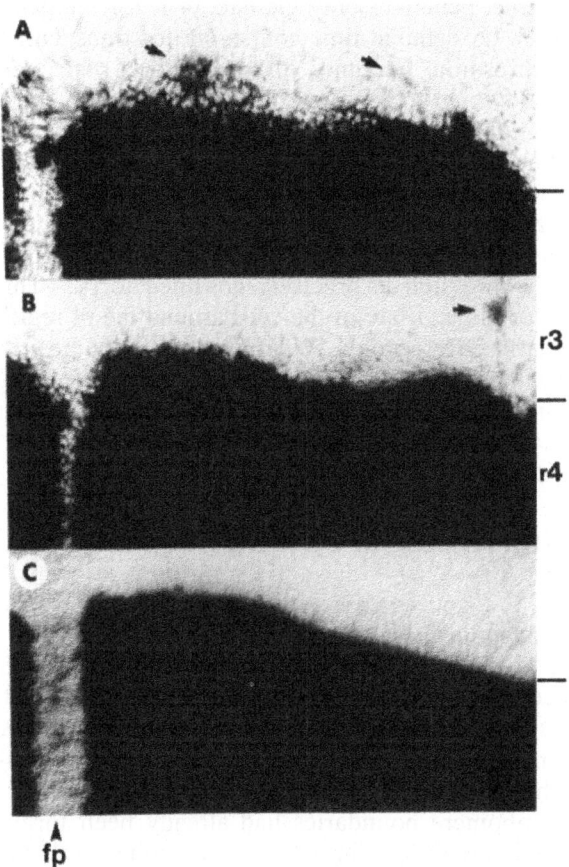

Figure 1. Flat mounts of (A) 8.5 dpc, (B) 9.5 dpc and (C) 10.5 dpc transgenic embryos showing how lacZ staining cells are progressively restricted to r4.

Reporter gene constructs in transgenic mice have been useful in reproducing the endogenous expression pattern of some *Hox* genes. This has been important in defining regulatory elements involved in spatially restricted domains of *Hox* expression (Boncinelli et al., 1988; Puschel et al., 1990; Whiting et al., 1991; Marshall et al., 1992; Sham et al., 1992) and identifying upstream factors responsible for their regulation (Marshall et al., 1992; Sham et al., 1993). An example is the segmental expression of *Hoxb-2* in r3 and r5, which has been found to be directly regulated by *Krox 20*, a zinc finger gene expressed in r3/5 restricted domains (Schneider-Maunoury et al., 1993; Sham et al., 1993; Swiatek and Gridley, 1993). Cells expressing lacZ can be easily followed in transgenic lines because of the high cell resolution and which can be a powerful tool in helping to understand the dynamic process of expression of *Hox* genes.

Because of its highly restricted expression in r4, *Hoxb-1* provides a good opportunity for studying the molecular and cellular mechanisms involved in segmental patterning and the processes that are related to the cellular formation and sharpening of rhombomeric boundaries. The early and late endogenous expression patterns of *Hoxb-1* including its ability to respond to RA have been reconstructed in transgenic mice (Marshall et al., 1992). An internal subfragment of this transgenic construct generates only the late, r4-restricted pattern of expression (Guthrie et al., 1992). A detailed time course on this transgenic line showed that the pattern of expression becomes progressively restricted to r4 during development and allows us to monitor the changes in the establishment of the expression pattern in detail. Initially the domain of expression in presumptive r4 has no clearly defined cellular boundaries and patches of lacZ staining cells are also present in presumptive r3 (Fig. 1A indicated by arrows) and r5. About half a day later this fuzzy pattern becomes progressively sharpened (Fig. 1B) until distinct boundaries of expression between r3/4 (Fig. 1C) and r4/5 are generated, and the domain of expression is precisely confined to r4. This process raised questions about the origin and fate of the cells around the r4 region at a time when rhombomere boundaries are forming. What happens to the positive cells in r3 that express the transgene early and not later? Do they die, migrate into r4 or downregulate their expression? These are fundamental questions concerning rhombomere formation in the mouse, therefore we examined the underlying molecular basis of this process.

TWO COMPONENTS ARE NECESSARY TO ACHIEVE R4 RESTRICTION

Deletion analysis in the 5' flanking region of the *Hoxb-1* gene mapped a fragment which functions as an enhancer on a heterologous promoter/lacZ reporter gene and reproduces the r4-restricted late pattern. Further deletions produced a domain of expression in the region of r4 and associated neural crest, but this pattern turned out not to be limited to r4 (Fig. 2B). A subpopulation of cells in r3 and r4, closest to rhombomere 4 boundaries, expressed lacZ markers at a time (9.0-9.5 dpc) when rhombomere boundaries had already been formed and when the endogenous gene was strictly confined to r4. We believed that cells in r3 and r5 were expressing *Hoxb-1* in an ectopic location because a negative control region required for restricting expression to r4, located upstream of the r4 enhancer was missing. We performed experiments by adding back flanking regions to map this repressor domain, and found a region which was able to restore proper r4-restricted expression (Fig. 2A). This negative region was termed 'r3/5 repressor', and on its own was unable to direct any segment-specific expression from a lacZ reporter. Hence its role appears to be to work cooperatively with the r4 enhancer to restrict *Hoxb-1* expression. Interestingly, as indicated by the arrows in Fig. 2B, the absence of the repressor region (-Rep.) did not allow all cells in r3 and r5 to express

Hoxb-1/lacZ transgene, suggesting that lacZ staining cells could be of r4 origin. Moreover the repressor was specifically able to block expression in r3 and r5, when *Hoxb-1*/lacZ was under the control of an r3/r5 enhancer region from the *Hoxb-2* gene [Fig. 3A] (Sham et al., 1992, 1993). Removal of the repressor (-Rep.) allowed the transgene to be expressed in r3/5 domains (Fig. 3B). These complementary experiments showed that a fragment present in the 5' flanking region of the *Hoxb-1* gene plays a negative role in blocking expression in r3 and r5 and as a consequence restricts *Hoxb-1* expression strictly to r4 (as summarized in Fig. 5B) with sharp boundaries at the r3/4 and r4/5 borders.

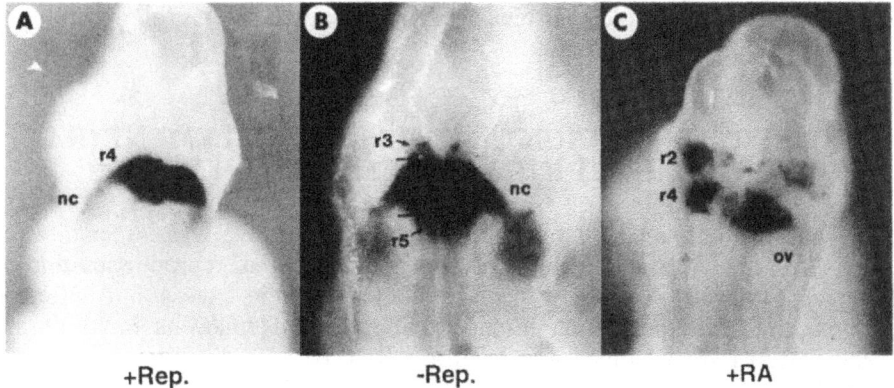

Figure 2. Dorsal views of 9.5 dpc transgenic embryos in the presence (A) and in the absence (B) of the r3/5 repressor region. In (B) lac Z staining cells spread into the adjacent rhombomeres. An ectopic r3 stripe appears in (C) after administration of RA.

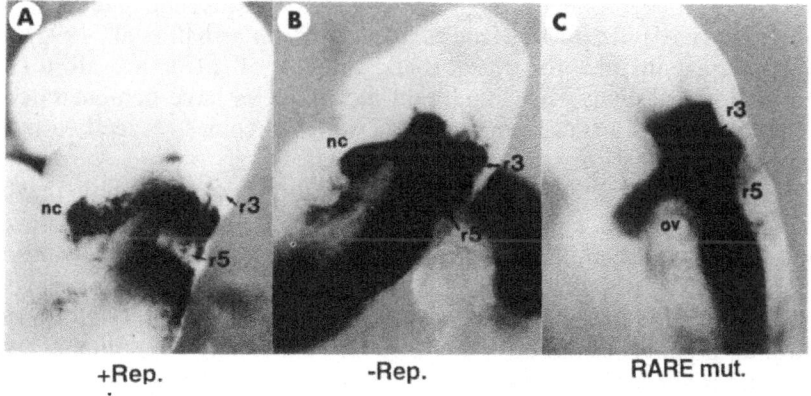

Figure 3. Transgenic 9.5 dpc embryos in which the r4 enhancer is expressed in r3 and r5 in the presence of the r3/5 repressor (A), in the absence of the r3/5 repressor (B) and in the presence of the r3/5 repressor which contains point mutations in the RARE.

By analogy to what had been found in *Drosophila* embryos, where stripes of gene expression are generated by a mechanism of gene activation and repression (Ip et al., 1992; Jackle et al., 1992), *Hoxb-1* expression in r4 requires the combination of a positive element, which generates a broad domain of expression, and a negative element, which confines the expression to r4. Negative regulation, such as that found in *Hoxb-1*, could play an important role in maintaining rhombomere identities during boundary formation. As mentioned above little is known about how *Hox* expression is linked to boundary formation. However the absence of a negative region in the 5' flanking region of *Hoxb-1* allows the *Hoxb-1*/lacZ transgene to be expressed in a subpopulation of r3/5 cells at a stage in rhombomere formation when the endogenous gene is not expressed in r3 and r5. Endogenous *Hoxb-1* is not expressed in the adjacent rhombomeres presumably because it has been downregulated in r3/5 in order to permit the cells with an r4 identity to be respecified to an r3/5 identity. We think that the repressor region could be one of the molecular mechanisms that play a crucial role in rhombomere-restricted expression of *Hoxb-1*.

A CONSERVED RETINOIC ACID RESPONSE ELEMENT (RARE) IS REQUIRED TO RESTRICT *Hoxb-1* EXPRESSION TO R4

It has long been established that vitamin A derivatives, especially retinoic acid (RA), play a crucial role in normal growth, vision, reproduction and overall survival. Moreover the teratogenic effects of retinoid excess and deficiency on developing embryos, which include dramatic malformations in hindbrain and craniofacial regions, limbs and axial skeleton, have markedly contributed to the belief that RA could act as a morphogen by conferring positional information during development (Lohnes et al., 1993). Ectopic doses of RA have shown to alter *Hox* gene expression in cell culture and subsequently in developing embryos of different species. In the hindbrain phenotypic changes suggest that RA is altering hindbrain organisation by altering *Hox* gene expression (Morriss-Kay et al., 1991; Conlon and Rossant, 1992; Marshall et al., 1992; Kessel, 1993). In RA treated mouse embryos *Hoxb-1* is expressed in more anterior regions, which at slightly later stages resolve into two stripes in r2 and r4 (Marshall et al., 1992). This process is reproduced in the *Hoxb-1*/lacZ transgene generating the late expression pattern of *Hoxb-1* (Guthrie et al., 1992; Marshall et al., 1992) and by the enhancer responsible for the broad domain of r4 expression (Fig. 2C). Neuroanatomical analysis of hindbrain motor nuclei have demonstrated that RA treatment produces a transformation of r2 which becomes respecified to an r4 fate (Marshall et al., 1992).

However despite the large body of information regarding the biological action of RA, little is known about its endogenous function during development. The discovery of a family of nuclear retinoic acid receptors (RAR) and retinoid X receptors (RXR), which act as ligand-activated transcription factors, have provided the first indication that RA might act at the molecular level through transcriptional control of target genes (Kliewer et al., 1992; Leid et al., 1992; Stunnenberg, 1993). RARs and RXRs have been shown to bind in vitro as heterodimers to retinoic acid response elements of RA regulated-target genes. RAREs consist of two directly repeated core motifs, as shown in Fig. 4A, separated by two or five base pairs [DR2 or DR5] (Kliewer et al., 1992; Leid et al., 1992; Stunnenberg, 1993). Interestingly RAR and RXR transcripts show overlapping, but distinct spatio-temporal expression patterns in the developing embryo and in various adult mouse tissues (Morriss-Kay, 1993). Certain receptors also have a segmental expression, which may suggest a role in *Hox* gene regulation.

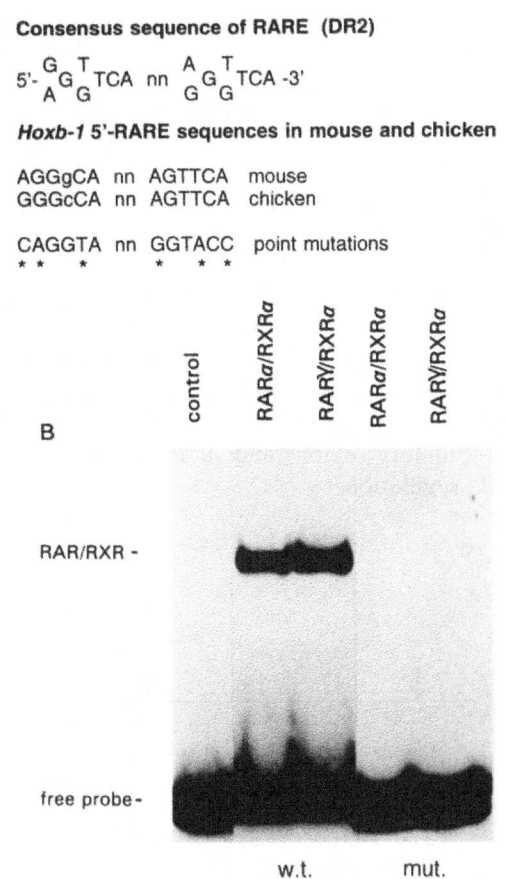

A   Consensus sequence of RARE (DR2)

$$5'- \begin{smallmatrix}G\\A\end{smallmatrix}\begin{smallmatrix}T\\G\end{smallmatrix}TCA \ nn \ \begin{smallmatrix}A\\G\end{smallmatrix}\begin{smallmatrix}T\\G\end{smallmatrix}TCA -3'$$

***Hoxb-1* 5'-RARE sequences in mouse and chicken**

AGGgCA nn AGTTCA   mouse
GGGcCA nn AGTTCA   chicken

CAGGTA nn GGTACC   point mutations
* *   *   *   * *

Figure 4. (A) A comparison at the sequence level of the consensus sequence of an RARE, as described in the literature, and the 5'-RAREs found in *Hoxb-1*. (B) An electrophoretic mobility shift assay shows that RAR/RXR heterodimers bind to the w.t. RARE but not to the mutated version.

As mentioned above the r4 enhancer of the *Hoxb-1* gene responds to RA by producing an ectopic stripe in r2 (Fig. 2C). In order to identify motifs responsible for the RA response of *Hoxb-1* we used an evolutionary approach and compared sequences with similar regulatory function of different species. In the chicken, *Hoxb-1* shows a similar pattern of expression as the mouse, which suggests that regulatory regions were conserved during vertebrate evolution. Indeed sequence comparison of the 5' flanking regions of the two species have identified conserved sequences that, when tested in transgenic mice, were able to reproduce the expected r4 domain of expression and the response to RA (Studer et al., 1994). Surprisingly no RARE could be found in the r4 enhancers, suggesting that the ectopic RA response might be mediated indirectly through another factor.

However a conserved RARE of the DR2 type had been found in the repressor region of the two species. The two repeats shown in Fig. 4A differ slightly in the consensus sequence of their first motif but were able to bind heterodimers of RAR/RXR in vitro (Fig. 4B). Point mutations in each half of the direct repeat eliminated not only in vitro binding, but also the repressor's biological function (Fig. 3C and Studer et al., 1994) in vivo, by releasing the repression in r3 and r5. This demonstrated that the *Hoxb-1* 5' RARE is an essential component involved in the repressor activity and that it is required during normal development in the restriction of *Hoxb-1* expression to r4 (summarized in Fig. 5B).

RA PLAYS MULTIPLE ROLES IN *Hoxb-1* EXPRESSION DURING DEVELOPMENT

As shown in Fig. 5A, during mouse development between headfold stage and early neurulation, *Hoxb-1* presents two distinct phases of expression and three different responses to retinoids and/or RA. Using reporter constructs in transgenic mice, we were able to reconstruct the whole pattern of expression of *Hoxb-1*, and separate the different phases of expression and different responses to RA by identifying regulatory regions and short motifs in the gene responsible for specific aspects of its regulation.

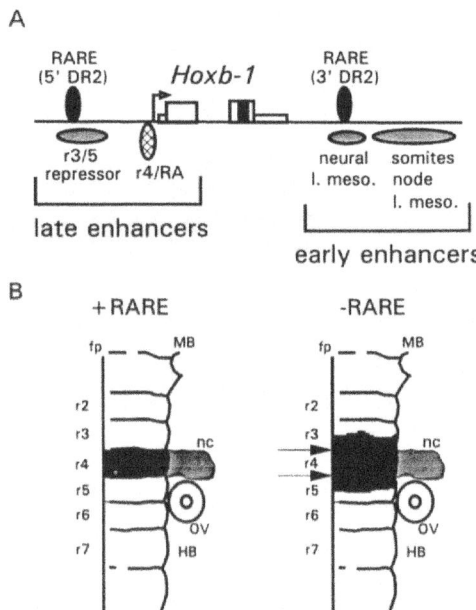

Figure 5. (A) summarizes all the regulatory regions including their RARE motifs found in the *Hoxb-1* gene important in regulating the early and late expression pattern. (B) The presence of a full functional RARE is necessary in the restriction of *Hoxb-1* to r4.

Recently two regulatory regions located downstream of the gene were identified and found to reproduce the early neural and mesoderm expression

(Marshall et al., 1994). The element responsible for the neuroectoderm and part of the lateral mesoderm expression responded to RA by shifting its expression in more anterior domains. Sequence analysis of the fragment identified an RARE of the DR2 type that in analogy with the one found in the repressor region is able to bind heterodimers of RAR/RXR in vitro. Point mutations in the two repeats or complete deletion of the whole motif abolished the full early neuroectoderm and part of the lateral mesoderm expression without influencing r4 expression. This demonstrates that the 3' RARE is not only necessary for the RA response but is essential in establishing a subset of the normal early expression of *Hoxb-1* during embryogenesis. Our data have shown that retinoids through the RAREs are directly required for the early establishment of *Hoxb-1* and its progressive restriction to r4. Indirectly RA also shifts the patterns of expression by mechanisms mediated through the r4 enhancer. We believe that the two types of regulation mediated through the RARE elements are important components of the normal control of the *Hoxb-1* gene because they have been found in more than one species. However we still do not understand the mechanism and the function of the indirect RA response in the r4 enhancer.

The 3' RARE has a positive role in establishing the early neuroectoderm expression, while the 5' RARE plays a negative role in refining the late r4 restricted expression. This raises the question of how specificity can be achieved during development. We think that specificity is achieved by the presence of other embryonic factors that are localized in defined regions and present only at certain time frames during development. Such factors might cooperate directly with the RAREs, which would play a positive or negative role depending on their characteristics. If this is a general mechanism of *Hox* regulation, we would expect to find other RAREs dispersed in the whole *HoxB* cluster as well as in other *Hox* clusters which are important for in vivo regulation. In fact, it has been already reported a DR5 RARE in the 3' region of *Hoxa-1*, a labial paralogue of *Hoxb-1*, which is necessary for RA response in cell culture (Langston and Gudas, 1992).

In conclusion these experiments have begun to dissect the complexity that establishes and maintains the expression of *Hoxb-1*. Multiple RAREs are clearly essential components for generating the *Hox* domains, and it will be interesting to determine what kind of factors interact with the RAREs. These data also confirm that there is a link between *Hox* regulation and rhombomere formation and provide support for the idea that RA has a normal role in regulating *Hox* genes during hindbrain regionalisation.

ACKNOWLEDGEMENTS

We thank Professor Chambon for the RAR and RXR proteins. AK, RK and HM were supported in part by an HFSP collaborative grant, MS by fellowships from SNF and EMBO and HP by an EMBO fellowship.

REFERENCES

Akam M (1987): The molecular basis for metameric pattern in the *Drosophila* embryo. Development 101: 1-22.
Awgulewitsch A and Jacobs D (1992): Deformed autoregulatory element from *Drosophila* functions in a conserved manner in transgenic mice. Nature 358: 341-345.

Birgbauer E and Fraser SE (1984): Violation of cell lineage restriction compartments in the chick hindbrain. Development 120: 1347-1356.

Boncinelli E, Somma R, Acampora D, Pannese M, D'Esposito M, Faiella A and Simeone A (1988): Organization of human homeobox genes. Hum Reprod 3: 880-886.

Boncinelli E, Simeone A, Acampora D and Mavilio F (1991): HOX gene activation by retinoic acid. TIG 7: 329-334.

Carpenter E, Goddard J, Chisaka O, Manley N and Capecchi M (1993): Loss of Hox-a1 (Hox-1.6) function results in reorganisation of the murine hindbrain. Development 118: 1063-1075.

Chisaka O, Musci T and Capecchi M (1992): Developmental defects of the ear, cranial nerves and hindbrain resulting from targeted disruption of the mouse homeobox gene Hox-1.6. Nature 355: 516-520.

Conlon R and Rossant J (1992): Exogenous retinoic acid rapidly induces anterior ectopic expression of murine Hox-2 genes in vivo. Development 116: 357-368.

Dekker E-J, Pannese M, Houtzager E, Boncinelli E and Durston A (1993): Colinearity in the Xenopus laevis Hox-2 complex. Mech Development 40: 3-12.

Dollé P, Lufkin T, Krumlauf R, Mark M, Duboule D and Chambon P (1993): Local alterations of Krox-20 and Hox gene expression in the hindbrain of Hox-a1 (Hox-1.6) homozygote null mutant embryos. Proc Natl Acad Sci USA 90: 7666-7670.

Duboule D and Dollé P (1989): The structural and functinal organization of the murine HOX gene family resembles that of *Drosophila* homeotic genes. EMBO J 8: 1497-1505.

Fraser S, Keynes R and Lumsden A (1990): Segmentation in the chick embryo hindbrain is defined by cell lineage restrictions. Nature 344: 431-435.

Frohman M, Boyle M and Martin G (1990): Isolation of the mouse Hox-2.9 gene; analysis of embryonic expression suggests that positional information along the anterior-posterior axis is specified by mesoderm. Development 110: 589-607.

Graham A, Papalopulu N and Krumlauf R (1989): The murine and *Drosophila* homeobox clusters have common features of organisation and expression. Cell 57: 367-378.

Guthrie S and Lumsden A (1991): Formation and regeneration of rhombomere boundaries in the developing chick hindbrain. Development 112: 221-229.

Guthrie S, Muchamore I, Marshall H, Kuroiwa A, Krumlauf R and Lumsden A (1992): Neuroectodermal autonomy of Hox-2.9 expression revealed by rhombomere transpositions. Nature 356: 157-159.

Guthrie S, Prince V and Lumsden A (1993): Selective dispersal of avian rhombomere cells in orthotopic and heterotopic grafts. Development 118: 527-538.

Hunt P, Gulisano M, Cook M, Sham MH, Faiella A, Wilkinson D, Boncinelli E and Krumlauf R (1991): A distinct Hox code for the branchial region of the head. Nature 353: 861-864.

Ingham P (1988): The molecular genetics of embryonic pattern formation in *Drosophila*. Nature 335: 25-34.

Ip YT, Levine M and Small SJ (1992): The bicoid and dorsal morphogens use a similar strategy to make stripes in the *Drosophila* embryo. J Cell Sci 16(Suppl): 33-38.

Izpisua-Belmonte J, Falkenstein H, Dolle P, Renucci A and Duboule D (1991): Murine genes related to the *Drosophila* AbdB homeotic gene are sequentially expressed during development of the posterior part of the body. Embo J 10: 2279-2289.

Jackle H, Hoch M, Pankratz MJ, Gerwin N, Sauer F and Bronner G (1992): Transcriptional control by *Drosophila* gap genes. J Cell Sci 16(Suppl): 39-51.

Kessel M (1993): Reversal of axonal pathways from rhombomere 3 correlates with extra Hox expression domains. Neuron 10: 379-393.

Kliewer SA, Umesono K, Mangelsdorf DJ and Evans RM (1992): Retinoid X receptor interacts with nuclear receptors in retinoic acid, thyroid hormone and vitamin D3 signalling. Nature 355: 446-449.

Krumlauf R (1992): Evolution of the vertebrate Hox homeobox genes. Bioessays 14: 245-252.

Kuratani SC and Eichele G (1993): Rhombomere transposition repatterns the segmental organization of cranial nerves and reveals cell-autonomous expression of a homeodomain protein. Development 117: 105-117.

Langston AW and Gudas LJ (1992): Identification of a retinoic acid responsive enhancer 3' of the murine homeobox gene Hox-1.6. Mech Dev 38: 217-228.

Leid M Kastner P and Chambon P (1992): Multiplicity generates diversity in the retinoic acid signalling pathways. Trends Biochem Sci 17: 427-433.

Lewis E (1978): A gene complex controlling segmentation in *Drosophila*. Nature 276: 565-570.

Lohnes D, Kastner P, Dierich A, Mark M, LeMeur M and Chambon P (1993): Function of retinoic acid receptor g in the mouse. Cell 73: 643-658.

Lufkin T, Dierich A, LeMeur M, Mark M and Chambon P (1991): Disruption of the Hox-1.6 homeobox gene results in defects in a region corresponding to its rostral domain of expression. Cell 66: 1105-1119.

Lumsden A (1990): The cellular basis of segmentation in the developing hindbrain. Trends Neurosci 13: 329-335.

Lumsden A and Keynes R (1989): Segmental patterns of neuronal development in the chick hindbrain. Nature 337: 424-428.

Malicki J, Cianetti L, Peschle C and McGinnis W (1992): A human HOX 4B regulatory element provides head-specific expression in *Drosophila* embryos. Nature 358: 345-347.

Mark M, Lufkin T, Vonesch J-L, Ruberte E, Olivo J-C, Dollé P, Gorry P, Lumsden A and Chambon P (1993): Two rhombomeres are altered in Hox-a1 null mutant mice. Development 119: 319-338.

Marshall H, Nonchev S, Sham M-H, Muchamore I, Lumsden A and Krumlauf R (1992): Retinoic acid alters the hindbrain Hox code and induces the transformation of rhombomeres 2/3 into a rhombomere 4/5 identity. Nature 360: 737-741.

Marshall H, Studer M, Pöpperl H, Aparicio S, Kuroiwa A, Brenner S and Krumlauf R (1994): A conserved retinoic acid response element required for early expression of the homeobox gene Hoxb-1. Nature 370: 567-571.

McGinnis W and Krumlauf R (1992): Homeobox genes and axial patterning. Cell 68: 283-302.

Morriss-Kay G (1993): Retinoic acid and craniofacial development: molecules and morphogenesis. Bioessays 15: 9-15.

Morriss-Kay G, Murphy P, Hill R and Davidson D (1991): Effects of retinoic acid on expression of Hox 2.9 and Krox 20 and on morphological segmentation in the hindbrain of mouse embryos. EMBO J 10: 2985-2996.

Murphy P and Hill R (1991): Expression of mouse labial-like homeobox-containing genes, Hox 2.9 and Hox 1.6, during segmentation of the hindbrain. Development 111: 61-74.

Murphy P, Davidson D and Hill R (1989): Segment-specific expression of a homeobox-containing gene in the mouse hindbrain. Nature 341: 156-159.

Papalopulu N, Lovell-Badge R and Krumlauf R (1991): The expression of murine Hox-2 genes is dependent on the differentiation pathway and displays collinear sensitivity to retinoic acid in F9 cells and Xenopus embryos. Nucleic Acid Res 19: 5497-5506.

Puschel A, Balling R and Gruss P (1990): Postion-specific activity of the Hox 1.1 promoter in transgenic mice. Development 108: 435-442.

Schneider-Maunoury S, Topilko P, Seitanidou T, Levi G, Cohen-Tannoudji M, Pournin S, Babinet C and Charnay P (1993): Disruption of Krox-20 results in elimination of rhombomeres 3 and 5 in the developing hindbrain. Cell 75: 1199-1214.

Scott MP (1992): Vertebrate Homeobox Gene Nomenclature. Cell 71: 551-553.

Scott MP, Tamkun JW and Hartzell GW 3rd (1989): The structure and function of the homeodomain. Biochim. Biophys. Acta 989: 25-48.

Sham M-H, Hunt P, Nonchev S, Papalopulu N, Graham A, Boncinelli E and Krumlauf R (1992): Analysis of the murine Hox-2.7 gene: conserved alternative transcripts with differential distributions in the nervous system and the potential for shared regulatory regions. EMBO J 11: 1825-1836.

Sham MH, Vesque C, Nonchev S, Marshall H, Frain M, Das Gupta R, Whiting J, Wilkinson D, Charnay P and Krumlauf R (1993): The zinc finger gene Krox-20 regulates Hox-b2 during hindbrain segmentation. Cell 72: 183-196.

Simeone A, Acampora D, Arcioni L, Andrews PW, Boncinelli E and Mavilio F (1990): Sequential activation of HOX2 homeobox genes by retinoic acid in human embryonal carcinoma cells. Nature 346: 763-766.

Simeone A, Acampora D, Nigro V, Faiella A, D'Esposito M ,Stornaiuolo A, Mavilio F and Boncinelli E (1991): Differential regulation by retinoic acid of the homeobox genes of the four HOX loci in human embryonal carcinoma cells. Mech Develop 33: 215-227.

Studer M, Pöpperl M, Marshall H, Kuroiwa A and Krumlauf R (1994): Role of conserved retinoic response element in rhombomeric restriction of Hoxb-1. Science 265: 1728-1732.

Stunnenberg H (1993): Mechanisms of transactivation by retinoic acid receptors. Bioessays 15: 309-315.

Sundin O and Eichele G (1990): A homeo domain protein reveals the metameric nature of the developing chick hindbrain. Genes Dev 4: 1267-1276.

Sundin O and Eichele G (1992): An early marker of axial pattern in the chick embryo and its respecification by retinoic acid. Development 114: 841-852.

Swiatek PJ and Gridley T (1993): Perinatal lethality and defects in hindbrain development in mice homozygous for a targeted mutation of the zinc finger gene Krox 20. Genes Dev 7: 2071-2084.

Vaage S (1969): The segmentation of the primitive neural tube in chick embryos (Gallus domesticus). Adv Anat Embryol Cell Biol 41: 1-88.

Whiting J, Marshall H, Cook M, Krumlauf R, Rigby P, Stott D and Allemann R (1991): Multiple spatially-specific enhancers are required to reconstruct the pattern of Hox-2.6 gene expression. Genes Dev 5: 2048-2059.

Wilkinson D and Krumlauf R (1990): Molecular approaches to the segmentation of the hindbrain. Trends Neurosci 13: 335-339.

Wilkinson D, Bhatt S, Chavrier P, Bravo R and Charnay P (1989a): Segment-specific expression of a zinc finger gene in the developing nervous system of the mouse. Nature 337: 461-464.

Wilkinson D, Bhatt S, Cook M, Boncinelli E and Krumlauf R (1989b): Segmental expression of Hox 2 homeobox-containing genes in the developing mouse hindbrain. Nature 341: 405-409.

PAX GENES AS PLEIOTROPIC REGULATORS OF EMBRYONIC DEVELOPMENT

Patrick Tremblay[1], Susanne Dietrich, Anastasia Stoykova, Edward T. Stuart and Peter Gruss.

Department of Molecular Cell Biology, Max Planck Institute for Biophysical Chemistry, Am Fassberg, 37077 Göttingen, GERMANY.

[1]Present Address: Department of Neurology, HSE 781, University of California, San Francisco, CA, U.S.A.

INTRODUCTION

Genes involved in the regulation of key cellular pathways have been, through strong evolutionary pressure, well-conserved. As a consequence, genes often harbour well-conserved domains evolved from duplication and recombination events. This process has led to the multiplication and diversification of basic functional units. This principle has been repeatedly confirmed by the isolation of multigene families whose significance transverses the species barrier (Beato, 1989; Bopp et al., 1986; Frigerio et al., 1986; Kessel and Gruss, 1990; Rosenfeld, 1991; Schubert et al., 1993).

DNA binding motifs of transcription factors often present a striking degree of conservation. For instance, the paired box, initially isolated from the genome of *Drosophila*, encodes a 128 amino acid domain shared between various transcription factors: *paired* (*prd*), *gooseberry-proximal* (*gsb-p*), *gooseberry-distal* (*gsb-d*), *Pox-neuro* and *Pox-meso* (Baumgartner et al., 1987; Bopp et al., 1986; Bopp et al., 1989). These genes function in *Drosophila* as developmental control switches; while *prd*, *gsb-d* and *gsb-p* are important for the segmentation process of the fly, *Pox-neuro* and *Pox-meso* appear to be involved in the specification of the neuronal and the mesodermal cell fates. Besides *Drosophila*, genes containing the paired box are present in nematodes, echinoderms (sea urchin) and vertebrates [Xenopus, zebrafish, chick, mouse, human] (Burri et al., 1989). The homology screen approach has been successfully used to isolate paired box containing genes (Pax) from the mouse genome (Deutsch et al., 1988; Burri et al., 1989; Dressler et al., 1990; Plachov et al., 1990; Goulding et al., 1991; 1993; Jostes et al., 1991; Walther et al., 1991; Asano et al., 1992). In addition to their structural conservation, all members of the Pax gene family are expressed

Neural Cell Specification: Molecular Mechanisms and Neurotherapeutic Implications
Edited by Juurlink *et al.*, Plenum Press, New York, 1995

along the antero-posterior axis, suggesting that they play an important role in the establishment of the main body plan.

The recent demonstration that various mutations within members of the murine Pax gene family correspond to previously characterized developmental mutants has confirmed the importance of Pax genes as crucial regulators of embryonic development. Firstly, various mutations within the *Pax1* gene have been shown to be responsible for the *undulated* (*un*) phenotype. Secondly, modifications within the *Pax3* gene give rise to the murine *splotch* (*Sp*) phenotype while the *small-eye* (*Sey*) phenotype is caused by alterations within the *Pax6* gene. In both homozygous *Sp* and *Sey* mice, the development of the central nervous system is seriously compromised. The observations made on these mutants underline the importance of Pax genes for the development of various structures such as the brain, the neural tube, neural crest derived organs, muscles or the skeleton.. Similar alterations within the human *PAX3* and *PAX6* genes have been identified in Waardenburg syndrome and aniridia patients respectively. Finally, inappropriate expression of these genes may be involved in the process of tumorigenesis, thereby demonstrating that both loss and gain of function of these genes can play an important role in the genesis of various pathologies.

STRUCTURE AND FUNCTION OF PAX GENES

All the information gathered on the structure and function of Pax genes indicate that Pax proteins act as transcription factors; their predicted structure, nuclear localisation, in vitro binding and transactivating abilities are all consistent with this notion.

Structural analysis of the Pax genes has revealed that their putative translation products vary between 361 and 479 amino acids. However, each gene may give rise to various protein isoforms as alternative splicing is a common feature of these genes. The characteristic structure of the Pax proteins consists of a 128 amino acid paired domain located very close to their amino terminus (Fig. 1). This domain appears to contain 3 α-helices, the first situated close to the aminoterminal part of the domain while the two last form a helix-turn-helix (HTH)-like motif near its carboxy terminus (Bopp et al., 1986; Frigerio et al., 1986; Treisman et al., 1989; Treisman et al., 1992; Walther et al., 1991). In addition, the paired domain may be accompanied by a 61 amino acid paired-type homeodomain, as in Pax3, 4, 6 and 7, also harbouring 3 α-helices (Fig. 1). Again, the last 2 α-helices of the homeodomain form a HTH motif similar to those shown to mediate DNA-binding activity in prokaryotes (Laughon and Scott, 1984; Pabo and Sauer, 1984). Interestingly, Pax2, 5 and 8 proteins contain a rudimentary paired-type homeodomain encoding only the first α-helix. All the murine Pax proteins except Pax4 and Pax6 include a well-conserved octapeptide between the paired domain and paired-type homeodomain, as previously documented for the *Drosophila* gooseberry and human PAX proteins HuP1 and HuP2 [Fig. 1] (Burri et al., 1989). Finally, the carboxy terminus constitutes a proline/serine/threonine-rich region reminiscent of the transcription activating domain of other transcription factors such as OCT-2 and CTF-1 (Mermod et al., 1989; Tanaka and Herr, 1990).

The potential of the Pax proteins to act as transcription factors has been tested in vivo and in vitro. First, their cellular localisation is nuclear (Pox neuro and Pox meso: Bopp et al., 1989; Pax1: Fritsch and Gruss, unpublished observations; Pax2: Dressler and Douglass, 1992; Pax6: Fritsch, Walther and Gruss, unpublished observations). Secondly, Pax proteins possess DNA binding abilities. Initially, the *Drosophila* prd protein was shown to bind the e5 sequence, found within the *even-skipped* gene promoter (Hoey and Levine, 1988). The *prd* gene product binds via its the paired domain and paired-type homeodomain, each recognising discrete sequences (Treisman et al., 1989; Treisman et al., 1991). Using permutations of the e5 sequences, many Pax proteins (Pax1, 2, 3, 5, 6, 8) demonstrate in vitro-binding abilities. As shown by interference experiments, the Pax1 protein recognizes a 24 bp sequence which includes a GTTCC binding core. The same results have been obtained with truncated protein forms encompassing only the paired domain indicating that the paired domain alone is able to confer DNA binding ability (Chalepakis et al., 1991). The ability of Pax proteins to act as transcription modulators has been further analysed for different members of the murine family (Chalepakis et al., 1991; Dressler and Douglass, 1992; Fickenscher et al., 1993; Goulding et al., 1991; Chalepakis, Asano and Gruss, unpublished results).

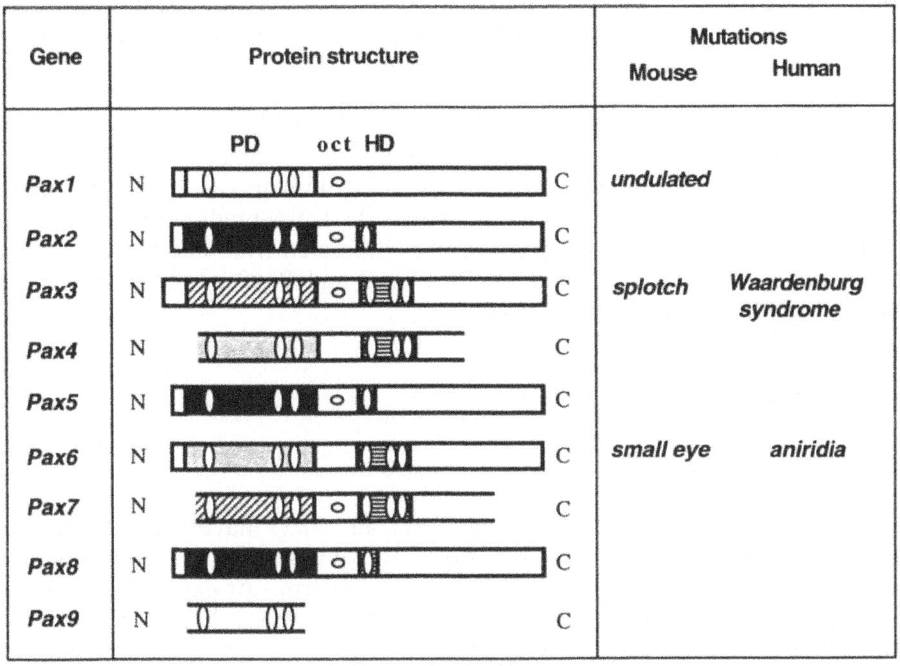

Figure 1. Structure of the members of the murine Pax gene family. Protein structure and corresponding murine and human mutations are shown. The patterns used to fill the paired boxes represent the different paired domain classes and link Pax genes in paralogous groups. Small white boxes represent putative α-helices. Open boxes indicate that the full sequence is not available. Abbreviations: C, carboxy terminus; HD, homeodomain; N, amino terminus; oct, octapeptide; PD, paired domain.

As a prototype for Pax proteins, the functional domains of the Pax3 protein have been characterized by gel shift and Gal-4 fusion experiments (Chalepakis et al., 1994; Goulding et al., 1991). The Pax3 protein binds via its paired domain to target sequences containing the core GTTCC sequence. This binding is enhanced by the presence of an ATTA motif recognised by the homeodomain. However, this motif is in itself insufficient to mediate binding (Chalepakis et al., 1994). This cooperative binding does not present any stereospecificity. The transactivation domain of the Pax3 protein has been mapped in its carboxy terminus while a transcription inhibiting activity has been unveiled in the amino terminus including part of the paired domain (Chalepakis and Gruss, unpublished observations). Meanwhile, many of the Pax proteins have been shown to activate transcription from minimal promoters fused to multimers of these binding sites, clearly underlining the function of the Pax proteins as transcription factors (Chalepakis et al., 1991; Fickenscher et al., 1993; Chalepakis and Gruss, unpublished observations).

The idea that Pax genes act as transcriptional modulators has been supported by the identification of natural sequences able to bind the Pax gene products. The *Pax5* gene is expressed in the B-cell lineage (Adams et al., 1992). This gene corresponds to the B-cell specific transcription factor BSAP which regulates the B-cell specific CD19 transmembrane receptor (Barberis et al., 1990; Kozmik et al., 1992). A BSAP binding site, located in the CD19 promoter region can be bound in vitro both by the Pax5 and Pax1 proteins (Adams et al., 1992). Sequences within thyroid-specific thyroglobulin and thyroperoxidase promoters are also bound by the Pax8 protein. Its gene product can also enhance transcription from these promoters (Zannini et al., 1992).

The different activities of the Pax genes, such as transactivation and transcription inhibition, are probably modulated by the synthesis of various protein forms arising from alternative splicing. For instance, an alternatively spliced 14 amino acid domain within the paired domain has been documented (Walther and Gruss, 1991) and is likely to alter the DNA-binding properties of the protein. Similarly, different protein isoforms generated from alternatively spliced messengers have been identified for the *Pax2* and *Pax3* genes (Dressler and Douglass, 1992; Tsukamoto et al., 1994).

PAX MUTANTS

*Pax*1 and *un*

All the murine Pax genes except *Pax1* and *Pax9* have been shown to be expressed in the developing nervous system. In the case of *Pax1*, its predominant expression domain is confined to the paraxial mesoderm. At day 8.25 p.c., briefly before the first somite de-ephithelialises, *Pax1* transcription begins in the ventral part, labelling the prospective sclerotome. The signal persists in the mesenchymal sclerotome during its condensation which gives rise to the neural arches, the intervertebral discs and parts of the vertebral bodies. As cartilage formation begins, *Pax1* expression vanishes, suggesting that *Pax1* plays a role during the mesenchymal phase of the vertebral column formation (Deutsch et al., 1988; Dietrich et al., 1993; Wallin et al., 1994).

Initially, the *Pax1* gene was mapped to chromosome 2 close to the *un* locus (Wright, 1947). This mutant which comprises several alleles, is characterised by malformations of the axial skeleton. (Grüneberg, 1950; Grüneberg, 1954; Wright, 1947). The identification in the *un* mutant of a point mutation at a highly conserved position within the *Pax1* paired box supported the idea that the *un* phenotype is due to alterations within that gene (Balling et al., 1988). Analysis of the DNA-binding properties of Pax1 indicated that the amino acid substitution within the Pax1[un] paired domain reduces or abolishes the affinity of the protein for its in vitro-defined binding sequences thereby obliterating the ability of Pax1 to activate transcription (Chalepakis et al., 1991). Furthermore, the genetic basis of two other *un* alleles has been established. Firstly, *Undulated short tail* (*Un^s*) mice (Blandova and Egorov, 1975) harbour a deletion of at least 48 kb, eliminating the whole *Pax1* gene as well as 38 kb of flanking sequences (Balling et al., 1992;

Dietrich and Gruss, in press). Thus, this allele can be considered as a null mutation, although it can not be excluded that the deletion of contiguous genes contributes to the Un^s phenotype. Secondly, the *undulated extensive* (*un^{ex}*) allele (Wallace, 1979, 1980) carries a deletion of at least 28 kb encompassing the fifth exon and the poly(A) signal (Dietrich and Gruss, in press). Consequently, the *un^{ex}/Pax1* transcripts lack the poly(A) tail, probably accounting for the low *Pax1* mRNA levels observed in these animals (Dietrich and Gruss, in press). The allelic series of *un* mutants constitute an important tool to investigate the role of the *Pax1* gene during development, as phenotypic aspects common to the three alleles should reflect the basic function of this gene.

In all homozygotes as well as in $Un^s/+$ heterozygotes, the most prominent phenotype is the short kinky tail resulting from vertebral malformations (for review see Balling et al., 1992). This aberration is most pronounced in the lumbar region as the vertebrae can lack vertebral bodies and intervertebral discs (Fig. 2)(Balling et al., 1992; Wallin et al., 1993; Dietrich and Gruss, in press). Similarly, the transverse processes are reduced or absent, while the neural arches are shortened. The strongest vertebral reductions are found in Un^s/Un^s homozygotes where the ventral halves of the lumbo-sacral vertebrae are missing while the neural arches fail to fuse dorsally. Since mild phenotypes can also be found in *un/+* and *un^{ex}/+*, all types of *un* mutations have to be considered as semidominant or haploid insufficient (Dietrich and Gruss, in press).

The vertebral phenotype can be traced back in development to the mesenchymal state of axial skeleton formation; in *un* mutants, the process of sclerotome condensation, normally giving rise to the neural arch anlage, appears delayed. In the perichordal region, the condensations forming the intervertebral discs and parts of the vertebral body are even more retarded (Grüneberg, 1963; Dietrich and Gruss, in press). As a critical cell density may not be reached, the formation of the ventro-medial vertebral column components may be compromised. These sclerotomal deficiencies observed in *un* mutants directly correlate with the *Pax1* expression pattern in the paraxial mesoderm (Fig. 2). These observations suggest that cell accumulation within the embryonic sclerotome is an intrinsic, *Pax1* dependent program.

The sternum, derived from lateral mesoderm, represents another *Pax1* expression domain (Chen, 1952; Deutsch et al., 1988). In *un* mutants, this structure is affected by sternebral fusions while the xiphoid process and anterior portions of the sternum may be split, a feature particularly prominent in the Un^s mutants (Grüneberg, 1963; Dietrich and Gruss, in press). On the other hand, the fused sternebrae phenotype may be secondary to the shortened ribs, preventing sternal ossification (Chen, 1953). Thus, the aberrant sternebrae may result from the sclerotome deficiency. On the other hand, the broadened and split xiphoid process may result from incompletely fused sternal bands, and be dependant on *Pax-1* function during elongation and unification of the contralateral sternal anlagen (Chen, 1953).

Another mesodermal expression domain is confined to the developing limbs where *Pax-1* activity starts at day 10.5 p.c. at the antero-dorsal margin of the forelimb (Dietrich et al., 1993), the region believed to give rise to the acromion and parts of the scapular blade [Fig. 2] (Beresford, 1983; Saunders, 1948). Around day 12.0 p.c., transient expression in the postero-dorsal margin appears, probably corresponding to the medial component of the scapula (Beresford, 1983; Saunders, 1948). Simultaneously, an identical pattern is observed in the hind limbs as well as in the carpal and tarsal regions of the paws at day 13.5 and 14.5 p.c. Consistent with these expression domains, *un* mutant mice exhibit reductions of the spina scapulae and acromion (Fig. 2). While *un/un*, *un^ex/un^ex* and *un/un^ex* mutant mice frequently lack the acromion, *Un^s* animals develop this structure but in close vicinity to the shoulder articulation (Dietrich and Gruss, in press). In the *Un^s* compound heterozygotes, a second, fork-like protrusion accompanies the acromion, both components fusing with the clavicle. While this phenotype illustrates the qualitative differences in the scapular phenotypes of *un* mutants, the reduction of the acromion in *un* and *un^ex* supports the idea that *Pax-1* is required for the outgrowth of this appendicular skeletal element.

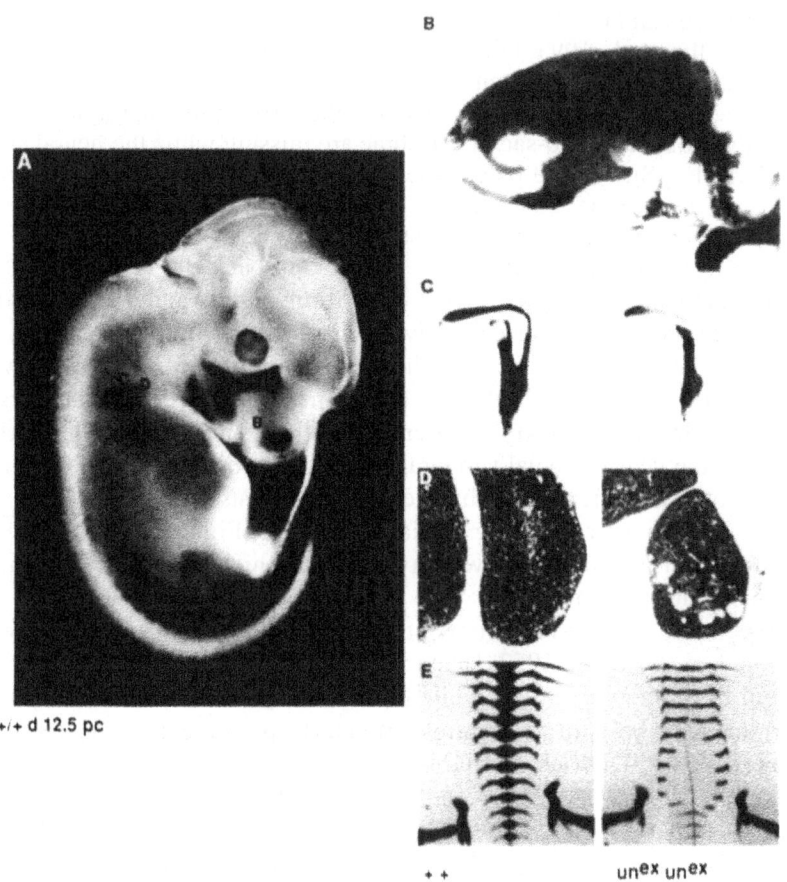

Figure 2. Correlation of *Pax1* expression pattern with aspects of the *un* phenotype in the mouse. (A) *Pax1* expression in normal embryo day 12.5 p.c. by whole mount in situ hybridization. Transcripts are localized in the face (b), the shoulder girdle (c), the thymus anlage (d) and the developing vertebral column. (B) Shortened facial skeleton of adult *un^ex/un^ex* animal. (C) Scapula of wildtype (left) and *un^ex/un^ex* animal (right). The mutant scapula lacks the acromion. (D) Wildtype (left) and *un^ex/un^ex* (right) thymus of newborn animals. The mutant thymus is reduced and interspersed with cysts. (E) Vertebral column of wildtype (left) and *un^ex/un^ex* (right) animals at day 13.5 p.c. The vertebral bodies are missing in the mutant animal.

From day 11.5 p.c. the *Pax-1* gene is expressed in the lateral portion of the maxillary prominence and, shortly thereafter, in the basis of the mandibular component of the first branchial arch/Meckel's cartilage, the maxillary edge of the naso-lacrimal groove, the nasal prominence and the developing eyelid (Fig. 2)(Deutsch, 1990; Dietrich and Gruss, in press). Consistent with these expression domains, *un* mutants display a remarkable size reduction of the facial skeleton, particularly evident in *un^ex/un^ex* adults where the snout appears snubbed [Fig. 2] (Dietrich and Gruss, in press). Moreover, their jaws may exhibit an extended curvature. This phenotype appears to be linked to a neural crest derived deficiency, illustrating the importance of *Pax-1* for the development of these neural crest derived skeletal elements (Couly et al., 1993).

Finally, an endodermal expression domain is represented by the pharyngeal pouches which start expressing *Pax-1* at day 8.25 p.c. [Fig. 2] (Dietrich et al., 1993). Here, the *Pax-1* signal persists until juvenile age in the third and fourth pouches from which the thymus is derived (Deutsch et al., 1988). In the mutants, the thymus is reduced in size (Dietrich and Gruss, in press). In addition, the presence of cysts within the developing thymus of *un* embryos, underscores the importance of *Pax-1* for the development of this structure (Fig. 2).

Pax3 and *Sp*

The *Sp* mutant is a well-recognised neural crest and neural tube defect model (Auerbach, 1954). The various *Sp* alleles (*Sp, Sp-delayed: Sp^d, Sp-retarded: Sp^r, Sp2H*) exhibit a semidominant phenotype associated with various *Pax3* expressing structures (Auerbach, 1954; Beechey and Searle, 1986; Dickie, 1964). The chromosomal localisation of the *Pax3* gene indicated that it may correspond to the *Sp* locus, both being located on band C4 of chromosome 1 (Evans et al., 1988; Goulding et al., 1993a). The *Pax3* gene has now been shown to be altered by deletions or point mutations in various *Sp* alleles (Epstein et al., 1991; Epstein et al., 1993; Goulding and Gruss, 1989; Vogan et al., 1993).

In the heterozygous state, *Sp* alleles are characterized by white spotting on the belly, the limbs and the tail, defects which originates from a neural crest deficiency. Conversely, in the homozygous state, most alleles result in embryonic death around day 13-14 p.c. (Auerbach, 1954). One allele, *Sp^r*, presents post-implantation lethality while another, *Sp^d*, permits development to proceed until birth (Dickie, 1964; Evans et al., 1988). Homozygous animals present various defects associated with neural crest: reduction or absence of dorsal root ganglia, Schwann cell deficiency, cardiovascular defects and thymic, parathyroid and thyroid malformations (Auerbach, 1954; Franz, 1989; Grim et al., 1992; Moase and Trasler, 1989). Cranial neural crest deficiencies have also been recently observed in this mutant (Tremblay and Gruss, unpublished observations).

The most recognizable feature of *Sp* homozygous embryos is the development of rachischisis (spina bifida) generally in the lumbo-sacral region and of cranioschisis at the mid-hindbrain level [exencephaly] (Fig. 3). Neuropore measurements have demonstrated a delay in anterior and posterior neuropore closure in *Sp* embryos (Dempsey and Trasler, 1983). Because of the expression pattern of the *Pax3* gene in the neural tube, this phenotypic feature appears to originate as an intrinsic deficiency within the developing tube. However, other observations suggest that this abnormality may be consecutive to other deficiencies.

It has been suggested by Schoenwolf and Smith that multiple factors act in concert to promote the closure of the neural tube: factors intrinsic to the neural tube such as changes in cell shape promote the wedging of the neural tube and somitic proliferation may also provide additional lateral forces to facilitate its closure (Schoenwolf and Smith, 1990). In that respect, somites appear to be of maximal size in the lumbo-sacral region and their normal proliferation may be mandatory to drive proper neural tube closure. In addition, it has recently been documented that the differential growth between the endoderm and the neural tube is responsible for the spina bifida and curly tail phenotypes in the curly tail mutant (Brook et al., 1991; van Straaten et al., 1993). Growth retardation within the endoderm appears to increase the curvature of the embryo therefore augmenting the constraints on the closure of the posterior neuropore. In this case, the curvature which is already maximum in the lumbo-sacral region would explain the predisposition of this region for the development of this phenotype. All these putative factors could be involved in the genesis of the *Sp* phenotype.

Pax3 is first expressed prior to neural tube closure at the tip of the neural folds. The juxtaposition of the folds resulting in neural tube closure is accompanied by the expansion of the *Pax3* expression domain down to the sulcus limitans delineating the ventral dorsal parts of the neural tube. From this stage, *Pax3* expression is maintained in the dorsal part of the neural tube and remains confined to the mitotically active ventricular zone. Neural crest deficiencies as well as a disorganized dorsal spinal cord have been documented in the *Pax3* mutant *Sp* (Auerbach, 1954; Yang and Trasler, 1991). These observations are consistent with an intrinsic deficiency at the level of the developing neural tube. Secondly, at the level of the segmentation plate *Pax3* expression is detected in the paraxial mesoderm. After segmentation of this structure and under the influence of the notochord, as *Pax1* is induced in the ventral part of the still epithelial somite, *Pax3* expression is maintained in the dorsal part of the somite, the dermomyotome (Bober et al., 1994). Since *Pax3* mutations appear to affect the organization of the developing dermomyotome (see below) as well as the generation, the differentiation, the proliferation and/or the migration of myoblasts [Fig. 3] (Bober et al., 1994), it can be envisaged that mutations within the *Pax3* gene indirectly affect neural tube closure by slowing down somitic proliferation. Finally, *Pax3* expression is observed at the tip of the hindgut between E9.5 and 12.5 p.c. (Tremblay and Gruss, unpublished observations). This endodermal expression domain is consistent with the idea of a proliferation defect in the hindgut of *Sp* embryos, thereby increasing the curvature of the embryo and hindering the proper closure of the neural tube. Thus, it appears likely that the development of spina bifida in *Sp* animals is a result of multiple deficiencies with the neural tube, the somites and the endoderm all contributing to the evolution of this phenotype and influencing its penetrance and expressivity.

Finally, mesoderm-derived deficiencies have also been documented in *Sp* mice. While the paraxial mesoderm exhibits widespread cellular disorganization as well a lower cell number (Yang and Trasler, 1991), the axial musculature appears to be reduced in size and the limb muscles are absent (Franz et al., 1993). Recent analysis of the *Sp* mutants indicated that the dermomyotome and myotome compartments present a generally disorganized appearance in the mutant. Moreover, *Pax3* mutations completely block the colonization of the limbs by a *Pax3*-expressing myoblast population (Fig. 3). While myogenesis can be followed at the level of the body wall muscles and the deep muscles of the back, the shoulders and the limbs of homozygous *Sp* animals are devoid of cells expressing myogenic factors (*myf-5*, *myoD*, *myogenin*) resulting in absence of muscle (Bober et al., 1994). Thus, *Pax3* appears essential for early dermomyotome-derived myoblast precursors destined to colonize the limbs and to a lesser extent for the body wall and back muscles.

The expression of *Pax3* in neural crest, the dorsal part of the spinal cord and the developing somites perfectly matches with the various *Sp*-associated deficiencies. Furthermore, the developmental abnormalities in *Sp* underlines the role of *Pax3* at three different levels. Firstly, *Pax3* is required for the release, migration and/or proliferation of neural crest cells and, consecutively, for the development of various structures and organs. Secondly, *Pax3* function is essential for the proper neural tube closure and/or neuroectoderm proliferation. Thirdly, *Pax3* appears to be directly or indirectly important for the specification, migration and/or proliferation of myoblast precursors destined to give rise to the limb musculature. A translocation t(2;13) involving the human *PAX3* gene in alveolar rhabdomyosarcomas (Barr et al., 1993) strongly supports the notion that *PAX3* plays an important function in muscle development and that its inactivation, its overexpression or some alterations of its structural integrity can yield diverse pathological effects.

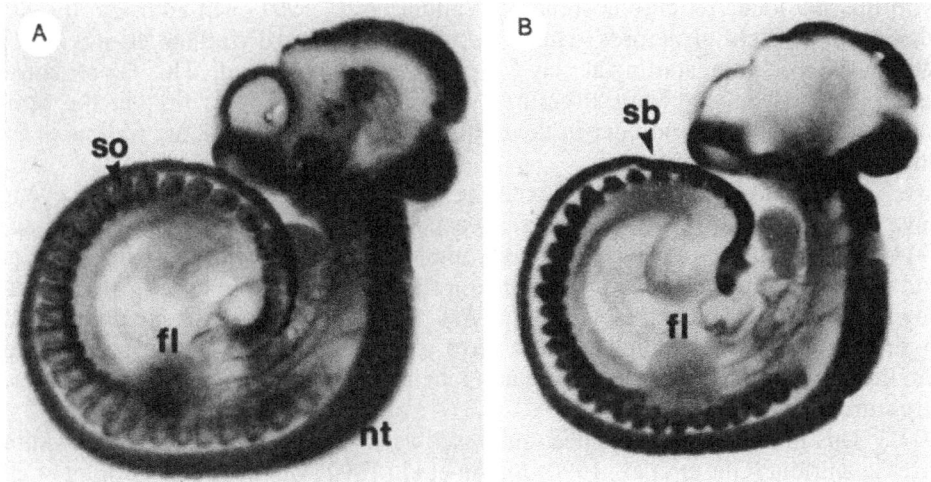

Figure 3. Expression at day 10.5 p.c. of the *Pax3* gene visualized by whole mount in situ hybridization in wildtype (A) and homozygous mouse *Sp* embryos (B). Expression can be observed in the developing brain, neural tube (nt), somites (so) and in myoblast precursors as they invade the forelimb bud (fl). In the mutant embryo, spina bifida (sb) can be observed in the lumbosacral region. No Pax3 expressing myoblasts colonize the limb bud while the somites present a disorganized *Pax3* signal.

Although the precise molecular cascade affected by the *Sp* mutations downstream from the *Pax3* gene is unknown, the composition of extracellular matrix molecules appears to be significantly in *Sp* and *Sp^d* mutants (O'Shea et al., 1987; for a review: Moase and Trasler, 1992). Such alterations could explain some of the morphological and/or migration aberrations documented in *Sp* embryos, but it is difficult to assess whether these extracellular matrix molecules constitute a direct target of Pax3.

Alterations in the human *PAX3* gene have also been linked with the development of the Waardenburg syndrome (Baldwin et al., 1992; Ishikiriyama et al., 1989; Tassabehji et al., 1993). Patients affected by this disorder mainly present cranio-facial abnormalities presumably originating from a cranial neural crest deficiency. In addition, musculo-skeletal hypoplasia sometimes accompanies this syndrome (Waardenburg, 1951).

Pax6 and Sey

The *Sey* mutations (*Sey^H*, *Sey^Dey*, *Sey^Neu*) are all of semidominant nature with variable expressivity and nearly complete penetrance (Roberts, 1967). These mutants present similar phenotypes but of various severities. The main defect observed in the heterozygous state, is a reduction of the diameter of the eye. In addition, other ocular abnormalities can be associated: smaller ocular orbit, absence of optic chiasma, thickened ocular extracellular membranes (especially of the lens), cellular degeneration, iris hypoplasia, corneal opacification, lens vacuolation and dislocation as well as overgrowth and invasiveness (Glaser et al., 1990; Hogan et al., 1988; Hogan et al., 1986; Jordan et al., 1992; Theiler et al., 1978). In homozygotes, the conditions can be readily recognized at day 10.5 p.c. of gestation when eye and nose placodes fail to develop. These embryos, which are devoid of ocular and nasal structures, die neonatally.

The chromosomal localization of *Pax6* indicated that it is closely associated with the *Sey* locus on chromosome 2 (Walther et al., 1991). In addition, the *Sey* phenotype affects structures where *Pax6* is expressed (Walther et al., 1991). Within the forebrain, starting at day 8.5 p.c., *Pax6* is expressed. This expression is remarkably associated with structures of the developing eye; first in the optic placode and optic pit and later in the optic sulcus, the optic stalk, as well as in the developing lens and cornea.

The nasal epithelium also expresses this gene. Various mutational events have been mapped to *Pax6* sequences to account for *Sey* alleles. In the *Sey^H* and *Sey^Dey* alleles the *Pax6* sequences are completely deleted, while a point mutation and an insertion of 116 bp cause premature translation termination upstream and downstream of the homeodomain in the *Sey* and *Sey^Neu* alleles respectively (Hill et al., 1991). These data indicate that *Pax6* is essential, within the neuroectoderm, for the complex local induction leading to the formation of the nose and eye structures.

This mouse mutant has been proposed as a model for the human congenital disease aniridia (Glaser et al., 1990; Jordan et al., 1992; van der Meer de Jong et al., 1990). In fact, several mutations have now been identified within the human *PAX6* gene in patients presenting aniridia as well as Peter's anomaly, affecting the anterior chamber of the eye and its associated structures (Glaser et al., 1992; Jordan et al., 1992; Ton et al., 1992).

PATTERNING BY PAX GENES

Pax Genes and Regional Specification in the Paraxial Mesoderm

The paraxial mesoderm which is destined to give rise to the axial skeleton, musculature and dermis is laid down during gastrulation along both sides of the neural tube. First organised as mesenchyme, this tissue epithelialises to give rise to the segmentally organised somites. The axial identity of the somites appears to be determined early during gastrulation and to be under the control of the homeobox gene family (Kieny et al., 1972; Kessel and Gruss, 1991). On the other hand, the establishment of the dorso-ventral polarity within the individual somites occurs later, after their formation from the segmental plate, under the influence of the notochord (Aoyoma and Asamoto, 1988; Pourquie et al., 1993). While the dorsal part of the somite gives rise to the dermomyotome (muscle and dermis), the ventral part of the somites are induced to form sclerotome i.e. axial skeleton (for a review see Christ and Wilting, 1992). The identity of the genes involved in establishment of the dorso-ventral polarity of differentiating somites is still unknown.

The onset of Pax gene expression coincides with the establishment of dorso-ventral polarity of the paraxial mesoderm and can thus be monitored by the expression of *Pax3* and *Pax1*; *Pax3* expression is initiated before somitogenesis, weakly labelling the entire anterior region of the presomitic mesoderm. Slightly prior to the de-epithelialization of the somite, the *Pax3* signal is maintained and strengthened dorsally in the dermomyotome (Goulding et al., 1991) while the *Pax1* signal is induced ventro-medially in the region giving rise to the sclerotome (Fig. 4). This suggests that an external signal is required to induce the ventral fate within the paraxial mesoderm, predisposed to a dorsal fate (Dietrich et al., 1993; Goulding et al., 1994; Williams and Ordahl, 1994). *Pax1* positive cells can be followed as they migrate and condense around the notochord. The expression of these genes indicates that they may constitute important determinants of the dorso-ventral polarity of the somites. This hypothesis is supported by the analysis of *Pax1* and *Pax3* mutants in which deficiencies in the corresponding compartments have been documented, as the homozygous *un* and *Sp* mutants suffer from sclerotomal and dermomyotomal derived defects respectively (Bober et al., 1994; Grüneberg, 1954).

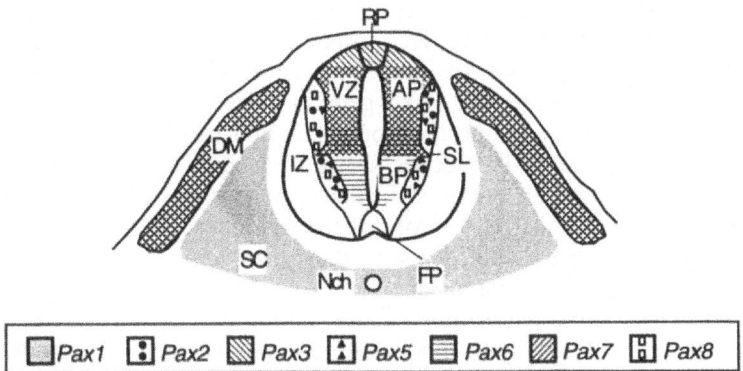

Figure 4. Schematic representation of the distribution of Pax gene expression as observed on a cross section of the spinal cord and paraxial mesoderm (day 11.5 p.c. mouse embryo). Abbreviations: AP, alar plate; BP, basal plate; DM, dermomyotome; FP, floor plate; IZ, intermediate zone; Nch, notochord; RP, roof plate; SC sclerotome ; SL, sulcus limitans; VZ ventricular zone.

The dynamics of this expression pattern also correlate with the presence of a ventralising signal emanating from the notochord and acting both on the neural tube and paraxial mesoderm. This signal appears to be required to induce the ventral properties and restrict dorsal properties within the paraxial mesoderm. This hypothesis is supported by analysis of murine notochord mutants and by transplantation experiments in the chick. In the mouse mutants *Brachyury curtailed* and *truncate*, notochord agenesis prevents the activation of *Pax1* and the dorsal restriction of *Pax3* activity (Dietrich et al., 1993). In *Danforth's short tail* and *Pintail* mutant mice the notochord degenerates during development. This deficiency results in the loss of *Pax1* expression while the *Pax3* signal occupies the previously *Pax1* expressing zone (Dietrich et al., 1993; Koseki et al., 1993). These findings indicate that the activity of the notochord is essential to initiate and maintain the ventral pathway within the developing somites.

Similar results have been obtained from notochord manipulation experiments in the chick where the notochord has been shown to be essential for the induction of sclerotome from the ventral somite (Pourquie et al., 1993). Upon removal of the notochord, the expression of *Pax9*, a sclerotomal marker coexpressed with *Pax1*, is absent while the domain of *Pax3* expression is expanded ventrally. Conversely, the implantation of a supernumerary notochord dorsally, inhibits the expression of *Pax3* dorsally while it promotes sclerotome differentiation as *Pax9* expression covers the entire paraxial mesoderm on the side of the transplantation (Brand-Saberi et al., 1993; Goulding et al., 1994). These results indicate that the notochord is required for the induction of the ventral fate in the paraxial mesoderm and that this induction process can be monitored by the Pax genes.

Pax Genes in the Establishment of Dorso-Ventral Polarity of the Neural Tube

Besides *Pax1* and *Pax9*, all Pax genes are expressed during the development of the neural tube. However, the Pax genes harbouring a complete paired type homeodomain (*Pax3*, *Pax6* and *Pax7*) appear to be expressed earlier during embryonic development (day 8 to 8.5 p.c.) and restricted to the mitotically active ventricular region of the neural tube [Fig. 4] (Goulding et al., 1991; Jostes et al., 1991; Walther et al., 1991). These neuroepithelial cells are destined to stop proliferating, migrate radially and settle at specific positions in the intermediate zone; their differentiation will give rise to both neuronal and glial cells. While the expression of the paralogous *Pax3* and *Pax7* genes encompasses the dorsal and alar plates of the neural tube, the *Pax6* gene is detected in the basal plate of the developing spinal cord, excluding the floor plate and its vicinity. The precocious expression of this group of Pax genes along with their dorso-ventral restriction indicates that they might be involved in the establishment of polarity within the neural tube. Recent studies on the role of the notochord in the patterning of the neural tube have supported the idea that these Pax genes are implicated in the dorso-ventral polarity of the neural tube.

The dorso-ventral patterning of the spinal cord has been shown to be dependent on specialised ventral structures, namely the floor plate and the notochord (Kitchin, 1949; van Straaten et al., 1985a; van Straaten et al., 1988; van Straaten et al., 1985b). The presence of the notochord induces the floor plate at the ventral midline of the neural tube, triggering the bilateral differentiation of motor neurons. The developmental process triggered at the ventral part of the neural tube proceeds as a wave, the differentiation events propagating toward the dorsal tube, where neurons are born and differentiate later into different cell fates, typically represented by the commissural and interneurons cell types (Altman and Bayer, 1984). Thus, determination of cell fate in the neural tube depends on extrinsic factors that act in a position-dependent manner to induce the differentiation of particular cell types in the appropriate dorso-ventral position. Grafting experiments indicate that the presence of a supernumerary notochord on the side of the neural tube induces the development of a second floor plate and of ectopic motor neurons, while its removal abrogates the differentiation of the floor plate and of motor neurons (Hirano and Iwakura, 1990; Placzek et al., 1990; van Straaten and Hekking, 1991; Yamada et al., 1991).

Transplantation studies reveal that the repatterning of the *Pax3* and *Pax6* expression shortly follows the repatterning of the neural tube resulting from the grafting or the removal of the notochord, but precedes the detectable ectopic differentiation of any cell types (Goulding et al., 1993b). While removal of the notochord results in the upregulation of the *Pax3* gene in the ventral part of the neural tube and the absence of *Pax6*, a second notochord juxtaposed to the lateral plate shifts the expression of the *Pax3* and *Pax6* genes more dorsally. This repatterning is observed as early as 6 hours post-transplantation and precedes the birth of ectopic neurons. These data indicate that *Pax3* and *Pax6* could be involved in the early establishment of the dorso-ventral polarity of the neural tube.

The other group of Pax genes (*Pax2, 5, 8*), containing only a rudimentary paired type homeobox, start to be expressed later (day 10 p.c.), in postmitotic cells of the intermediate zone located on both sides of dorso-ventral midline, the sulcus limitans (Fig. 4). These cells migrate out and form the intermediate grey zone of the spinal cord where *Pax2* continues to be expressed into adult stages while *Pax8* expression fades away around day 13.5 p.c. These *Pax2* cells have recently been identified in the zebrafish as interneurons of the grey matter (Krauss et al., 1991; Mikkola et al., 1992).

Pax Genes as Local Regulators in the Developing Brain

Development of the vertebrate nervous system begins by neural plate induction from an uncommitted neuroectoderm under the influence of factors released by the axial mesoderm. This process triggers the folding of the neural plate and leads to the formation of the neural tube while the neural crest cells are specified at the tips of the neural folds. Positional identity along the antero-posterior axis of the neural tube, from the hindbrain to the tail, appears to be specified early in the primitive streak and to be controlled, directly or indirectly, by the homeobox gene family (Boncinelli et al., 1991; Kessel, 1991; McGinnis and Krumlauf, 1992).

Conversely, the mechanisms by which the region anterior to the hindbrain is patterned to give rise to the various functional regions of the brain remain obscure. Many genes, initially isolated in *Drosophila* as major players controlling the development of the head of the fly, have permitted the isolation of related vertebrate genes whose expression pattern suggests their involvement in the specification of domains of the brain. Such genes include members of the Otx, Emx, Dlx, Wnt and Nkx families (Boncinelli et al., 1993; McMahon et al., 1992; Price et al., 1992; Price et al., 1991). In that respect, the temporally and spatially restricted distribution of Pax gene expression in the developing brain strongly suggests that they are important for the regionalization of the neuroepithelium to give rise to specific brain structures as their expression pattern delineates specific domains within the developing brain [Fig. 5] (Stoykova and Gruss, 1994). In addition, some experimental evidence provides us with more insight into the role of these genes during brain development.

As in the neural tube, Pax genes which contain a complete paired type homeobox (*Pax3, Pax6, Pax7*) are expressed early (E8.0 - 8.5) in the mitotically active cells of the ventricular neuroepithelium. The expression of *Pax3* and *Pax7* is observed in the entire neuroepithelium of the prosencephalon, later giving rise to the telencephalon and diencephalon, as well as in the midbrain and hindbrain regions. At this early stage, *Pax6* expression is confined to the prosencephalon and the anterior rhombencephalon, excluding the mesencephalic roof and the region of the future metencephalon (Stoykova and Gruss, 1994; Walther and Gruss, 1991). Pax gene transcript distribution coincides with the establishment of various anatomical boundaries. The mid-hindbrain boundary, established early during development, is delineated by the rhombencephalic isthmus and the fovea isthmi. All Pax gene transcripts are detected with their specific dorso-ventral expression domains in the spinal cord and hindbrain up to the rostral region of the hindbrain. In more anterior regions, three Pax genes display restricted expression along longitudinal and transverse planes which relate to former neuromeric domains. *Pax3* and *Pax7* transcripts are observed in the diencephalic region posterior to the fasciculus retroflexus, including the epithalamus and the pretectum. The axonal tracts of the posterior commissure mark the boundary between the mesencephalon and the pretectum (the former synencephalon). At day 10.5 p.c., this boundary delineates the rostral limit of *Pax3* and *Pax7* expression domains. The same boundary applies for the caudal expression limit of *Pax6*, extending from the telencephalic cortex over the epithalamus and pretectum, excluding the roof of the mesencephalon (Stoykova and Gruss, 1994; Walther and Gruss, 1991). Furthermore, *Pax6* appears to have a region specific expression within the developing ventral thalamus (corresponding to the parencephalon anterior), strictly respecting the boundary between the dorsal and ventral thalamus at the level of the zona limitans intrathalamica. Within the dorsal thalamus (former parencephalon posterior), *Pax6* transcripts are detected in the internal germinative layer and epithalamus (dorso-lateral domain within the parencephalon posterior). With regard to the recently proposed subdivisions of the pretectal area into neuromeric territories (D3, D4: Figdor and Stern, 1993), additional experiments will be needed to define whether the *Pax3*, *6* and *7* expression domains delineate these putative compartments.

Comparative in situ analysis at midgestation and in adult mouse brains revealed a similar distribution of Pax gene transcripts along the antero-posterior axis. Rostrally, in the telencephalon and diencephalon, *Pax6* is expressed. In the midbrain, *Pax3, 5, 6 and 7* expression is observed while transcripts of *Pax2, 3, 5, 6, 7* and *8* are all present in the hindbrain. In many cases, the expression of a given Pax gene in various adult brain nuclei, correlates adequately with its expression in the presumptive region of origin within the midgestation brain. Such examples are, firstly, the expression of *Pax6* in the adult diencephalon (zona incerta, entopeduncular nucleus, reticular nucleus) and its expression in the embryonic ventral thalamu, the cerebellar granular layer and embryonal external granular layer; secondly, *Pax6* expression in the adult and embryonic septal areas; *Pax7* expression in the colliculus superioris and the mesencephalic roof; finally various Pax genes are expressed in isthmic nuclei corresponding their expression in the most rostral part of the embryonic hindbrain (Stoykova and Gruss, 1994).

The spatially-restricted expression of the Pax genes within the developing brain and the young adult brain strongly suggests their involvement in the regionalization and in the functional maintenance of various brain regions. This hypothesis has been supported by recent experimental data. Observations made on the *Pax6* mutant *Sey* confirm the importance of *Pax6* for the generation, the organisation and/or the maintenance of the telencephalic and diencephalic areas of the brain. As recently reported, animals homozygous for the chemically induced *Sey^neu* allele present widespread telencephalic defects: deficient neuronal migration, abnormal cortical lamination, expanded ventricular/subventricular zones, cortical heterotopias and leptomeningeal ectopias, improper compartmentalization of the striatum and reduction of various axonal tracts (Schmal et al., 1993). In addition, an abnormal development of various *Pax6* expression domains within the diencephalon is also evident in this mutant (Stoykova et al, unpublished observations).

The other paralogous group of Pax genes, *Pax2, 5* and *8* appears to be equally important for the patterning of another region of the brain. These genes are expressed along the neural tube up to the mid-hindbrain boundary while *Pax5* expression extends in the caudal part of the mesencephalic tegmentum. Interestingly, an antibody directed against the zebrafish homologue of the *Pax2* gene, *Pax[zf-b]*, affects the mid-hindbrain region by inhibiting the formation of the isthmus (Krauss et al., 1992). Thus, these groups of genes appear to be important determinants for the patterning of the mid-hindbrain region. These observations are reminiscent of the phenotypic effects resulting from the inactivation of *En-1*, *En-2* and *Wnt-1* (McMahon and Bradley, 1990; Millen et al., 1994; Thomas et al., 1991; Wurst et al., 1994) and suggest some functional interactions between Pax, En and Wnt genes in the generation of boundaries within the developing nervous system.

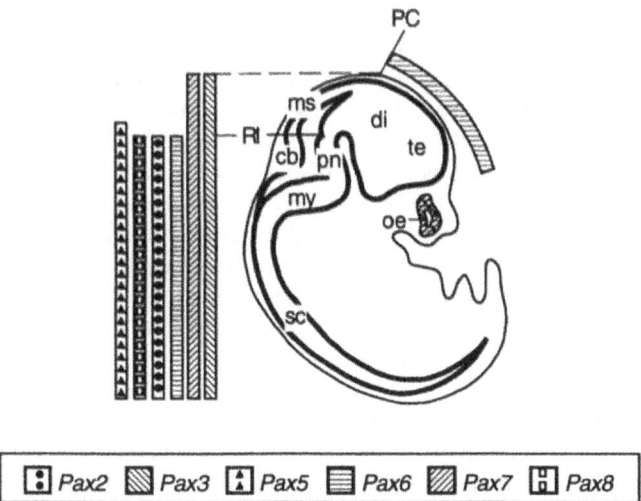

Figure 5. Schematic representation of the distribution of Pax gene expression in the nervous system as observed on a sagittal section (day 12.5 p.c. mouse embryo). The filled bars represent the limit of expression of the various Pax genes. Abbreviations: cb, cerebellum; di, diencephalon; ms, mesencephalon; my, myencephalon; oe, olfactory epithelium; PC, posterior commissure; pn, pons; RI, rhombencephalic isthmus; sc, spinal cord; te, telencephalon.

PAX GENES IN ONCOGENESIS

As already described, the spatial and temporal expression of Pax genes is tightly regulated during development. However, recent experimental data concerning the overexpression or aberrant regulation of Pax genes both in vitro and in vivo have given credence to the theory that Pax genes can act as protooncogenes.

The transforming ability of the *Pax1, -2, -3, -6, -8* has been tested by transfection experiments on NIH 3T3 or 208 cells (Maulbecker and Gruss, 1993). Foci of transformed cells were frequently observed and the subcutaneous injection of these cells into nude mice resulted in the rapid development of tumors. This underlines the transforming potential of intact Pax genes. This tumorigenic potential is dependent on the presence of a functional paired box, as the *Pax1[un]* allele, harbouring a point mutation in a well conserved residue of the paired domain, was unable to transform. On the other hand, this potential is independent of the presence of a homeobox as all Pax genes tested, including the *Pax1* gene devoid of such a box, demonstrated transforming abilities. This in vitro transformation potential is corroborated by in vivo evidence. Firstly, the aberrant expression of *PAX2* and *PAX8* appears to be involved in the genesis of kidney tumors. *Pax2* and *Pax8* transcripts are during murine kidney development (Dressler et al., 1990; Plachov et al., 1991). Their expression normally disappears upon terminal differentiation but persists in human Wilms' tumors (Dressler and

43

Douglass, 1992; Poleev et al., 1992). In addition, *Pax2* blocks the terminal differentiation of renal tubules and glomeruli when overexpressed in transgenics under the control of the CMV promoter (Dressler et al., 1993). The pathological state observed is reminiscent of some precancerous lesions. Thus, improper *Pax2* expression may inhibit the differentiation of certain cell types and, therefore, promote tumorigenesis. Secondly, alterations in the human *PAX3* AND *PAX7* genes have been identified in alveolar rhabdomyosarcomas, a type of cancer involving a t(2;13) and more rarely a t(1;13) translocation (Barr et al., 1993; Davis et al., 1994). Characterization of the translocated regions indicated that the *PAX3* and *PAX7* genes are rearranged, resulting in the replacement of their carboxy terminal sequences by a portion of chromosome 13. These translocations leave the *PAX* paired domain and homeoboxes intact, fused with the carboxy terminal half of the same forkhead domain from a novel gene, *FKHR* (Galili et al., 1993; Shapiro et al., 1993; Davis et al., 1994). Thirdly, the *PAX5* locus which is located on 9p13, may be involved in the most malignant form of astrocytomas, glioblastoma multiforme. Increased *PAX5* expression has been observed in highly malignant astrocytomas (Stuart et al., unpublished observations), indicating that *PAX5* may play a role in the progression of astrocytic tumors.

Taken together, these data demonstrate that disregulation and/or rearrangement of Pax genes can promote tumorigenesis in vivo. Although the link between uncontrolled cell growth and the altered expression of genes involved in the control of development appears both appealing and likely, it has yet to be determined what role these genes play in the multistep nature of cancer. Since the development of cancer appears to require the accumulation of multiple genetic events involving both the activation of protooncogenes as well as the inactivation of tumor suppressor genes, it will be of interest to identify the role of Pax genes in this process. In this regard, the identification of the target genes of Pax proteins will offer a good advantage in understanding better not only the molecular basis of cancer but also the molecular mechanisms behind the function of Pax genes.

REFERENCES

Adams B, Dörfler PAA., Kozmik Z, Urbánek P, Maurer-Fogy I and Busslinger M (1992): *Pax-5* encodes the transcription factor BSAP and is expressed in B lymphocytes, the developing CNS, and adult testis. Genes Dev 6: 1589-1607.

Altman J and Bayer SA (1984): "The Development of the Rat Spinal Cord," Springer: Berlin.

Aoyoma H and Asamoto K (1988): Determination of somite cells: independence of cell differentiation and morphogenesis. Development 104: 15-28.

Asano M and Gruss P (1992): *Pax-5* is expressed at the midbrain-hindbrain boundary during mouse development. Mech Dev 33: 27-38.

Auerbach R (1954): Analysis of the developmental effects of a lethal mutation in the house mouse. J Exp Zool 127: 305-329.

Baldwin CT, Hoth CF, Amos JA, da-Silva EO and Milunski A (1992): An exonic mutation in the HuP2 paired domain gene causes Waardenburg's syndrome. Nature 355: 637-638.

Balling R, Deutsch U and Gruss P (1988): *Undulated*, a mutation affecting the development of the mouse skeleton, has a point mutation in the paired box of *Pax-1*. Cell 55: 531-535.

Balling R, Lau CF, Dietrich S, Wallin J and Gruss P (1992): Development of the skeletal system In 'Postimplantation Development in the Mouse', Wiley, Chichester, pp 132-143.

Barr FG, Galili N, Holick J, Biegel JA, Rovera G and Emanuel BS: (1993): Rearrangement of the PAX3 paired box gene in the pediatric solid tumor alveolar rhabdomyosarcoma. Nature Genet 3: 113-117.

Baumgartner S, Bopp D, Burri M and Noll M (1987): Structure of two genes at the *gooseberry* locus related to the *paired* gene and their spatial expression during *Drosophila* embryogenesis. Genes Dev 1: 1247-1267.

Beato M (1989): Gene regulation by steroid hormone. *Cell* 56, 335-344.

Beechey CV and Searle AG (1986): Mutations at the Sp locus. Mouse News Lett 75: 28.

Beresford B (1983): Brachial muscles in the chick embryo: the fate of individual cells. J Embryol exp Morph 77: 99-116.

Blandova YR and Egorov IU (1975): Sut allelic with *un* Mouse News Lett 52: 43.

Bober E, Franz T, Arnold H-H, Gruss P and Tremblay P (1994): *Pax-3* is required for the development of limb muscles: a possible role for the migration of dermomyotomal muscle progenitor cells. Development 120: 603-612.

Boncinelli E, Gulisano M and Broccoli V (1993): Emx and Otx homeobox genes in the developing mouse brain. J Neurobiol 2: 1356-1366.

Boncinelli E, Simeone A, Acampora D and Mavilio F (1991): HOX gene activation by retinoic acid. Trends Genet 7: 329-334.

Bopp D, Burri M, Baumgartner S, Frigerio G and Noll M (1986): Conservation of a large protein domain in the segmentation gene *paired* and in functionally related genes of *Drosophila*. Cell 47: 1033-1040.

Bopp D, Jamet E, Baumgartner S, Burri M and Noll M (1989): Isolation of two tissue-specific *Drosophila* paired box genes, *Pox meso* and *Pox neuro*. EMBO J 8: 3447-3457.

Brand-Saberi B, Ebensperger C, Wilting J, Balling R and Christ B (1993): The ventralizing effect of the notochord on somite differentiation in chick embryos. Anat Embryol 188: 239-245.

Brook FA, Shum AS W, van Straaten HWM and Copp AJ (1991): Curvature of the caudal region is responsible for failure of neural tube closure in the curly tail (ct) mouse embryo. Development 113: 671-678.

Burri M, Tromvoukis Y, Bopp D, Frigerio G and Noll M (1989): Conservation of the paired domain in metazoans and its structure in three isolated human genes. EMBO J 8: 1183-1190.

Chalepakis G, Fritsch R, Fickenscher H, Deutsch U, Goulding M and Gruss P (1991): The molecular basis of the *undulated/Pax-1* mutation. Cell 66: 873-884.

Chalepakis G, Goulding M, Read A, Strachan T and Gruss P (1994): Molecular basis of splotch and Waardenburg *Pax-3* mutations. Proc Natl Acad Sci USA 91: 3685-3689.

Chen JM (1952): Studies on the morphogenesis of the mouse sternum. I. Normal development. J. Anat 86: 373-386.

Chen JM (1953): Studies on the morphogenesis of the mouse sternum. III. Experiments on the closure and segmentation of the sternal bands. J. Anat 87: 130-149.

Christ B and Wilting J (1992): From somites to vertebral column. Ann. Anat 174: 23-32.

Couly GF, Coltey PM and Le Douarin NM (1993): The triple origin of the skull in higher vertebrates: a study in quail-chick chimeras. Development 117: 409-429.

Davis RJ, D'Cruz CM, Lovell MA, Biegel JA and Barr FG (1994): Fusion of *PAX7* to *FKHR* by the variant t(1;13)(p36;q14) translocation in alveolar rhabdomyosarcoma. Cancer Res 54: 2869-2872.

Dempsey EE and Trasler DG (1983): Early morphological abnormalities in Splotch mouse embryos and predisposition to gene- and retinoic acid-induced neural tube defects. Teratology 28: 461-472.

Deutsch U, Dressler GR and Gruss P (1988): Pax1, a member of a paired box homologous murine gene family, is expressed in segmented structures during development. Cell

53: 617-625.

Dickie MM (1964): New *splotch* alleles in the mouse. J Hered 55: 97-101.

Dietrich S, Schubert FR and Gruss P (1993): Altered *Pax* gene expression in murine notochord mutants: the notochord is required to initiatew and maintain ventral identity in the somite. Mech Dev 44: 189-207.

Dressler GR, Deutsch U, Chowdhury K, Nornes HO and Gruss P (1990): *Pax2*, a new murine paired-box-containing gene and its expression in the developing excretory system. Development 109: 787-795.

Dressler GR and Douglass EC (1992): Pax-2 is a DNA-binding protein expressed in embryonic kidney and Wilms tumor. Proc Natl Acad Sci USA 89: 1179-1183.

Dressler GR, Wilkinson JE, Rothenspieler UW, Patterson LT, Williams-Simons L and Westphal H (1993): Deregulation of *Pax-2* expression in transgenic mice generates severe kidney abnormalities. Nature 362: 65-67.

Epstein DJ, Vekemans M and Gros P (1991): splotch (Sp^{2H}), a mutation affecting development of the mouse neural tube, shows a deletion within the paired homeodomain of *Pax-3*. Cell 67: 767-774.

Epstein DJ, Vogan KJ, Trasler DG and Gros P (1993): A mutation within intron 3 of the *Pax-3* gene produces aberrantly spliced mRNA transcripts in the splotch (*Sp*) mouse mutant. Proc Natl Acad Sci USA 90: 532-536.

Evans EP, Burtenshaw MD, Beechey CV and Searle AG (1988): A splotch locus deletion visible by Giemsa banding. Mouse News Lett 81: 66.

Fickenscher HR, Chalepakis G and Gruss P (1993): Murine Pax-2 protein is a sequence specific transactivator with expression in the genital system. DNA Cell Biol 12: 381-391.

Figdor MC and Stern CD (1993): Segmental organization of embryonic diencephalon. Nature 363: 630-634.

Franz T (1989): Persistent truncus arteriosus in the splotch mutant mouse. Anat Embryol 180: 457-464.

Franz T, Kothary R, Surani MAH and Grim M (1993): The Splotch mutation interferes with muscle development in the limbs. Anat Embryol 187: 153-160.

Frigerio G, Burri M, Bopp D, Baumgartner S and Noll M (1986): Structure of the segmentation gene *paired* and the Drosophila *PRD* gene set as part of a gene network. Cell 47: 735-746.

Galili N, Davis RJ, Fredericks WJ, Mukhopadhyay S, Rauscher,FR, Emanuel BS, Rovera G and Barr FG (1993): Fusion of a fork head domain gene to *PAX3* in the solid tumour alveolar rhabdomyosarcoma. Nature Genet 5: 230-235.

Glaser T, Lane J and Housman D (1990): A mouse model of the aniridia-Wilms tumor deletion syndrome. Science 250: 823-827.

Glaser T, Walton DS and Maas RL (1992): Genomic structure, evolutionary conservation and aniridia mutations in the human *PAX6* gene. Nature Genet 2: 232-239.

Goulding M, Lumsden A and Paquette AJ (1994): Regulation of *Pax-3* expression in the dermomyotome and its role in muscle development. Development 120: 957-971.

Goulding M, Sterrer S, Fleming J, Balling R, Nadeau J, Moore K, Brown SDM, Steel KP and Gruss P (1993a): Analysis of the *Pax-3* gene in the mouse mutant *splotch*. Genomics 17: 355-363.

Goulding MD, Chalepkis G, Deutsch U, Erselius JR and Gruss P (1991): Pax-3, a novel murine DNA binding protein expressed during early neurogenesis. EMBO J 10: 1135-1147.

Goulding MD and Gruss P (1989): The homeobox in vertebrate development. Curr Opin Cell Biol 1: 1088-1093.

Goulding MD, Lumsden A and Gruss P (1993b): Signals from the notochord and floor plate regulate the region-specific expression of two Pax genes in the developing spinal cord. Development 117: 1001-1016.

Grim M, Halata Z and Franz T (1992): Schwann cells are not required for guidance of motor nerves in the hindlimb in Splotch mutant mouse embryos. Anat Embryol 186: 311-318.

Grüneberg H (1950): Genetical studies on the skeleton of the mouse. I. Minor variations of the vertebral coloumn. J Genet 50: 111-141.

Grüneberg H (1954): Genetical studies on the skeleton of the mouse. XV. Tail kinks. J Genet 53: 536-550.

Grüneberg H (1963): 'The Pathology of Development', Blackwell Scientific: Oxford.

Hill RE, Favor J, Hogan BLM, Ton CCT, Saunders GF, Hanson IM, Prosser J, Jordan T, Hastie ND and van Heyningen V (1991): Mouse *Small eye* results from mutations in a paired-like homeobox-containing gene. Nature 354: 522-525.

Hirano T and Iwakura Y (1990): A novel transcriptional regulatory factor that binds to the polyoma virus enhancer in a developmental stage-specific manner. Biochimie 72: 327-36.

Hoey T and Levine M (1988): Divergent homeobox proteins recognize similar DNA sequences in Drosophila. Nature 332: 858-861.

Hogan BLM, Hirst EMA, Horsburgh G and Hetherington CM (1988): *Small eye (Sey)*: a mouse model for the genetic analysis of cranofacial abnormalities. Development 103(Supplement): 115-119.

Hogan BLM, Horsburgh G, Cohen J, Hetherington CM, Fisher G and Lyon MF (1986): *Small eyes (Sey)*: a homozygous lethal mutation on chromosome 2 which affects the differentiation of both lens and nasal placodes in the mouse. J Embryol exp Morph 97: 95-110.

Ishikiriyama S, Tonoki H, Shibuya Y, Chin C, Harado N, Abe K and Niikawa N (1989): Waardenburg syndrome type I in a child with de novo inversion (2) (q35q37.3). Am J Hum Genet 33: 505-507.

Jordan T, Hanson I, Zaletayev D, Hodgson S, Prosser J, Seawright A, Hastie N and van Heyningen V (1992): The human *PAX6* gene is mutated in two patients with aniridia. Nature Genet 1: 328-332.

Jostes B, Walther C and Gruss P (1991): The murine paired box gene, *Pax7*, is expressed specifically during the development of the nervous and muscular system. Mech Dev 33: 27-38.

Kessel M (1991): Molecular coding of axial positions by *Hox* genes. Sem Dev Biol 2: 367-373.

Kessel M and Gruss P (1990): Murine developmental control genes. Science 249: 374-379.

Kessel M and Gruss P (1991): Homeotic transformations of murine vertebrae and concomitant alteration of *Hox* codes induced by retinoic acid. Cell 67: 89-104.

Kieny M, Mauger A and Sengel P (1972): Early regionalization of the somitic mesoderm as studied by the development of the axial skeleton of the chick embryo. Dev Biol 28: 142-161.

Kitchin IC (1949): The effects of notochordectomy in Amblystoma mexicanum. J Exp Zool 112: 393-415.

Koseki H, Wallin J, Wilting J, Mizutani Y, Kispert A, Ebensperger C, Herrmann BG, Christ B and Balling R (1993): The basis of a genetic interaction between *Danforth's short tail (Sd)* and *undulated;* a role for *Pax-1* as a possible mediator of the notochord signals in the dorsoventral specification of vertebrae. Development 119: 649-660.

Krauss S, Johansen T, Korzh V and Fjose A (1991): Expression of the zebrafish paired box gene *pax[zf-b]* during early neurogenesis. Development 113: 1193-1206.

Krauss S, Maden M, Holder N and Wilson SW (1992): Zebrafish *pax[b]* is involved in the formation of the midbrain-hindbrain boundary. Nature 360: 87-89.

Laughon A and Scott MP (1984): Sequence of a Drosophila segmentation gene: Protein structure homology with DNA-binding proteins. Nature 310: 25-31.

Maulbecker CC and Gruss P (1993): The oncogenic potential of Pax genes. EMBO J 12: 2361-2367.

McGinnis W and Krumlauf R (1992): Homeobox genes and axial patterning. Cell 68: 283-302.

McMahon AP and Bradley A (1990): The *Wnt-1 (int-1)* proto-oncogene is required for development of a large region of the mouse brain. Cell 62: 1073-1085.

McMahon A. P, Joyner AL, Bradley A and McMahon JA (1992): The midbrain-hindbrain phenotype of Wnt-1⁻/Wnt-1⁻ mice results from stepwise deletion of engrailed-expressing cells by 9.5 days postcoitum. Cell 69: 581-595.

Mermod N, O'Neill EA, Kelly TJ and Tjian R (1989): The proline-rich transcriptional activator of CTF/NF-1 is distinct from the replication and DNA-binding domain. Cell 58: 741-753.

Mikkola J, Fjose A, Kuwada JY, Wilson S, Guddal PH and Krauss S (1992): The paired domain-containing factor *pax[b]* is expressed in specific commissural interneurons in zebrafish embryos. J Neurobiol 23: 933-946.

Millen K, Wurst W, Herrup K and Joyner AL (1994): Abnormal embryonic cerebelar development and patterning of post-natal foliation in two *Engrailed-2* mutants. Development 120: 695-706.

Moase CE and Trasler DG (1989): Spinal ganglia reduction in the Splotch-delayed mouse neural tube defect mutant. Teratology 40: 67-75.

Moase CE and Trasler DG (1992): Splotch locus mouse mutants: models for neural tube defects and Waardenburg syndrome type I in humans. J Med Genet 29: 145-151.

O'Shea KS, Rheinheimer JST and O'Shea JM (1987): Morphometric analysis of the forebrain anomalies in the delayed splotch mutant embryo. J Craniofacial Genet Dev Bio. 7: 357-369.

Pabo CO and Sauer RT (1984): Protein-DNA recognition. Ann Rev Biochem 53: 293-321.

Placzek M, Tessier-Lavigne M, Yamada T, Jessel T and Dodd D (1990): Mesodermal control of neural cell identity: Floor plate induction by the notochord. Science 250: 985-988.

Poleev A, Fickenscher H, Mundlos S, Winterpacht A, Zabel B, Fidler A, Gruss P and Plachov D (1992): *PAX8*, a human paired box gene: isolation and expression in developing thyroid, kidney and Wilms' tumor. Development 116: 611-623.

Pourquie O, Coltey M, Teillet M-A, Ordahl C and Le Douarin NM (1993): Control of dorsoventral patterning of somitic derivatives by notochord and floorplate. Proc Natl Acad Sci USA 90: 5242-5246.

Price M, Lazzaro D, Pohl T, Mattei M-G, Rüther U, Olivo J-C, Duboule D and Di Lauro R (1992): Regional expression of the homeobox gene *Nkx-2.2* in the developing mam malian forebrain. Neuron 8: 241-255.

Price M, Lemaistre M, Pischetola M, Di Lauro R and Duboule D (1991): A mouse gene related to *Distal-less* shows a restricted expression in the developing forebrain. Nature 351: 748-751.

Roberts RC (1967): *Small-eyes*, a new dominant mutant in the mouse. Genet Res 9: 121-122.

Rosenfeld MG (1991): POU-domain transcription factors: POU-er-full developmental regulators. Genes Dev 5: 897-907.

Saunders JW (1948): The proximo-distal sequence of origin of the parts of the chick wing and the role of the ectoderm. J Exp Zool 108: 363-403.

Schoenwolf GC and Smith JL (1990): Mechanisms of neurulation: traditional viewpoint and recent advances. Development 109: 243-270.

Schubert FR, Nieselt-Struwe K and Gruss P (1993): The Antennapedia-type homeobox genes have evolved from three precursors separated early in metazoan evolution. Proc Natl Acad Sci *USA* 90: 143-147.

Shapiro DN, Sublett JE, Li B, Downing JR and Naeve CW (1993): Fusion of *PAX3* to a member of the Forkhead Family of transcription factors in human alveolar rhabdomyosarcoma. Cancer Res 53: 5108-5112.

Stoykova A and Gruss P (1994): Roles of *Pax*-genes in developing and adult brain as suggested by expression patterns. J Neurosci 14: 1395-1412.

Tanaka M and Herr W (1990): Differential transcriptional activation by Oct-1 and Oct-2: Interdependent activation domains induce Oct-2 phosphorylation. Cell 60: 375-386.

Tassabehji M, Read AP, Newton VE, Patton M, Gruss P, Harris R and Strachan T (1993): Mutations in the *PAX3* gene causing Waardenburg syndrome type 1 and type 2. Nature Genet 3: 26-30.

Theiler K, Varnum DS and Stevens LC (1978): Development of Dickie's Small eye, a mutation in the house mouse. Anat Embryol 155: 81-86.

Thomas KR, Musci TS, Neumann PE and Capecchi MR (1991): *Swaying* is a mutant allele of the proto-oncogene *Wnt-1*. Cell 67: 969-976.

Ton CCT, Miwa H and Saunders GF (1992): *Small eye (Sey)*: Cloning and characterization of the murine homolog of the human Aniridia gene. Genomics 13: 251-256.

Treisman J, Gönczy P, Vashishita M, Harris E and Desplan C (1989): A single amino acid can determine the DNA binding specificity of homeodomain proteins. Cell 59: 553-562.

Treisman J, Harris E and Desplan C (1991): The paired box encodes a second DNA-binding domain in the Paired homeo domain protein. Genes Dev 5: 594-604.

Treisman J, Harris E, Wilson D and Desplan C (1992): The homeodomain: a new face for the helix-turn-helix? BioEssays 14: 145-150.

Tsukamoto K, Nakamura Y and Niikawa N (1994): Isolation of two isoforms of the PAX3 gene transcripts and their tissue-specific alternative expression in human adult tissue. Hum Gene . 93: 270-274.

van der Meer de Jong R, Dickinson ME, Woychik RP, Stubbs L, Hetherington C and Hogan BL (1990): Location of the gene involving the *small eye* mutation on mouse chromosome 2 suggests homology with human aniridia 2 (AN2). Genomics 7: 270-275.

van Straaten HMW and Hekking JWM (1991): Development of floor plate, neurons and axonal outgrowth pattern in the early spinal cord of the notochord-deficient chick embryo. Anat Embryol 184: 55-63.

van Straaten HW, Hekking JW, Thors F, Wiertz-Hoessels EL and Drukker J (1985a): Induction of an additional floor plate in the neural tube. Acta Morphol Neerl Scand 23: 91-97.

van Straaten HW, Hekking JW, Wiertz-Hoessels EJ, Thors F and Drukker J (1988): Effect of the notochord on the differentiation of a floor plate area in the neural tube of the chick embryo. Anat Embryol 177: 317-324.

van Straaten HWM, Hekking JWM, Consten C and Copp AJ (1993): Intrinsic and extrinsic factors in the mechanism of neurulation: effect of curvature of the body axis on closure of the posterior neuropore. Development 117: 1163-1172.

van Straaten HWM, Thors F, Wiertz-Hoessels EL and Drukker J (1985b): Effect of a notochordal implant on the early morphogenesis of the neural tube and neuroblasts: histometrical and histological results. Dev Biol 110: 247-254.

Vogan KJ, Epstein DJ, Trasler DG and Gros P (1993): The *splotch-delayed (Sp^d)* mouse mutant carries a point mutation within the paired box of the *Pax-3* gene. Genomics 17: 364-369.

Waardenburg P (1951): A new syndrome combining developmental anomalies of the eyelids, eyebrows, and nose root with congenital deafness. Am J Hum Genet 3: 195-253.

Wallin J, Mizutani Y, Imai K, Miyashita N, Moriwaki K, Taniguchi M, Koseki H and

Balling R (1993): A new Pax gene, Pax-9, maps to mouse chromosome 12. Mammalian Genome 4: 354-358.

Wallin J, Wilting J, Koseki H, Fritsch R, Christ B and Balling R (1994) The role of Pax-1 in axial skeleton development. Development 120: 1109-1121.

Walther C (1989): Isolierung und Charakterisierung Paired Box homologer Gene der Maus: Die Pax-Familie.

Walther C and Gruss P (1991): *Pax-6*, a murine paired box gene, is expressed in the developing CNS. Development 113: 1435-1449.

Walther C, Guénet J-L, Simon D, Deutsch U, Jostes B, Goulding M, Plachov D, Balling R and Gruss P (1991): Pax: a murine multigene family of paired box containing genes. Genomics 11: 424-434.

Williams BA and Ordahl CP (1994): Pax-3 expressionin segmental mesoderm marks early stages in myogenic cell specification. Development 120: 785-796.

Wright ME (1947): *Undulated*: A new genetic factor in *Mus musculus* affecting the spine and tail. Heredity 1: 137-141.

Wurst W, Auerbach AB and Joyner AL (1994): Multiple developmental defects in *Engrailed-1* mutant mice: and early mid-hindbrain deletion and patterning defects in forelimbs and sternum. Development 120: 2065-2075.

Yamada T, Placzek M, Tanaka H, Dodd J and Jessell TM (1991): Control of cell pattern in the developing nervous system: Polarizing activity of the floor plate and notochord. Cell 64: 635-647.

Yang X-M and Trasler DG (1991): Abnormalities of neural tube formation in pre-spina bifida Splotch-delayed mouse embryos. Teratology 43: 643-657.

Zannini M, Francis-Lang H, Plachov D and Di Lauro R (1992): Pax-8, a paired domain-containing protein, binds to a sequence overlapping the recognition site of a homeodomain and activates transcription from two thyroid-specific promoters. Mol Cell Biol 12: 4230-4241.

INDUCTION AND THE GENERATION OF REGIONAL AND CELLULAR DIVERSITY IN THE DEVELOPING MAMMALIAN BRAIN

Anthony-Samuel LaMantia[1], Melissa C. Colbert[2,3], and Elwood Linney[2]

Departments of Neurobiology[1] and Microbiology[2], Duke University Medical Center, Durham, NC, 27712, U.S.A.
[3]Present Address: The Children's Hospital Medical Center, Division of Molecular Cardiovascular Biology, The Children's Hospital Research Foundation, 3333 Burnett Ave, Cincinnati, OH 45229, U.S.A.

INTRODUCTION

Local inductive interactions, in contrast to a rigid mosaic of transcription factor expression, define domains and distinguish subsets of cells that prefigure the establishment of regional and cellular diversity in the developing central nervous system. The olfactory bulb in the forebrain and the cervical and lumbar enlargements of the spinal cord differentiate from regions of the neural tube where, at an earlier time, local retinoid signals activate a subset of more widely distributed retinoid sensitive transcription factors (retinoid receptors) leading to local differences in gene expression. In addition, within these domains of differential gene expression, subset of cells respond to the inductive signal (in this case, retinoic acid) while others do not. These, differences in gene expression that prefigure the division of the brain into distinct functional subdivisions and cell types may reflect the coincidence of inductive signals and relevant signal transduction molecules, including transcription factors at both the regional and cellular level. This relationship may contribute both to the establishment of functionally distinct regions and cell classes in the developing central nervous system.

TRANSCRIPTION FACTORS, INDUCTIVE SIGNALS AND REGIONAL AND CELLULAR DIVERSITY IN THE DEVELOPING BRAIN

Over the past five years, the developing neural tube has been divided into a complex checkerboard of expression domains for various developmentally

Neural Cell Specification: Molecular Mechanisms and Neurotherapeutic Implications
Edited by Juurlink *et al.*, Plenum Press, New York, 1995

regulated genes (for review see Holland and Hogan, 1988; Wilkinson et al., 1989; Lumsden and Keynes, 1989; Kesssel and Gruss, 1990; Rubenstein and Puelles, 1993). These domains have been compared to the segments and compartments that play a role in earlier embryonic pattern formation in a number of invertebrate and vertebrate species (for review see Ingham, 1988; McGinnis and Krumlauff, 1992). In the developing vertebrate brain, many of the genes that define these domains code for putative transcription factors with homologies to known DNA binding proteins in *Drosophila* (Wilkinson et al., 1989; McGinnis and Krumlauf, 1992), others are for secreted signaling molecules of the peptide or steroid hormone family (Jessel and Melton, 1992; Linney and LaMantia, 1994), and others are for cell surface receptors, adhesion and recognition molecules (Jessel, 1988; Lumsden and Keynes, 1989; Hynes and Lander, 1992).

The mosaic distribution of these molecules, particularly transcription factors, might directly predict the subdivision of the neural tube into distinct structural or functional regions of the mature brain. This suggestion seems sensible based upon comparisons with the coincidence of segmental boundaries and domains of transcription factor expression in invertebrate embryos (reviewed by Ingham, 1988). Occasionally, expression domains of several developmentally regulated molecules prefigure or coincide with rudimentary brain regions; just as often, however, they do not. Furthermore, when some of these molecules are inactivated, either by naturally occurring or targeted mutations, the effects do not easily accord with their pattern of expression (McMahon and Bradley, 1988; Thomas and Capecchi, 1990; Chisaka et al., 1992; Erickson, 1992; Lohnes et al., 1993; Lufkin et al., 1993; Tomasiewicz et al., 1993). Thus, the establishment of regional identity in the brain cannot be simply explained by expression patterns of either transcription factors or signaling molecules.

If the expression patterns of developmentally regulated molecules are insufficient to explain the accquisition of positional information and cellular identity in the developing brain, what other sorts of cellular and molecular mechanisms might contribute to these essential steps of neural development? One possibility is cell-cell interactions that depend upon the concerted activity of signaling molecules, receptors and transcription factors during embryonic development. This sort of signaling is now generally considered the cellular and molecular equivalent of embryonic induction. Embryonic induction - where one cell or tissue signals the differentiation of another adjacent cell or tissue - has been recently rediscovered as a tractable problem for cell and molecular biology as well as neurobiology (reviewed by Gurdon, 1987; Slack, 1991; Jessel and Melton, 1992). This renaissance is based on observations that placed inductive interactions squarely in the realm of cell-cell signaling (reviewed by Evans, 1988; Schlessinger and Uhlrich, 1992). Much of the recent interest in inductive interactions in the nervous system has been focused upon the first steps of neural plate and neural tube formation (Amaya et al., 1990; Hemmati-Brivanlou and Melton, 1992; Hogan et al., 1992; Lamb et al., 1993; Echelard et al., 1993). Less attention has been paid to inductive interactions that occur while the neural tube is differentiating into a rudimentary brain. These interactions may lead to local changes in gene expression via direct or indirect interactions with transcription factors, or they may precipitate other cellular changes that influence regional or cellular differentiation of the developing brain.

It is tempting to assume that the singular presence of either a transcription factor or a signaling molecule in a particular region or cell is equivalent to differentiation in that same region or cell. Functionally significant domains, however, might arise as a consequence of the modulation of regional differences in gene expression that rely upon the interactions of local inductive signals and transcription factors. These interactions might then establish a transcriptional history that biases a region or group of cells toward a particular fate. This seems plausible since some of the intracellular targets of fairly well characterized inductive signaling molecules are transcription factors, either directly, as in the case of steroid/thyroid-like molecules (Evans, 1989) or indirectly, as for peptide growth factors (Cantley et al., 1991).

Local modulation of transcription factor activity by inductive signals might provide a more flexible way to subdivide the neural tube into forerunners of functionally or structurally distinct brain regions. Furthermore, variability in inductive signaling between individual cells might lead to local differences in cell fate. Inductive interactions require two distinct embryonic tissues or cells to be in the right place at the right time, and to be competent to interact with one another. This sort of requirement provides increased specificity as well as reliability in the establishment of positional information and cellular identity. A major challenge, therefore, is to assess the relationship between inductive signals, transcription factors, regional and cellular differences in gene expression during neural development. We have approached this challenge by studying the role of retinoid signaling in establishing regional and cellular diversity in the developing brain and spinal cord.

STUDYING RETINOID SIGNALING IN THE DEVELOPING BRAIN AND SPINAL CORD

Several lines of evidence support a role for retinoid-mediated induction in the regional and cellular differentiation of the developing brain and spinal cord. Retinoic acid (RA), the acidic form of vitamin A, has been known to be a powerful teratogen since the mid-1930's (reviewed by Schardein, 1993; Linney and LaMantia, 1994). Exogenously provided RA has specific effects on morphogenesis in vertebrate embryos (Tickle et al., 1982) and cellular differentiation in cultured cell lines (Strickland and Mahadavi, 1978; Jones-Villeneuve et al., 1982). Many of the teratogenic effects of various retinoids are focused upon the brain and spinal cord, particularly when retinoid exposure happens during early to mid-gestation (Schenefelt, 1972), the time when distinct regions and cell types begin to be established. Endogenously produced retinoids can act to influence morphogenesis in developing embryos via specific nuclear receptor/transcription factors (Thaller and Eichele, 1987; Giguere et al., 1987; reviewed by Linney, 1992). In fact, there is a broad family of retinoid signaling molecules (Fig. 1) that influence both retinoid metabolism (reviewed by Giguere, 1994) and retinoid dependent gene expression via interactions with specific response elements (de The et al., 1990; Sucov et al., 1990). Thus, the available evidence suggests that the details of RA signaling might provide insight into the role that inductive interactions play in the regional and cellular differentiation of the brain.

To investigate the relationship between retinoid signaling, regional differences in gene expression, and differentiation of functionally significant subdivisions in the developing brain we took advantage of the basic molecular mechanism by which RA is thought to modulate gene expression (Fig. 1). Retinoic acid (RA) binds to retinoic acid receptors or retinoid X receptors (RARs and RXRs) which are found in cell nuclei. Activated RARs dimerize, bind to specific response elements found in the regulatory regions of genomic DNA (retinoic acid response elements, or RAREs), and thereby modulate gene expression. Thus, the molecular mechanisms of RA signaling offer an opportunity to investigate the activity of the signaling molecule, target transcription factors, and downstream gene activation.

To take advantage of these features of the retinoid signaling pathway we used a retinoid sensitive indicator transgene to detect retinoid-mediated gene expression either in transgenic indicator mouse embryos (Balkan et al., 1992; see also Mendelsohn et al., 1992; Rossant et al., 1991) or in an indicator cell line (Colbert et al., 1993; see also Wagner et al., 1992; Chen et al., 1992). The transgene consists of a contrived promoter with three RAREs followed by a viral thymidine kinase promoter and bacterial β-galactosidase as the reporter (Fig. 1b; see also Balkan et al., 1992). We used mouse embryos as well as mouse L cells (a fibroblast line) carrying this indicator transgene to determine the relationship between where RA is available and where RA-mediated gene expression occurs in the developing brain and spinal cord. The transgenic embryos demonstrate where and when a specific type of RA mediated-gene expression occurs during development since transgene expression relies upon the coincidence of RA, plus RARs or RXRs. The cell line provides a uniform population of RA-responsive cells to evaluate the production of RA from distinct tissues in the developing embryo. Finally, we used in situ hybridization to determine where and when RARs, RXRs and other RA signaling molecular messages are actually present in the embryo.

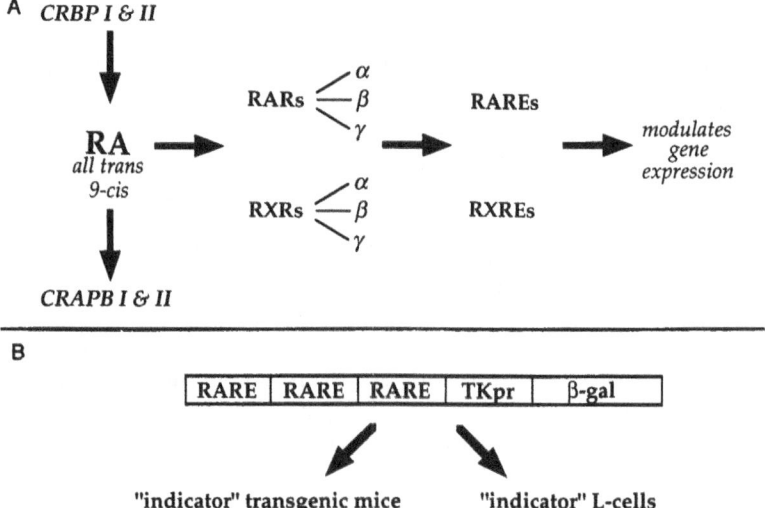

Figure 1. (A) Schematic of the retinoid signaling pathway showing the endogneous activating ligands *all trans* retinoic acid and *9-cis* retinoic acid (RA). In addition the several retinoic acid receptor (RARs) and retinoid X receptors (RXRs) subtypes are indicated as well as the catabolic and metabolic cellular retinoic acid and cellular retinol binding proteins. The schematic from left to right suggests the basic sequence of events for retinoid signaling in several classes of cells including neurons. (B) Schematic of the retinoid sensitive transgene used to detect RA-mediated gene expression in cell lines and mouse embryos.

Retinoid-mediated gene expression prefigures regional differentiation in the forebrain and spinal cord

First, one must determine whether RA-mediated signaling leads to region specific gene expression that is significant for the development of distinct brain regions. To examine this question, we took advantage of the transgenic mouse line carrying the indicator transgene whose expression is RA-regulated (Fig. 1B). One assumes that this contrived RA-responsive promoter should drive β-gal expression in cells where RA receptors are present and RA is available. In the brain of mouse embryos at mid gestation, there are two patterns of RA-dependent transgene expression that prefigure the differentiation of functionally significant regions and pathways (Fig. 2, see also LaMantia et al., 1993; Colbert et al., 1993; 1994).

A distinct pattern is seen in the forebrain vesicle and olfactory placode between the 8th and 11th day of gestation (E8-E11). Initially, in the head there is a crescent shaped domain of RA-dependent transgene expression in the ventrolateral aspect of the prosencephalic vesicle, and a patch of transgene expression in a thickened region of the lateral surface ectoderm that defines the olfactory placode. The ventrolateral expression domain persists once the prosencephalic vesicle has differentiated into two paired telencephalic vesicles. The olfactory pit, which differentiates from the olfactory placode, also remains transgene positive (Fig. 2A). Subsequently, the earliest axons from the retinoid activated olfactory epithelium grow specifically into the retinoid activated domain in the forebrain, and it is highly likely that this region goes onto differentiate into the olfactory bulb and other olfactory-recipient regions of the basal forebrain (LaMantia et al., 1993; Whitesides and LaMantia, 1995; Drake et al., 1994). Thus, the coordination of RA-mediated gene expression in the forebrain and placode apparently prefigures the establishment of distinct forebrain regions and some of their major connections.

The second pattern of retinoid activated gene expression is found in the spinal cord from E12.5 until birth. These retinoid activated domains correspond to the cervical and lumbar enlargements. The thoracic and sacral regions are devoid of transgene activity. This observation can be confirmed quantitatively by counting sensory ganglia that correspond to cervical, thoracic, lumbar and sacral regions and then correlating this data with the pattern of transgene expression (Fig. 2B; see also Colbert et al., 1994). The final numbers and types of spinal cord motor and sensory neurons vary in register with these four regions. Thus, our evidence suggests that retinoid mediated gene expression in the cervical and lumbar domains might accompany the establishment of these cytological differences in the spinal cord.

RA-activated domains appear at an appropriate time, just before or during the establishment of the relevant rudimentary functional subdivisions, to be considered as a potential contributor to the process of regional differentiation. Furthermore, the consistency of the boundaries of these domains across hundreds of embryos in at least three independent transgenic lines reinforces the conclusion that these transgene expression patterns represent highly regulated inductive fields. Thus, understanding how these fields are defined should offer insight into the process by which the undifferentiated neuroepithelium becomes divided into rudiments of major functional subdivisions of the mature brain.

How are domains established in the developing forebrain and spinal cord?

Domains that prefigure regional differentiation in the brain could result primarily from a mosaic pattern of transcription factor expression (in this case, RARs and RXRs), local activation of transcription factors by a discrete inductive signalling source, or the close spatial and temporal coordination of transcription factor expression and signalling molecule availability. Thus, there are at least three combinations of retinoid signals, RARs and RXRs that might explain the patterns of RA mediated gene expression in the forebrain and spinal cord (Fig. 3). First, RA might be available widely and sensitive transcription factors - in this case the RARs - expressed in a region-specific pattern. Second, sources of RA may be localized and activate only a subset of receptors from a more widely distributed population. Third, both retinoid sources and receptors might be localized. To answer which of these three combinations explains the domains of RA-mediated gene expression we have observed, we first asked whether RA comes from local sources in the developing embryo, and we then evaluated the distribution of RARs and RXRs in the developing brain.

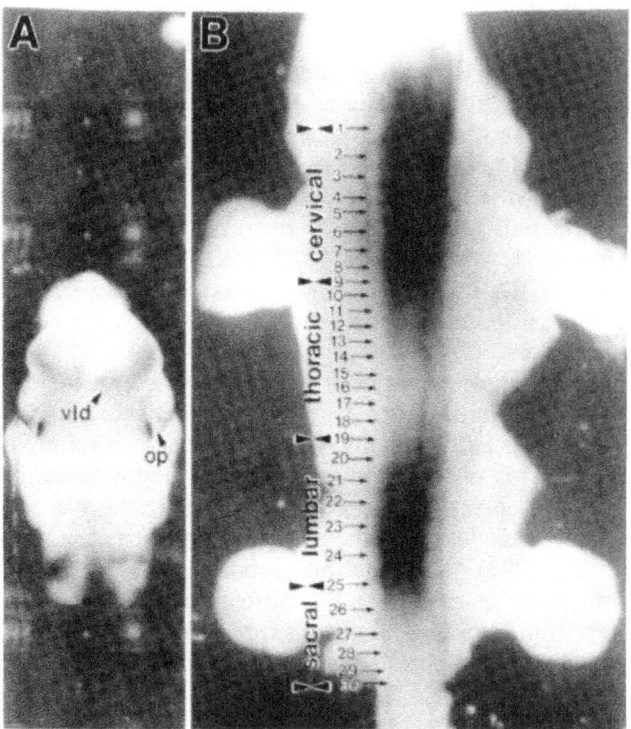

Figure 2. Patterns of retinoid activated gene expression prefigure the differentiation of major functional domains in the forebrain and spinal cord. (A) In the forebrain on embryonic day 10.5 (day of breeding = embryonic day 0.5), a crescent of transgene expression is seen in the ventrolateral telencephalon as well as in the olfactory placode. The initial axons that extend from the placode grow specifically into this retinoid activated domain, prior to the differentiation of a recognizable rudiment of the olfactory bulb. (B) On embryonic day 12.5, two domains of retinoid activated gene expression are seen in the spinal cord. When one counts sensory ganglia and compares the segmental levels with the labeling pattern, it is clear that this labeling corresponds to regions of the cord where the cervical and lumbar enlargements will emerge.

We used the transgenic indicator cell line to determine whether RA comes from local sources or is available widely in the embryo. If the pattern of RA-mediated gene expression in the forebrain and spinal cord reflects local sources of RA, then subregions of the forebrain, spinal cord or adjacent surrounding tissues should activate transgene expression when cocultured with our uniformly RA responsive indicator cells whereas non adjacent regions should not. Our results show that RA sources co-localize with domains of retinoid-mediated gene expression. For the forebrain and olfactory placode, the mesenchyme sandwiched in between the two retinoid responsive domains is the apparent source of RA (Fig. 4A), thus providing a common signal for these two regions. Surrounding tissues did not elicit transgene expression in the indicator cells. Similarly, the spinal cord induces transgene expression in indicator cells that are adjacent only to the cervical and lumbar cord (Fig. 4B). This suggests that at the mid-point of neural development, when few neurons have been generated and functional regions have yet to differentiate, retinoid sources are already localized in register with the pattern of retinoid mediated gene expression in the developing forebrain and spinal cord.

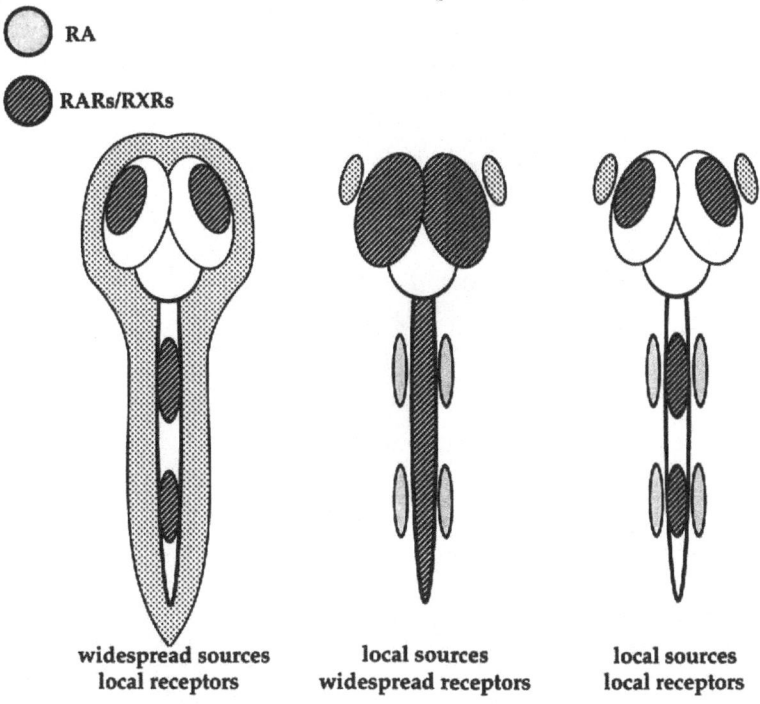

Figure 3. Three possible explanations for domain specific inductive activation of retinoid mediated gene expression. (A) Widely available signaling molecule, domains are shaped by limited expression of responsive transcription factors. (B) Localized signal that activates gene expression via a subset of widely distributed responsive transcription factor. (C) Coincidence of local sources and responsive transcription factors.

Despite the discrete localization of RA sources, it is possible that the expression of the transcription factors which respond directly to RA signals, RARs and RXRs, are similarly patterned. This would make it difficult to determine whether local inductive signaling or a transcription factor mosaic was primarily responible for the patterns of RA-mediated gene expression seen in the forebrain and spinal cord. Therefore, we examined the expression patterns of retinoid signal transduction molecules—particularly the RA-activated transcription factors—in a developing embryos using digoxygenin-labeled RNA probes for the relevant mRNAs. In the forebrain, RARα and RXRγ are seen throughout the prosencephalic or telencephalic vesicles, up to the diencephalic/mesencephalic junction (Fig. 5A,B). These two retinoid receptors are also seen throughout the olfactory placode and epithelium. The expression of RARα and RXRγ in the spinal cord is similarly widespread (Fig. 5D,E); mRNAs for these receptors are found in the thoracic and sacral spinal cord, where RA-mediated transgene expression is absent, as well as in the cervical and lumbar cord. In contrast, the distribution of the RA inducible RARβ (Fig. 5C,F), as well as the RA inducible metabolic binding protein CRBPI and the RA inducible catabolic binding protein CRABPII follows more closely (but not precisely) the localized sources of RA in both the forebrain and spinal cord (Colbert et al., 1995).

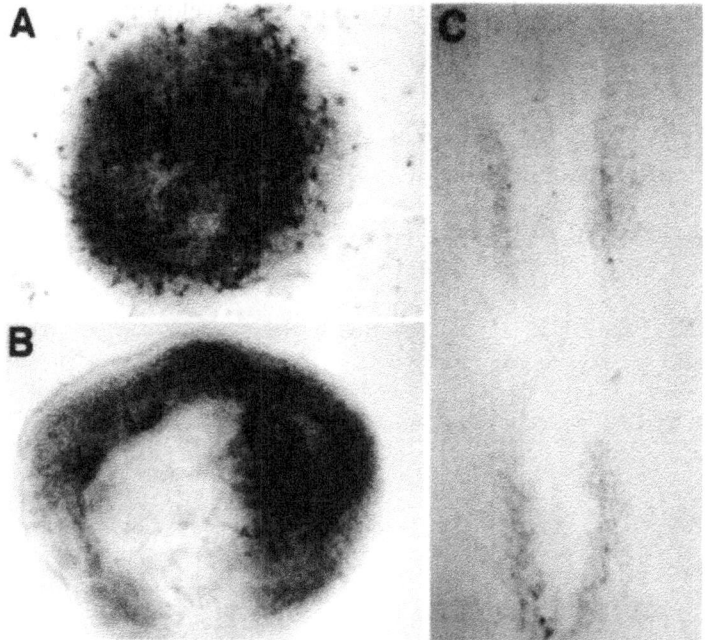

Figure 4. Retinoids that activate the indicator transgene come from localized sources in the E10.5 and 12.5 mouse embryo. (A) lateral cranial mesenchyme underlying the olfactory pit induces transgene expression in L-C2A5 indicator cells. (B). Olfactory pit neuroepithelium fails to induce transgene expression. (C). Transgene expression in indicator cells co-cultured with whole spinal cord is limited to the cervical and lumbar regions of the cord. (A) and (B) from LaMantia et al., 1993; (C) from Colbert et al., 1993.

The widespread distribution of at least two RA receptors - as well as the local expression of several RA-inducible RA signalling molecules - suggests that the establishment of limited domains of RA-dependent gene expression is primarily dependent upon the localization of RA sources. The local expression of RARβ, as well as that of the RA-inducible catabolic and metabolic RA binding proteins might help to delineate further the boundaries of RA-mediated gene expression, or provide a wider variety of receptors for dimerization and thus enrich possibilities for transactivation of genes within RA-specified inductive domains. The distribution of RARα and RXRγ in the forebrain and spinal cord suggests that some RA responsive transcription factors are distributed widely; accordingly, their ability to activate gene expression may be determined by local sources of RA. This also suggests that some of these retinoid receptors may remain "silent" during distinct stages of development, since they apparently do not activate retinoid-dependent gene expression as measured by the RA-sensitive transgene.

Are there "silent" transcription factors?

If the "silence" of some retinoid receptors results from the fact that they never receive a retinoid signal, then the RARs and RXRs in the medial forebrain or thoracic and sacral spinal cord should retain the ability to activate RA-mediated transgene expression, if provided with RA. It is also possible, however, that accessory factors repress RA-mediated gene expression in these regions or that the transgene is only activated via an RARβ-dependent mechanism. If this is the case, exogenous RA might be ineffective in eliciting RA-dependent gene expression outside of the ventrolateral forebrain and the cervical and lumbar spinal cord. To answer this question in the forebrain and spinal cord, pregnant mother mice were fed large doses of RA either on E9.5 when the RA activated forebrain domains are first recognizable, or on E12.5 shortly after the cervical and lumbar spinal cord domains are first seen. The resulting pattern of transgene expression in these embryos was examined twenty-four hours later.

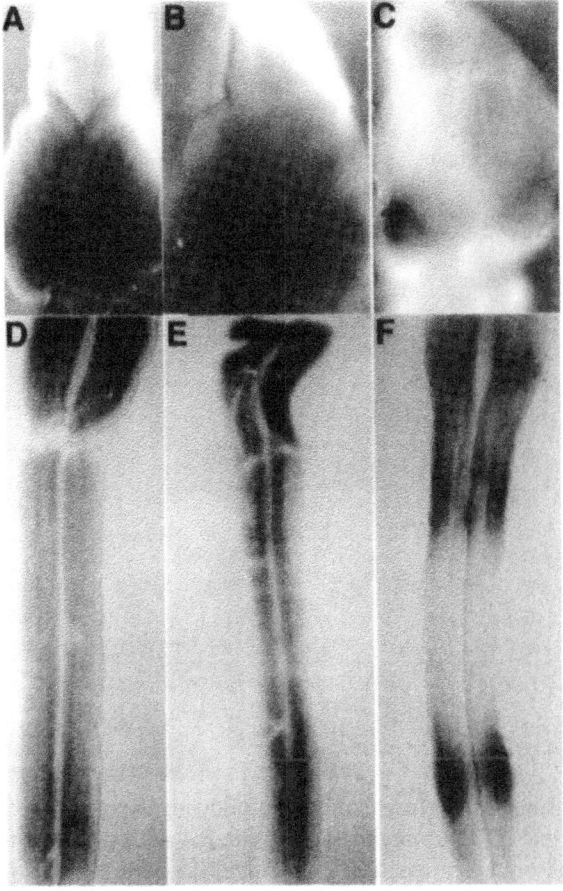

Figure 5. Widespread distribution of RARα and RXRγ in the forebrain and spinal cord accompanied by limited distribution of RARβ. (A) RARα in E 10.5 forebrain (B) RXRγ in E 10.5 forebrain (C) RARβ in E10.5 forebrain. (D) RARα in E 12.5 spinal cord. (E) RXRγ in E 12.5 spinal cord. (F) RARβ in E 12.5 spinal cord.

In the forebrain there was a consistent expansion of transgene expression to include all of the medial prosencephalon and diencephalon (Fig. 6A). This expanded expression domain stopped quite cleanly at the diencephalic/mesencephalic boundary, in register with the limits of expression of RARα and RXRγ. Thus, receptors found beyond the normal limits of an RA induced domain are able to activate ectopic expression of RA sensitive genes. "Silent" RARs and RXRs in the thoracic and sacral spinal cord are similarly able to activate ectopic RA-mediated transgene expression in the presence of exogenous RA. When E12.5 embryos received RA via maternal circulation or when the spinal cords from E12.5 embryos were cultured for 24 hours in the presence of 10^{-7} M RA, there is transgene activity in the the alar region of the thoracic and sacral cord (Fig. 6B). Thus, retinoid receptors outside the normal retinoid activated domains retain the ability to drive transgene expression when provided with RA.

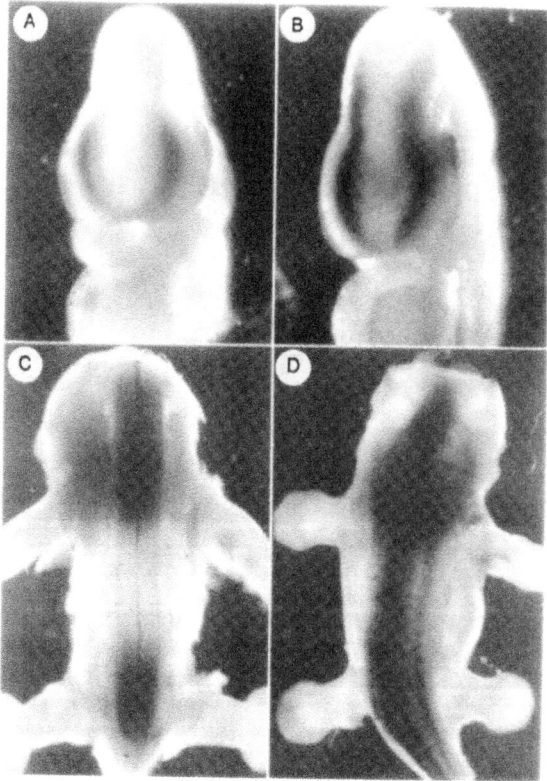

Figure 6. Expansion of transgene expression following exposure to exogenous retinoic acid in vivo (A) Normal E10.5 embryo showing the RA-activated ventrolateral domain in the forebrain and the olfactory placode. (B) An E10.5 embryos showing expansion of the ventrolateral domain following maternal administration of RA at E9.5. Note also that the eye, which is normally not activated at this time has been activated. (C) Spinal cord in an E12.5 embryo showing the normal pattern of cervical and lumbar domains. (D) Spinal cord in an E12.5 embryo following maternal administration of RA at E11.5.

Induction, normal development, and teratogenesis

Regional specification in the nervous system, at least that which uses RA as a signal, apparently involves the following relationship between induction and transcription factor expression: local sources of inductive signal activate a subset of more widely distributed transcription factors (see Fig. 3). These activated transcription factors then go on to influence gene expression in those areas. Its important to note, however, that overlayed on this general plan, perhaps as a second amplification step, some transcription factors (in this case, those whose expression is retinoid-modulated, like RARβ) are expressed in register with local sources of RA. Exogenous RA leads not only to ectopic transgene expression but ectopic RARβ expression as well (Conlon and Rossant, 1992; Mendelsohn et al., 1992; Colbert et al., 1995). This suggests that the normal expression of RA-sensitive genes depends primarily upon the location of RA sources.

There is apparently a field of "silent" transcription factors—including RARα and RXRγ. When given the opportunity, via exogenously provided RA, these silent transcription factors are competent to respond to inductive signals. It is possible that this opportunity arises during teratogenesis. Some of the deleterious effects of early retinoid exposure might be explained by the ectopic stimulation of RARs and RXRs leading to subsequent inappropriate gene expression and differentiation (see also Morriss-Kay et al., 1991; Conlon and Rossant, 1992; Zimmer and Zimmer, 1992; Kessel, 1993). In such cases, the developmental anomalies might bear little or no relation to the endogenous sites of RA-signaling and activation. This observation raises a flag of caution in interpreting teratogenic observations. The teratogenic effects of retinoids or other inductive signaling molecules might reflect ectopic activation of gene expression via otherwise silent transcription factors rather than overstimulation of normal signaling in appropriate locations.

Silent transcription factors might also be relevant to interpreting embryological experiments where bits of tissue are transplanted from one position in the embryo to another to assess developmental potential. The relocated tissues, which might never have the chance to respond to certain inductive signals in their original location, may retain the ability to do so due to the presence of silent transcription factors that are sensitive to inductive signals. Thus, if the tissue is transplanted to a location that provides appropriate signals, region-appropriate trans-differentiation might occur. Accordingly, the embryological phenomenon of competence might have as its basis the normal mismatch between local inductive sources and widespread responsive transcription factors. The widespread distribution of certain ligand activatable transcription factors like the RARs and RXRs might confer competence to fairly extensive regions of the neuroepithelium to respond to retinoid signals. For example, it is possible that experiments in which the transplanted olfactory placode (perhaps with RA-producing mesenchyme attached) elicits forebrain differentiation at several sites in the neural tube (Stout and Graziadei, 1980), might be explained by this observation.

Retinoid-activated transcriptional regulation provides a particularly clear and compelling case for the local sculpting of domains of gene expression from a more widespread responsive field. The eventual coincidence of these domains with distinct functional regions supports the idea that inductive interactions are employed at an early time perhaps to shape significant aspects of brain regional organization and connectivity. These principles of regional differentiation, however, need not operate only through ligand activated transcription factors like RARs and RXRs. The activity of several other classes of developmentally regulated transcription factors including HOX and hox-like proteins might be influenced by peptide hormone inductive signals that exert their influence via second messenger mediated post-translational modifications rather than direct ligand binding. The activation of gene expression via local induction - which we have shown for retinoids - should be considered when interpreting the signficance of patterns of transcription factor expression in the developing brain. The simple presence of a transcription factor in a particular field may not be sufficient to suggest a contribution by that factor to establishing boundaries of gene expression that prefigure regional differentiation.

INDUCTION AND THE GENERATION OF CELLULAR DIVERSITY IN THE DEVELOPING BRAIN

A great deal of attention has been paid to understanding inductive interactions and transcriptional regulation at the level of entire fields of embryonic cells. It is not always clear, however, whether or not every cell within a particular field responds uniformly to a particular signal, or uses a particular transcription factor to modulate its gene expression. It is entirely possible that embryonic fields may be defined grossly based upon macroscopic limits of activation of a particular molecule. It is also possible that individual cells within those boundaries exhibit heterogeneity of expression or activation of that molecule. This heterogeneity might contribute to the generation of diverse cell types that occurs in any differentiating embryonic tissue.

We have begun to examine the relationship between heterogeneity for inductive interactions and transcriptional activation and the generation of cellular diversity within retinoid-activated forebrain and spinal cord domains. The most important observation to emerge thus far is that within any of the RA-activated domains not all cells respond to RA-signals. Within the neuroepithelium of the ventrolateral forebrain, cervical or lumbar spinal cord or olfactory pit there is a mosaic of RA-responding cells (Fig. 7), most of which are proliferative cells based upon location, morphology, and their lack of expression of neuron or glial specific molecules (LaMantia et al., 1993; Colbert et al., 1995). This mosaicism may be correlated with discontinuities in the cellular distribution of a number of retinoid receptors and binding proteins (Colbert et al., 1995). The macroscopic limits of an inductive field are probably established by local inductive sources acting upon responsive signal transduction molecules. Cellular responses to the signal, however, may be determined on a cell by cell basis. This mosaicism may contribute to the establishment of subsets of precursor cells that are biased toward particular cellular fates within newly emerging, functionally distinct domains.

CONCLUSION

For an undifferentiated neural precursor cell or neuroblast, the establishment of a regional and cellular fate is the ultimate goal. The molecular mechanisms that facilitate this goal include the regulation of gene expression by inductive signals and transcription factors at very early stages in the life histories of these cells. The evidence reviewed here suggests that the signaling molecule retinoic acid and the transcription factors that respond to this signal—the RARs and RXRs—interact to assist subsets of neural precursors to know where they are, and perhaps, what they are to become. The importance of these early events for biasing undifferentiated cells toward a specific fate that reflects both position and cell type cannot be understated. Current efforts to recapitulate the differentiation of distinct classes of neurons in vitro from neuronal stem cell lines might indeed be even more successful if one could manipulate the full complement of inductive signals and transcriptional regulation that helps to define their position and fate long before the final facets of their neuronal phenotype are established. This sort of manipulation can be imagined with the help of a detailed knowledge of the molecular and cellular mechanisms that bias neural precursors toward their regional and cellular fates early in embryonic development.

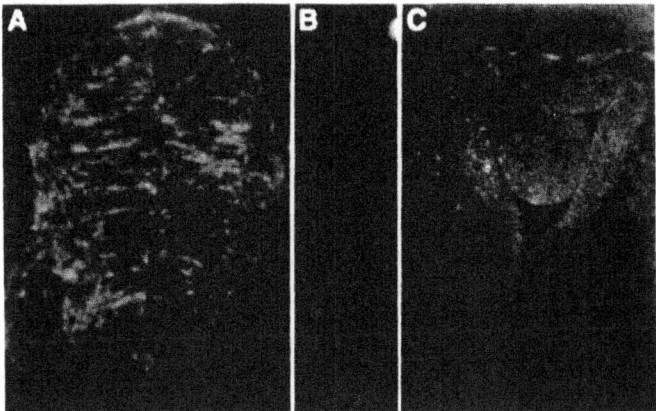

Figure 7. Cellular heterogeneity of RA-responsiveness in the neuroepithelium of the ventrolateral forebrain (A), the cervical spinal cord (B) and the olfactory epithelium (C). In the forebrain at E10.5 and spinal cord at E12.5,the majority of the responsive cells are apparently proliferative cells. In contrast, by E13.5 in olfactory epithelium many of the responsive cells are neurons, based upon double labeling with neurons specific markers like NCAM and GAP43 (not shown).

ACKNOWLEDGEMENTS

We thank Beth Gerwe for excellent technical assistance. Tiffany Jurgens contributed to preliminary experiments in the forebrain, and Will Rubin to recent work on retinoid-mediated inductive events in the spinal cord. A.-S. L. is supported by NIH HD 29178, a National Down Syndrome Society Science Scholar Award and a Sloan Foundation Fellowship. M.C.C. is supported by HD07637 and E.L. by CA39066, HD 28855, and HD 24130.

REFERENCES

Amaya E, Musci TJ and Kirschneer M (1991): Expression of a dominant negative mutant of the FGF receptor disrupts mesoderm formation in Xenopus embryos. Cell 66: 257-270.

Balkan W, Colbert MC, Bock C and Linney E (1992): Transgenic indicator mice for studying activated retinoic acid receptors during development. Proc Natl Acad Sci USA 89: 3347-3351.

Chen Y, Huang L, Russo AF and Solursh M (1992): Retinoic acid is enriched in Hensen's node and is developmentally regulated in the early chicken embryo. Proc Natl Acad Sci USA 89: 10056-10059.

Chisaka O, Musci T and Capecchi (1992): Regionally restricted developmental defects resulting from targetted disruption of the mouse homeobox gene hox 1.5. Nature 350: 473-479.

Colbert MC, Linney E and LaMantia A-S (1993): Local sources of retinoic acid coincide with retinoid-mediated trangene activity during embryonic development. Proc Natl Acad Sci USA 90: 6572-6576.

Colbert MC, Rubin W.W, Linney E and LaMantia A-S (1995): Retinoid signaling and the generation of regional and cellular diversity in the embryonic mouse spinal cord. Dev Dynamics (submitted).

Conlon RA and Rossant J (1992): Exogenous retinoic acid rapidly induces anterior ectopic expression of murine Hox-2 genes in vivo. Development 116: 357-368.

de The H, Vivanco-Ruiz MDM Tiollais P, Stunnenberg H and Dejean A (1990a) : Identification of a retinoic acid responsive element in the retinoic acid receptor b gene. Nature 343: 177-180.

Drake DP, Gerwe EA, Linney E and LaMantia A-S (1994): Role of retinoid signalling in the developing mutant mouse *small eye*. Soc. Neurosci. Abstr. (in press).

Echelard Y, Epstein DJ, St-Jacque B, Shen L, Mohler J, McMahon JA and McMahon AP (1993): Sonic hedgehog, a member of a family of putative signaling molecules, is implicated in the regulation of CNS polarity. Cell 75: 1417-1430.

Giguère V (1994): Retinoic acid receptors and cellular retinoid binding proteins: complex interplay in retinoid signaling. Endocrine Rev 15: 61-79.

Giguère V, Ong ES, Sequi P and Evans RM (1987): Identification of a receptor for the morphogen retinoic acid. Nature 330: 624-628.

Gurdon JB (1987): Embryonic induction - molecular prospects. Development 99: 285-306.

Hemmati-Brivanlou A and Melton DA (1992): A truncated activin receptor inhibits mesoderm induction and formation of axial structures in Xenopus embryos. Nature 359: 609-614.

Hogan BLM, Thaller C and Eichele G (1992): Evidence that Hensen's node is a site of retinoic acid synthesis. Nature 359: 237-241.

Holland PWH and Hogan BLM (1988): Expression of homeo box genes during mouse development: a review. Genes and Develop 2: 773-782.

Hynes RO and Lander AD (1992): Contact and adhesive specificities in the associations, migrations, and targeting of cells and axons. Cell 68: 303-322.

Ingham PW (1988): The molecular genetics of embryonic pattern formation in Drosophila. Nature 335: 25-34.

Jessell TM (1988): Adhesion molecules and the hierarchy of neural development. Neuron 1: 3-13.

Jessell,T.M and Melton DA (1992): Diffusible factors in vertebrate embryonic induction. Cell 68: 257-270.

Jones-Villeneuve EMV, Rudnicki MA, Harris JF and McBurney MW (1983): Retinoic acid-induced neural differentiation of embryonal carcinoma cells. Mol Cell Biol 3: 2271-2279.

Kessel M (1993): Reversal of axonal pathways from rhombomere 3 correlates with extra *Hox* expression domains. Neuron 10: 379-393.

LaMantia A-S, Colbert MC and Linney E (1993): Retinoic acid induction and regional differentiation prefigure olfactory pathway formation in the mammalian forebrain. Neuron 10: 1035-1048.

Lamb TM, Knecht AK, Smith WC, Stachel SE, Economides AN, Stahl N, Yancopoulos GD and Harland RW (1993): Neural induction in the secreted polypeptide noggin. Scienc*e* 262: 713-718.

Linney E (1992): Retinoic acid receptors: Transcription factors modulating gene regulation, development, and differentiation. Curr Top Devl Biol 27: 309-350.

Linney E and LaMantia A-S (1994): Retinoid signalling in mouse embryos. Adv Develop Biol 3: 73-114.

Lohnes D, Kaster P, Dierich A, Mark M, LeMeur M and Chambon P (1993): Function of retinoic acid receptor g in the mouse. Cell 73: 643-658.

McGinnis W and Krumlauf R (1992): Homeobox genes and axial patterning. Cell 68: 283-302.

McMahon AP and Bradley A (1988): The *Wnt*-1 (*int*-1) proto-oncogene is required for development of a large region of the mouse brain. Cell 62: 1073-1085.

McMahon AP and Moon RT (1989): Ectopic expression of the proto-oncogene *int*-1 in Xenopus embryos leads to duplication of the embryonic acids. Cell 58: 1075-1084.

Mendelsohn C, Ruberte E, LeMeur M, Morriss-Kay G and Chambon P (1991): Developmental analysis of the retinoic acid-inducible RAR-β2 promotor in transgenic animals. Development 113: 723-734.

Morriss-Kay G, Murphy P, Hill RE and Davidson DR (1991): Effects of retinoic acid excess on expression of Hox-2.9 and Krox-20 on morphological segmentation in the hindbrain of mouse embryos. EMBO J 10: 2985-2995.

Petkovich M, Brand N, Krust A and Chambon P (1987): A human retinoic acid receptor which belongs to the family of nuclear receptors. Nature 330: 444-450.

Puelles L and Rubenstein JL (1993): Expression patterns of homeobox and other putative regulatory genes in the embryonic mouse forebrain suggest a neuromeric organization. Trends Neurosci 16: 472-479.

Rossant J, Zirngibl R, Cado D, Shago M and Giguère V (1991): Expression of a retinoic response element-hsplacZ transgene defines specific domains of transcriptional activity during mouse embryogenesis. Gene Develop 5: 1333-1344.

Schardein JL (1993): 'Chemically Induced Birth defects. Second Edition, Revised and Expanded'. Marcel Dekker, New York.

Schlessinger J and Ullrich A (1992): Growth factor signaling by receptor tyrosine kinases. Neuron 9: 383-391.

Shenefelt RE (1972): Morphogenesis of malformations in hamsters caused by retinoic acid: Relation to dose and stage at treatment. Teratol 5: 103-118.

Stout RP and Graziadiei PPC (1980): Influence of the olfactory placode on the development of the brain in *Xenopus laevis* (Daudin). I. Axonal growth and connections of the transplanted olfactory placode. Neuroscience 5: 2175-2186.

Strickland S, and Mahdavi V (1978): The induction of differentiation in teratocarcinoma stem cells by retinoic acid. Cell 15: 393-403.

Sucov HM, Murakami KK and Evans RM (1990): Characterization of an autoregulated response element in the mouse retinoic acid receptor type β gene. Proc Natl Acad Sci USA 87: 5392-5396.

Thaller C and Eichele G (1987): Identification and spatial distribution of retinoids in the developing chick limb bud. Nature 327: 625-628.

Tickle C, Alberts B, Wolpert L and Lee J (1982): Local application of retinoic acid to the limb bond mimics the action of the polarizing region. Nature 296: 564-565.

Wagner M, Han B and Jessell TM (1992): Regional differences in retinoid release from embryonic neural tissue detected by an in vitro reporter assay. Development 116: 55-66.

Whitesides JG III and LaMantia A-S (1994): Differential adhesion of neurons and neural precursor cells from a distinct domain in the developing mammalian forebrain. Dev Biol (submitted).

Zimmer A and Zimmer A (1992): Induction of a RARβ2-β-gal transgene by retinoic acid reflects the neuromeric organization of the central nervous system. Development 116: 977-983.

GENETIC DETERMINANTS OF NEURAL CELL FATE

POTENTIAL ROLE OF HOMEOBOX GENES IN NEURAL CELL DIFFERENTIATION

Massimo Gulisano[1,2], Vania Broccoli[1,3], Fabio Spada[1], Edoardo Boncinelli[1,4]

[1]DIBIT, Istituto Scientifico H.S.Raffaele, Via Olgettina 60, 20132 Milano; [2]Istituto di Biologia Generale, University of Catania, Via Androne, 81, 95124 Catania; [3]Istituto di Genetica, Dipartimento di Biologia Evoluzionistica Sperimentale, Via Belmeloro 8, 40126 Bologna, [4]Centro Infrastrutture Cellulari, Via Vanvitelli 32, 20129 Milano, ITALY

INTRODUCTION

Homeobox Genes in Development

A fascinating question in developmental biology is how different regions of the animal embryo are specified. In particular, the uncovering of the molecular rules governing head development, has been the focus of studies by investigators working in a wide range of developmental systems.

During last fifteen years, new families of developmentally regulated genes have been extensively studied in different animal systems. The characteristic of these genes is to contain a small DNA binding conserved sequence, named homeobox. It is now well established that genes containing a homeobox are regulatory genes coding for nuclear proteins that act as transcription factors (Levine and Hoey, 1988). Through the recognition properties of their homeodomain (Gehring et al., 1994 for a review), homeoproteins encoded by homeobox genes are thought to regulate the expression of batteries of target genes both in vertebrates (Guazzi et al., 1994) and in flies (Capovilla et al., 1994).

Several homeobox genes are believed to control cell identity with a regional or segmental pattern both in flies and vertebrates (McGinnis and Krumlauf, 1992). The study of vertebrate homologues of regulatory genes operating in the *Drosophila* trunk has provided invaluable information about the genetic control of the system of positional values operating during development. Hox genes (Boncinelli et al., 1991; Kessel and Gruss, 1991; McGinnis and Krumlauf, 1992; Krumlauf, 1994) stand out among the various homeobox gene families so far identified as the vertebrate relatives of *Drosophila* homeotic selector genes of the

Neural Cell Specification: Molecular Mechanisms and Neurotherapeutic Implications
Edited by Juurlink *et al.*, Plenum Press, New York, 1995

69

Homeotic complex HOM-C. They are required to control the correct vertebrate axial specification and to provide positional cues in the developing neural tube from hindbrain to tail (Hunt et al., 1991).

The molecular mechanisms underlying the development of most anterior regions in both flies and vertebrates, to the contrary, remain to be clarified. Recently, some candidate genes have been isolated that play a role in patterning the *Drosophila* head (Finkelstein and Perrimon, 1991; Cohen and Jürgens, 1991, *btd*). Two of them, *empty spiracle* [*ems*] (Dalton *et al.*, 1989; Walldorf and Gehring, 1992) and *orthodenticle* [*otd*] (Finkelstein et al. 1990), have recently been identified and shown to contain a homeobox. *ems* (Dalton *et al.*, 1989; Walldorf and Gehring, 1992) and *otd* (Cohen and Jürgens, 1991; Finkelstein and Perrimon, 1991) mutations result in the deletion of specific anterior head structures in developing fly embryos. Products of both genes are required for proper development of these regions and they function at least in part as gap genes in defining antennal and pre-antennal segments in the head. These segments include antennal sense organs which are the main olfactory sensory structures of the *Drosophila* larva. At the blastoderm stage, gene products of both genes are expressed in two fairly anterior circumferential stripes, partially overlapping (Finkelstein and Perrimon, 1991). A group of mutations exist that fail to complement *otd* mutations. Flies homozygous for these mutations, *ocelliless*, show deletions of sense organs, including ocelli, in the adult head (Wieschaus et al., 1992).

We have used *Drosophila* sequences to identify vertebrate homologues and cloned Emx1 and Emx2 (Simeone et al., 1992b), related to *ems*, and Otx1 and Otx2 (Simeone et al., 1992a, 1993), related to *otd*. In E10 mouse embryos the four genes show a pattern of "nested" expression domains in the sequence Emx1<Emx2<Otx1<Otx2, both in dorsal and ventral brain regions defining a rostral pre-isthmic, brain as opposed to hindbrain and spinal cord (Simeone et al. 1992a).

The first appearance of the four genes is also sequential: Otx2 is already expressed at E5.5, followed by Otx1 and Emx2 at E8-8.5 and finally by Emx1 at E9.5. The four homeobox genes might play a role in establishing or maintaining the limits and identity of the various embryonic brain regions. The specification of the various regions of the rostral brain seems, then, to be a discrete process with its centre in the dorsal telencephalon.

EMX AND OTX ARE VERTEBRATE HOMOLOGS OF EMS AND OTD

Homeodomain proteins have been found not only in metazoan but also in fungi and plants, leading to the idea that they arose early during the evolution of eukaryotes. In the course of evolution, the amino acid sequence of the homeodomain has been conserved to a high degree. For example the human Hox-A7 homeodomain differs in only 1 out of 60 amino acids from that of the Antennapedia (Antp) homeodomain, which is its putative homolog in *Drosophila*, even though vertebrates and insects separated more than 500 millon years ago. This indicates that there is strong evolutionary pressure to conserve the amino acid sequence of the homeodomain.

A striking sequence homology is found between Otx genes and their *Drosophila* cognates in the homeodomain. In fact, the homeodomains of the predicted mouse Otx1 and Otx2 proteins are strikingly similar to that of the fly protein, differing by only 3 and 2 amino acid residues respectively (out of 60). The only other sequence similarity between *otd* and the Otx genes however is limited to residues immediately flanking the homeodomain. More differences are found between Emx and ems genes: there are eleven and eight amino acid residues different between ems and Emx 1 and Emx2 respectively.

Table 1. Comparison of Emx and Otx homeodomains with *Antp*, *otd*, *bcd* and *gsc* homeodomains, using the one-letter amino acid code. Five amino acid residues following the homeodomain are also shown. Dashes indicate amino acid identity with *Antp*. An intron is present between the residues 44 and 45 in the Emx homeodomains and between the residues 46 and 47 in Otx and *otd* homeodomains (not shown). No evidence for a similar intron has been reported for *ems* (Dalton et al., 1989). Amino acids in bold indicate differences between Xenopus and mouse genes. Human genes are identical in the homeodomain to the mouse. M: mouse; H: human

```
              10         20         30         40         50         60
Antp      RKRGRQTYTR YQTLELEKEF HFNRYLTRRR RIEIAHALCL TERQIKIWFQ NRRMKWKKENK TKGEP

bcd       PR-T-T-F-S S-IA---QH- LQG----AP- LADLSAK-A- GTA-V----K ---RRH-IQSD QHKDQ

gsc       KR-H-TIF-D E-LEA--NL- QETQ-PDVGT -EQL-RRVH- R-EKVEV--K ---A--RRQKR SSS-E

otd       QR-E-T-F-- A-LDV--AL- GKT--PDIFM -E-V-LKIN- P-SRVQV--K ---A-CRQQLQ QQQQS 46

otx1      QR-E-T-F-- S-LDV--AL- AKT--PDIFM -E-V-LKIN- P-SRVQV--K ---A-CRQQQQ SGSGT 46   M/H
Xotx1     QR-E-T-F-- S-LDV--TL- AKT--PDIFM -E-V-LKIN- P-SRVQV--K ---A-CRQQQQ SSNNS 46

otx2      QR-E-T-F-- A-LDV--AL- AKT--PDIFM -E-V-LKIN- P-SRVQV--K ---A-CRQQQQ QQQNG 46   M/H
Xotx2     QR-E-T-F-- A-LDI--AL- AKT--PDIFM -E-V-LKIN- P-SRVQV--K ---A-CRQQQQ QQQNG 46

ems       P--I-TAFSP S-L-K--HA- ES-Q-VVGAE -KAL-QN-N- S-T-V-V--- ---T-H-RMQQ EDEKG

emx1      P--I-TAFSP S-L-R--RA- EK-H-VVGAE -KQL-GS-S- S-T-V-V--- ---T-Y-RQKL EEEG- 44   M/H
Xemx1     P--I-TAFSP S-L-R--RA- EK-H-VVGAE -KQL-SS-S- S-T-V-V--- ---T-Y-RQKL EEEG- 44

emx2      P--I-TAFSP S-L-R--HA- EK-H-VVGAE -KQL--S-S- --T-V-V--- ---T-F-RQKL EEEGS 44   M/H
Xemx2     P--I-TAFSP S-L-R--HA- EK-H-VVGAE -KQL--T-S- --T-V-V--- ---T-F-RQKL EEEGS 44
```

Emx and Otx genes are highly conserved among the vertebrate species examined so far. In the case of Otx2, for example, between mouse and man there is a single conservative amino acid substitution (outside the homeodomain) out of approximately 300 residues (Simeone et al., 1993). Between the *Xenopus* and murine Otx2 proteins there is an overall sequence similarity of 95% (our unpublished results). There are no amino acid residue differences between mouse and human within the homeodomain, either for the Emx or for the Otx genes (Table 1). Instead, between the *Xenopus* and the mouse cognates there is a single amino acid substitution in the homeodomain of all the four genes; the *Xenopus* and murine Otx2 proteins have an overall sequence similarity of 95% (Pannese et al., 1994). *otd*- and *ems*-related genes have also been isolated from chick (Bally-Cuif et al., 1994) and zebrafish (our unpublished results). In particular, the otd/Otx homeodomains share the lysine residue at position 9 of the recognition helix only found so far in *Drosophila bicoid* (*bcd*) (Driever and Nüsslein-Volhard, 1988) and vertebrate *goosecoid* (Blumberg et al., 1991; Blum et al., 1992; Izpisua-Belmonte et al., 1993) genes (Table 1).

Recently we mapped the four human genes. OTX2 maps to chromosome region 14q22; the OTX1 locus maps to 2p13; EMX2 maps to 10q26.1 and finally EMX1 maps to 2p14-p13 (Kastury et al., 1994). The OTX1 and EMX1 loci may be closely linked on or near 2p13, prompting speculation that a clustered gene structure could have functional significance, as is the case for the Hox clusters (Boncinelli et al., 1991)

OTX2 IS INVOLVED IN GASTRULATION

The possibility that one of these genes, Otx2, was involved in the mechanisms of determination of anterior regions during the process of gastrulation was extremely plausible from the whole mount in situ experiments performed in the mouse (Fig. 1). Otx2 transcripts were detectable in mouse embryonic structures since the prestreak stage at about day 5.5-5.7 post coitus (i.e. E5.5-E5.7) in the embryonic ectoderm, or epiblast, and not in extraembryonic ectoderm. The epiblast of the gastrulating mouse embryo is believed to be the sole source of all definitive tissues in the fetus. A variety of grafting and transplant experiments demonstrate the potency of the epiblast to form derivatives of all three germ layers in gastrulating as well as in prestreak embryos (Tam, 1989; Lawson et al., 1991). The same expression pattern was observed in E6 embryos and in early-streak E6.5 embryos. Between day 7 and 7.5 Otx2 expression progressively receded to anterior regions (Fig. 1a), where it remain confined (Fig. 1b, c, d) . These regions correspond to the neuroectoderm of the prosencephalon and mesencephalon. Since at E9.0, Otx2 appears to demarcate with a sharp boundary the division between mesencephalon and metencephalon (Fig. 1c, d); Otx2 will remain expressed in these regions until late in gestation. The progressive confinement of Otx2 expression from the entire epiblast to presumptive fore- and mid-brain neuroectoderm occurs concomitantly with progressive regionalization of cell fate within the epiblast. This Otx2 progressive confinement is certainly correlated with the expression of other early developmental genes. For example, what is known about the evolution of expression patterns of early Hox genes (McGinnis and Krumlauf, 1992), in particular Hox-2.9 [i.e. Hoxb-1] (Wilkinson et al., 1989; Frohman et al., 1990), could suggest a relationship between the progressive displacement towards anterior of the anterior border of the Hox expression domain and the posterior border of the Otx2 expression domain.

The role of Otx2 in gastrulation was recently analysed in different systems such as *Xenopus* (Pannese et al., 1995), chick (Bally-Cuif et al., 1994) and zebrafish (our unpublished results). These systems permit manipulations, such as microinjection and embryo culture in the presence of retinoic acid, to answer some of the questions raised on the role of this gene during vertebrate gastrulation.

EMX AND OTX GENES EXPRESSION DURING MOUSE EMBRYO DEVELOPMENT

Early Embryo Development

Let us consider early expression of these genes. Emx1 is first expressed in E9 mouse embryos in the anterior dorsal region of the neural tube (Simeone et al., 1992b), an area fated to give rise to cortical telencephalic regions, at a stage when most of brain regionalization is probably already specified and cortical neurogenesis is just starting (Kuhlenbeck, 1973; Luskin et al., 1988). An Emx2 hybridisation signal is already detectable in anterior dorsal neuroectodermal regions in E8.5 embryos (Simeone et al., 1992b). In E9.5 embryos its expression domain in rostral brain clearly contains that of Emx1.

Otx1 is first expressed in a large region of the anterior neural tube of E8.25-8.5 embryos (Simeone et al., 1992a). Anterior-posterior delimitation of the Otx1 expression in the rostral neural tube is defined between E9 and E9.5 embryos. Dorsally, its expression domain is comprised of a continuous region including part of the telencephalon, the diencephalon and the mesencephalon. That the posterior boundary of this domain coincides with that of the mesencephalon is apparent in paramedian sections even if in more median sections strong hybridisation signal extends only half way along the mesencephalon. Ventrally, the Otx1 expression domain includes contiguous regions of both diencephalon and mesencephalon with sharp anterior and posterior boundaries.

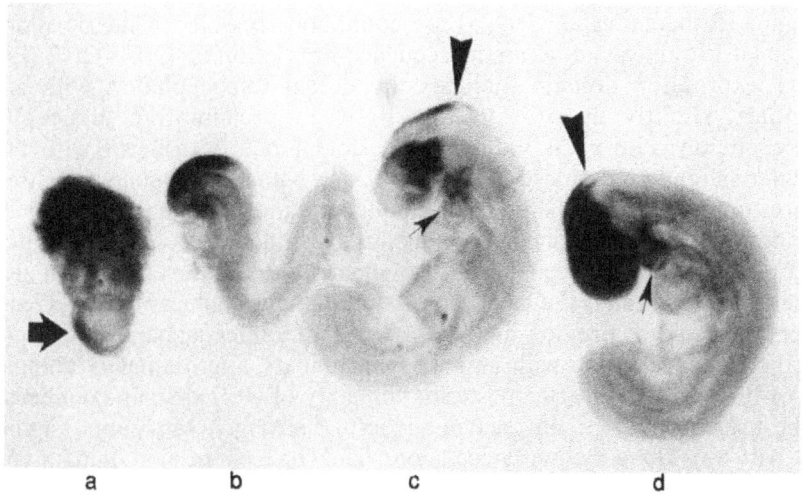

a   b   c   d

Figure 1. Otx2 expression in early mouse embryos, revealed by whole-mount in situ hybridisation with digoxigenin-labeled riboprobes. Anterior is to the left. **a)** E7.2 gastrulating embryo. Otx2 expression is already confined to the most anterior portion of neuroectoderm (arrow). **b)** E8 embryo. Otx2 expression is localised in presumptive fore- and mid-brain regions. **c)** and **d)** E8.8 and E9.2 embryos, respectively. Otx2 expression is localised in fore- and mid-brain with a sharp posterior boundary (arrowheads). Expression is also detectable in the first branchial arch (arrows).

We have previously seen that Otx2 is expressed in all the epiblast very early. At E9 the Otx2 expression domain contains that of Otx1 and includes the entire forebrain. Anterior-posterior delimitation of the Otx2 expression is clear in E9.75 embryos. Dorsally, it includes the entire telencephalon, the diencephalon and the mesencephalon. The anterior portion of this domain includes lamina terminalis and presumptive basimedial striatum, whereas the posterior boundary coincides with that of the mesencephalon. Ventrally, the Otx2 expression domain includes contiguous regions of diencephalon and mesencephalon with an anterior boundary just posterior to the optic chiasma.

Temporal sequence of these events may be put in relation to proposed patterns of brain regionalisation (Sakai, 1987; Puelles et al., 1987). For example, Emx2 first expression roughly coincides with appearance of a subdivision between forebrain and midbrain, corresponding to total level 4 of Sakai (1987), and precedes the appearance of total level 2 dividing telencephalon from diencephalon. After the appearance of this subdivision, Emx1 expression begins to be detectable in dorsal telencephalon. Emx2 is expressed in dorsal telencephalon but also in restricted anterior regions of diencephalon, both before and after the appearance of total level 3 within diencephalon. The significance of these correspondences is difficult to assess in the absence of morphogenetic data obtained through experimental manipulations. Furthermore, some authors disagree about this temporal sequence of events. According to some of them, for example, subdivision between dorsal and ventral thalamus even precedes that between telencephalon and diencephalon. Nevertheless, at least the case for the Otx genes defining the posterior mesencephalic boundary may be made (Fig.1c, d).

Expression at E10

In E10 mouse embryos the four genes are all expressed. Their expression domains (Simeone et al., 1992a) are continuous regions of the developing brain contained within each other in the sequence Emx1<Emx2<Otx1<Otx2 (Fig.2). The Emx1 expression domain includes the dorsal telencephalon with a posterior boundary slightly anterior to that between presumptive diencephalon and telencephalon. Emx2 is expressed in dorsal neuroectoderm with an anterior boundary slightly anterior to that of Emx1 and a posterior boundary within the roof of presumptive diencephalon. This boundary most probably coincides with the boundary between first and second thalamic segment. Its expression in restricted ventral regions of diencephalon is also apparent at this developmental stage. Both dorsally and ventrally, the Otx1 expression domain contains the Emx2 domain. It covers a continuous region including part of the telencephalon, the diencephalon and the mesencephalon with an anterior boundary approximately coincident with that of Emx2. Laterally, the posterior boundary of Otx1 domain coincides with that of the mesencephalon. In median sections a strong hybridisation signal extends only half way along the mesencephalon. The Otx2 expression domain contains the Otx1 domain both dorsally and ventrally. It covers practically the entire fore- and mid-brain. It is interesting to note that an extremely anterior region, that includes the optic chiasma and the optic recess, is excluded from the Otx2 expression domains. This region, with the exception of the optic recess, is the site of expression of several *Distal-less*-related (Dlx) genes (Price et al., 1991; Porteus et al., 1991; Robinson et al., 1991; Simeone et al., 1994; reviewed in Boncinelli, 1994). This anterior exclusion of Otx2 expression is in fact reminiscent of *otd* expression of the fruitfly embryo, which also retracts from the anterior pole. It suggests that, as in *Drosophila*, development of the extreme anterior terminus may be governed by a distinct genetic mechanism.

Expression of Emx and Otx genes identifies several regions in forebrain (Fig. 2). Some of these may correspond to presumptive anatomical subdivisions, whereas the significance of others remains to be assessed. Dorsally, for example, it is clear that the two Emx genes identify a presumptive cortical region, part of which will be neocortex and archicortex (area 4 in Fig. 2). Emx2 expression also defines the boundary between future dorsal (DT) and ventral thalamus (VT). On the other hand, expression of these genes does not offer an unambiguous cue for the boundary between presumptive ventral thalamus and area 5 of telencephalon (broken line in Fig. 2). In this light, it is interesting to consider that in recent experiments in the chick embryo (Martinez et al., 1991) ventral thalamus has been shown to share with telencephalon the inability to express *Engrailed-2* (*En-2*) upon grafting of metencephalic neuroepithelium. The potential to express *En-2* and to exhibit mesencephalic cell fate stops at the zona limitans intrathalamica, the boundary between dorsal and ventral thalamus. We could consider the existence of two territories within the early vertebrate forebrain: an anterior one, encompassing telencephalon and ventral thalamus, and a posterior one, including the rest of diencephalon (see also Figdor and Stern, 1993).

Just anterior to presumptive cortical region there is a region of dorsal telencephalon (area 2 in Fig. 2) where expression of Emx2 and Otx1 is less defined, both dorsally and ventrally. This region generates a further subdivision of the forebrain whose ventral floor probably corresponds to mammillary area. It is conceivable that presumptive territory generated by this subdivision corresponds to caudal portions of basal internal grisea. Between E10 and E10.25 the expression of Emx2, Otx1 and Otx2, begins to shift towards slightly more posterior brain areas.

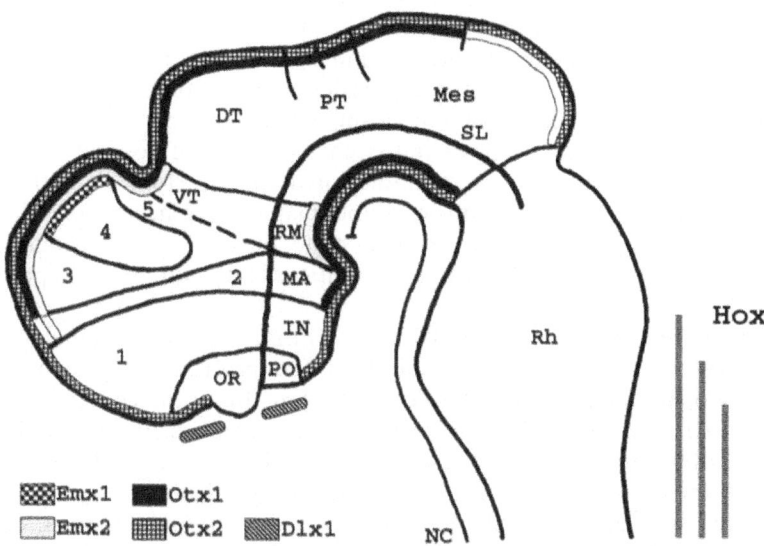

Figure 2. Summary of the expression domains of the four genes and Dlx1 in the developing central nervous system of E10 mouse embryos. Expression of members of the Hox gene family is also indicated. Expression of Emx and Otx genes identifies a number of regions in forebrain. Some of these correspond to anatomical subdivisions proposed by various authors, whereas the significance of others, indicated with numbers from 1 to 5, remains to be assessed. The broken line antero-ventral to presumptive ventral thalamus indicates a boundary not related to the expression of these genes. A fainter stippling designates an anterior region of dorsal telencephalon (area 2) where expression limits of Emx2 and Otx1 are less defined. This region generates a subdivision of the forebrain whose ventral floor corresponds to mammillary area. A fainter stippling in the Otx1 domain in posterior mesencephalon designates the region where Otx1 expression is weak along the mid-line and stronger in more lateral sections. DT, dorsal thalamus; IN, infundibular or tuberal region; MA, mammillary area; Mes, mesencephalon; NC, notochord; OR, optic recess; PO, post-optic region; PT, pretectum; Rh, rhomboencephalon; RM, retro-mammillary area; SL, sulcus limitans; VT, ventral thalamus.

EMX AND OTX GENES EXPRESSION DURING MIDGESTATION EMBRYO DEVELOPMENT

Emx Gene Expression

Both Emx1 and Emx2 are expressed in presumptive cerebral cortex in a developmental period, starting at day 9.5, corresponding to major events in cortical neurogenesis (Simeone et al., 1992b). During postnatal cortex development Emx2 remain confined to late germinal layers, i.e. hippocampus. Emx1 is instead clearly expressed in the hexalaminar cortex with a specific pattern (Gulisano M, Broccoli V and Boncinelli E, manuscript in preparation). It seems reasonable to hypothesize a role of Emx1 and Emx2 in establishing the limits and identity of the cerebral cortex and in cortical neurogenesis.

Temporal patterns of neurogenesis are believed to be important prerequisites for the establishment of precise anatomical interactions in the developing brain (Kuhlenbeck, 1973; Luskin et al., 1988). The two genes are expressed in most cortical regions with a precise pattern in space and time. The temporal profiles of expression of the two genes are very similar but not coincident. There is a shift between Emx1 and Emx2 expression, the latter being expressed earlier and declining earlier in anterior and lateral cortical regions. In its full extension, E12.5 to E13.5, the Emx1 expression domain comprises cortical regions including primordia of neopallium, hippocampal and parahippocampal archipallium (Kuhlenbeck, 1973). Emx1 expression is characteristic of cortical regions being mainly, but not exclusively, hexalaminar in nature. Hybridisation signal is uniformly distributed across the cortex without major differences. Emx1 is found to be expressed throughout all the cortical plate throughout embryogenesis and postnatal development. A role of this gene in cortical neurogenesis appears quite conceivable.

In its full extension, E12.5 to E13.5, the Emx2 expression domain comprises presumptive cortical regions including neopallium, hippocampal and parahippocampal archipallium and selected paleopallial localisations, but no basal internal grisea. In this period the hybridisation signal is uniformly distributed across the cortex without major differences but starting from E13.75 it appears to be confined to the germinal neuroepithelium of the ventricular zone, excluding the intermediate zone and cortical plate. From day E14.5 on, Emx2 cortical expression progressively declines in anterior and ventrolateral regions and at E17 is confined to germinal hippocampal layers. Emx2 results then to be correlated with neuroproliferative layers in the telencephalic cortex throughout development (Gulisano M, Broccoli V and Boncinelli E, manuscript in preparation).

Emx2 is also expressed in some neuroectodermal regions of the embryo including olfactory placodes. Particularly interesting is its expression in olfactory placodes, olfactory bulbs, olfactory epithelia of nasal chambers and in several cerebral locations related to olfaction. In fact, Emx2 is also expressed in specific sites of hippocampal and parahippocampal cortex, amygdala, specific areas of basal cortex, hypothalamus, ventral and dorsal thalamus, habenula, presumptive mammillary body, septal and tegmental regions. All these regions contain areas related to olfaction even if it remains to be seen whether Emx2 expression sites really coincide with primordia of these areas as is the case for olfactory epithelia in nasal pits and chambers. In flies, *ems* is involved in the regulation of olfactory sense organs during development. Mutant *ems* flies lack primordia of antennal sense organs (Dalton et al., 1989), the main olfactory sensory structures of the *Drosophila* larva. The idea that homeobox genes of the *ems* family might be already involved in the specification of the proto-olfactory system seems intriguing.

Otx Gene Expression

In midgestation embryos, the two Otx genes are expressed in specific regions of the brain (Simeone et al., 1993). Both are expressed in basal telencephalon, in diencephalon and mesencephalon but not in spinal cord of E12.5

embryos. Their expression domains in mesencephalon show a sharp posterior boundary, both dorsally and ventrally at the level of rhombic isthmus, already shown in earlier stages. From E9.5 onward, the expression of both genes clearly marks the posterior boundary of mesencephalon to the exclusion of presumptive anterior cerebellar domain. Ventrally, however, Otx1 expression re-appears posteriorly, in the anterior metencephalon, after a gap just posterior to the IV cranial nerve. Otx1 is also expressed in dorsal telencephalon, whereas Otx2 expression has disappeared from this region at E11.75. A specific Otx2 localisation is in choroid plexuses, both in the lateral ventricules and in the fourth ventricle.

In the telencephalon, Otx1 expression is detectable in the presumptive cerebral cortex from its anterior boundary to its posterior limit. The hybridisation signal is remarkably uniform across the cortex, without major variation. Sagittal sections reveal expression in the olfactory bulbs. Frontal sections show that the Otx1 domain includes neopallium, hippocampal and parahippocampal archipallium and selected paleopallial and septal localisations (Simeone et al., 1992a). Otx1 expression is also detectable in some noncortical basal telencephalic regions, namely in the germinal layer of the most lateral portion of lateral ganglionic eminence and in part of superior basimedial region.

Otx1 is also expressed in regions of diencephalon including epithalamus, dorsal thalamus and mammillary region of posterior hypothalamus. Its expression domain does not include the ventral thalamus. A two-layered narrow stripe of expression is detectable at the level of the boundary between dorsal and ventral thalamus, this is the zona limitans intrathalamica, the precursor of lamina medullaris externa and mammillo-thalamic tract. Other localisations are fasciculus retroflexus, the precursor of habenulo-interpenduncular tract, stria medullaris, including the region surrounding the posterior commissure, primordium of mammillotegmental tract, epiphysis, fornix and sulcus lateralis hypothalami posterioris. Frontal sections define its expression in the germinal layer of diencephalon, particularly at the level of two sulci, namely the diencephalicus dorsalis and medius. Posterior to diencephalon, it is expressed in mesencephalic regions of tectum and tegmentum, possibly at the level of presumptive dorsal periventricular bundle. Nonbrain Otx1 localisations are in auricular and ocular regions, nasal cavities including external ducts and pharynx (Simeone et al., 1993). Later in development Otx1 is found to be expressed in deeper layers of telencephalic cortex. When the cortex is already organized in layers (i.e. P9-P16) Otx1 is expressed in a subpopulation of neurons in layer 5 and overall layer 6 except in the frontal area (Frantz et al., 1994; our unpublished results).

Between E10.75 and E11.5, Otx2 expression disappears from the telencephalic cortex. In E12.5 embryos Otx2 is no longer expressed in cortical telencephalon but only in a subset of presumptive noncortical basal ganglia, in a quite complementary fashion to Otx1. It is not expressed in anterior septal region but very well expressed in septal regions contiguous to diencephalon and in germinal layer of anterior basimedial regions. Its expression in lamina terminalis confirms its very anterior expression shown earlier in development. Otx2 is expressed in regions of diencephalon (Fig.3) and mesencephalon with a pattern very similar to Otx1. Otx2 is also expressed in the anlage of neurohypophysis and in choroid plexuses. Expression in developing choroid plexuses, both in outer ventriculi and myelencephalon, precedes and accompanies the various morphogenetic events leading to their formation and appears to be confined to cells of the neuroepithelial layer and excluded from underlying mesenchyme (Boncinelli et al., 1993).

Otx genes show an interesting pattern in developing cerebellum during early postnatal mouse life. A code of Otx gene expression seems to identify three distinct population of granule cells progenitors in the developing cerebellum between P5 and P16: a first population of granule cells progenitors expressing Otx1 only (lobules 1-5, simple lobule); a second population, expressing both Otx1, with a gradient from medium to low levels and Otx2 with an opposite gradient, from low to medium levels (lobule 6-9); finally a third population with high levels of Otx2 expression (lobule 10 and flocculus) (Frantz et al., 1994; our unpublished results). Otx genes show then opposite, ovelapping A/P gradients during cerebellar development: Otx1, anterior to posterior, and Otx2, posterior to anterior, thus suggesting a specific role in establishing limits and identity during cell differentiation pathways in the cerebellar cortex (Frantz et al., 1994; our unpublished observations).

Both Otx genes are also expressed in the olfactory epithelium, as well as in the developing inner ear from early expression in the otic vesicle to epithelia in auricular ducts of sacculus and cochlea and in the developing eye (Simeone et al., 1993). Here the two genes are both expressed but with specific patterns maintained throughout eye development from their appearance in the optic stalk at E10. In E12.5 embryos Otx1 is expressed in the iris, in a peripheral region including ciliary bodies, sclera, external ectoderm and the external sheath of the optic nerve, whereas in E17 embryos it is only expressed in the iris. At E12.5, Otx2 is expressed in sclera, retinal pigmented layer and leptomeninges intimately surrounding the `optic nerve, whereas in E17 embryos it is expressed in the pigmented epithelium of retina and in neurosensory retina. In flies *otd* is required for the development of the eye-antennal imaginal discs (Cohen and Jürgens, 1991; Finkelstein and Perrimon, 1991; Wieschaus et al., 1992). Flies homozygous for *ocelliless* mutations show deletions of sense organs, including ocelli, in the adult head. We may conclude that developmental genes encoding homeoproteins containing a *otd*-like homeodomain play a major role in the development of sense organs from a very remote past in evolution.

Areas and Boundaries

Dlx1 is a murine homeobox gene (Price et al., 1991, 1992) containing a homeodomain related to that of the *Drosophila Distal-less* gene and is expressed in ventral thalamus and basal telencephalic regions from day 10 of mouse development. There is a certain degree of complementarity between localisations of Dlx1 expression (Price at al., 1991, 1992; Simeone et al., 1993) and that of the Otx genes both in diencephalon, striatum and basimedial region. This can be clearly observed hybridising Dlx1, Otx1 and Otx2 on contiguous sections (Simeone et al., 1993). In sagittal sections of diencephalon the dorsal border of Dlx1 expression abuts the ventral border of Otx expression. This border is comprised of two layers of cells orthogonal to the frontal surface of diencephalon surrounding the zona limitans intrathalamica present at the boundary between dorsal and ventral thalamus (Boncinelli et al., 1993). Otx genes appear in general to be expressed in germinal layers and boundary regions whereas Dlx1 expression extends to more internal regions, such as paraventricular and internal areas of ventral thalamus (Fig. 3).

Much the same is observable in frontal sections of diencephalon where the four longitudinal columns of Herrick, i. e. epithalamus, dorsal and ventral thalamus and hypothalamus, are easily identified, being separated by sulci termed sulcus diencephalicus dorsalis, medius and ventralis, respectively, from dorsal to ventral. Otx1 and Otx2 are expressed (Simeone et al., 1993) in ependymal layers of epithalamus and dorsal thalamus and in orthogonal layers of cells immediately above and below the sulcus diencephalicus medius. Dlx1 is expressed in deeper regions of ventral thalamus. Complementarity of Otx2 and Dlx1 expression patterns is also apparent in anterior basal regions where again Otx2 expression is more superficial and Dlx1 expression is distributed throughout deeper regions suggesting for these genes a role in different groups of neuroblasts. The exclusion of cells expressing Dlx1 from ventricular germinal layers suggests that these cells represent early differentiated neurons. Alternatively, Dlx1 expression might be transient in proliferating neuroblasts (Price et al., 1992). In general, Dlx1 expression appears to respond to already established boundaries. Conversely, the Otx genes could be involved in the early subdivision of the forebrain.

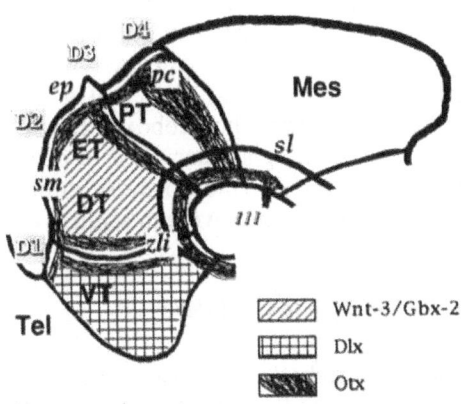

Figure 3. Schematic representation of the expression domains of *Otx1* and *Otx2* (Boncinelli et al., 1993), *Wnt3* (Salinas and Nusse, 1992), *Gbx-2* (Bulfone et al., 1993) and *Dlx* genes (Bulfone et al., 1993; Simeone et al., 1993) in diencephalon of E12.5 embryos. Within the diencephalic regions bold letters designate the columnar nomenclature: DT, dorsal thalamus; ET, epithalamus; PT, pretectum and VT, ventral thalamus. Outside the profile, the proposed new subdivision into four neuromeres, D1 to D4 (Figdor and Stern, 1993), is indicated. *ep*, epiphysis; Mes, mesencephalon; *pc*, posterior commissure; *sl*, sulcus limitans; *sm*, stria medullaris; Tel, telencephalon; *zli*, zona limitans intrathalamica; *III*, 3rd cranial nerve.

Expression of Otx genes in diencephalon and mesencephalon of E12.5-14.5 embryos colocalizes with boundary regions and presumptive axon tracts (Fig.3), including anterior and posterior commissure. This expression appears to be confined to precursor cells surrounding these structures as if these cells could be used as borders of pathways for the pioneer axon tracts. This is particularly evident in posterior commissure and along the zona limitans intrathalamica. Otx gene expression in posterior commissure is limited to cells of ventricular epithelium, whereas primary fibers running on its surface are not labelled.

Expression of Otx genes along the zona limitans intrathalamica might constitute a framework for the axon patterning of lamina medullaris and other structures physically separating dorsal thalamus from ventral thalamus. In sagittal and frontal sections of diencephalon the dorsal border of Dlx1 expression abuts the ventral border of Otx expression (Simeone et al., 1993). Expression of Otx genes in two layers of cells running parallel to the zona limitans intrathalamica present at the boundary between dorsal and ventral thalamus might account for the sharp dorsal boundary of the Dlx1 expression domain in ventral thalamus. This is schematically shown in Fig.3, where the expression domain of Wnt3 (Salinas and Nusse, 1992) is also represented, as an example. The ventral boundary of this domain also abuts the zona limitans intrathalamica. Expression of both genes around the optic nerve appears similar to that along the zona limitans intrathalamica in providing clues to axon pathfinding and patterning. Expression of Otx genes might provide a global framework for the primary scaffold of specific axon pathways in the early neuroepithelium of the forebrain.

It is of interest to consider the possibility that Otx genes play a determinant role in the development of the head in at least two stages. Homeobox genes do not necessarily have a single function and domain throughout development, for there are many instances to the contrary (Hoppler and Bienz, 1994; Jones and McGinnis, 1993)

Probably, Otx genes are used in development to specify territories or areas in rostral brain of E8-E10 mouse embryos and to provide later on positional cues required for axons to grow along specific pathways within the embryonic central nervous system (Figdor and Stern, 1993) even if it is not clear whether the two functions are independent. Vertebrate homeobox genes of the Emx, Dlx and Hox families do not appear to do anything similar. It is also of interest to consider that in *otd* mutant flies, pioneer axons of the posterior commissures fail to develop normally as if appropriate positional cues were missing (Tessier-Lavigne, 1992).

EVOLUTIONARY CONSIDERATIONS

Different families of vertebrate homeobox genes related to regulatory genes controlling the development of the fly are expressed in different body regions. The structural organisation and the regulation of these genes appear remarkably conserved in evolution. In fact, both Emx and Otx genes are expressed in the developing brain and in a few additional cephalic localisations. Emx1 and Otx2 expression is confined to the head. Emx2 and Otx1 are mainly but not exclusively expressed in the head, whereas the 38 homeobox genes of the Hox family are expressed in the brain stem and trunk only (Boncinelli et al., 1991). The Otx expression domain covers the first four cranial nerves and the Hox domain covers all others, leaving possibly only the fifth, stemming from rhombomere 2, out of this inventory.

A separation between head and trunk structures occurs very early in development of both flies and mammals. In flies a cephalic furrow forms in gastrulating embryos at the posterior boundary of presumptive head at about one third of the embryo length. This process is genetically controlled, for example by *bicoid* (Driever and Nüsslein-Volhard, 1988), and there is increasing evidence that the rules governing head formation may differ from the paradigm established for the central region of the body (Finkelstein and Perrimon, 1991; Cohen and Jürgens, 1991). In the mouse egg cylinder the head derives from regions anterior to the primitive streak whereas most of trunk paraxial mesoderm derives from growing cells stemming from the most anterior portion of the developing primitive streak (Tam, 1989). The embryonic axis lengthens up to the neural plate stage essentially by two processes: elongation of the primitive streak and expansion of the region of epiblast immediately cranial to the anterior end of the primitive streak (Lawson et al., 1991). This regional specification together with germ layer determination is achieved in steps.

Cephalization is thought to have occurred independently in the evolutionary lineages leading to insects and vertebrates (Kuhlenbeck, 1973; McGinnis and Krumlauf, 1992). Nevertheless, *ems*-related and *otd*-related genes are expressed in anterior cephalic regions both in flies and mammals. A couple of hypotheses can be advanced to explain this evolutionary paradox. First, these genes might be related to anteriority. Otx2 is one of the earliest genes expressed in the epiblast and straight afterwards in anterior neuroectoderm, demarcating rostral brain regions even prior to any sign of headfold formation. Its gene product contains a homeodomain of the *bicoid* class and is able to recognize and transactivate a *bicoid* target sequence (Simeone et al., 1993). Otx2 expression appears intimately linked to anterior specification and head formation. We can speculate that an *otd*-like gene already specified anteriority in early metazoans, too. From this ancestral gene probably derived both *otd* and vertebrate Otx genes. On the contrary, the maternal *bicoid* gene might have subsequently evolved in the evolutionary lineage leading to flies and other long germ-band insects. In common ancestors of flies and vertebrates Otx and Emx head genes might then exert their function in anterior neuroectoderm. When more and more anterior structures were added in evolution, the expression domain of these genes automatically shifted anteriorly. According to a second hypothesis, expression of these genes is related to major sense organs and follows their localisation in the body. The two classes of explanations need obviously not to be mutually exclusive.

As a first step toward answering these evolutionary questions we cloned Emx and Otx genes homologue in frog, chick and zebrafish and studied its expression. Despite the fact that the developmental brain anatomy of all these species is rather different, preliminary data show for example that the Otx2 expression domain is very similar in frogs, chick and mice throughout embryogenesis. The significance of these data with regards to the evolutionary story of brain and especially of forebrain remains to be assessed.

ACKNOWLEDGEMENTS

We wish to thank all components of our lab for comments and helpful suggestions, and M. Sottocorno for skilful secretarial assistance. This work was supported by grants from Progetti Finalizzati CNR, "Ingegneria Genetica" and "ACRO", the V AIDS Project of the Ministero della Sanita', the EC BIOTECH Programme, the Telethon-Italia Programme and the Italian Association for Cancer Research (AIRC).

REFERENCES

Bulfone A, Puelles L, Porteus MH, Frohman MA, Martin GR and Rubenstein JLR (1993): Spatially restricted expression of *Dlx-1*, *Dlx-2* (*Tes-1*), Gbx-2 and Wnt-3 in the embryonic day 12.5 mouse forebrain defines potential transverse and longitudinal segmental boundaries. J Neurosci 13: 3155-3172.

Bally-Cuif L, Gulisano M, Broccoli V and Boncinelli E (1994): *c-otx2* is expressed in two different phases of gastrulation and is sensitive to retinoic acid treatment in chick embryo. Mech Dev (in press).

Blum M, Gaunt SJ, Cho KWY, Steinbeisser H, Blumberg B, Bittner D and De Robertis EM (1992): Gastrulation in the mouse: the role of the homeobox gene *goosecoid*. Cell 69: 1097-1106

Blumberg B , Wright CVE, De Robertis EM and Cho KWY (1991): Organizer-specific homeobox genes in *Xenopus* laevis embryo. Science 253: 194-196

Boncinelli E, Simeone A, Acampora D, Mavilio F (1991) HOX gene activation by retinoic acid. Trends Genet 7: 329-334

Boncinelli E, Gulisano M and Broccoli V (1993): *Emx* and *Otx* homeobox genes in the developing mouse brain. J. Neurobiology 24: 1356-1366

Boncinelli E (1994): Early CNS development: *Distal-less* related genes and forebrain development. Current .Opinion. in Neurobiology 4: 29-36

Capovilla M, Brandt M and Botas J (1994): Direct regulation of *decapentaplegic* by *Ultrabithorax* and its role in Drosophila midgut morphogenesis. Cell 76: 461-475

Cohen S and Jürgens G (1991): *Drosophila* headlines. Trends Genet 7: 267-272

Dalton D, Chadwick R and McGinnis W (1989): Expression and embryonic function of empty spiracles: a *Drosophila* homeobox gene with two patterning functions on the anterior-posterior axis of the embryo. Genes Dev 3: 1940-1956

Driever W and Nüsslein-Volhard C (1988): The bicoid protein determines position in the *Drosophila* embryo in a concentration-dependent manner. Cell 54: 95-104

Figdor M and Stern C (1993): Segmental organization of embryonic diencephalon. Nature 363, 630-634.

Finkelstein R, Smouse D, Capaci T, Spradling AC and Perrimon,N (1990): The *orthodenticle* gene encodes a novel homeo domain protein involved in the development of the *Drosophila* nervous system and ocellar visual structures. Genes Dev 4: 1516-1527

Finkelstein R and Perrimon N (1991): The molecular genetics of head development in *Drosophila melanogaster*. Development 112: 899-912

Frantz GD, Weimann JM, Levin ME and McConnell SK (1994): *Otx1* and *Otx2* define layers and regions in developing cerebral cortex. J Neurosci 14: 5725-5740.

Frohman MA, Boyle M and Martin GR (1990): Isolation of the mouse *Hox-2.9* gene; analysis of embryonic expression suggests that positional information along the anterior-posterior axis is specified by mesoderm. Development 110: 589-607

Gehring WJ, Qiu Qian Y, Billeter M, Furukubo-Tokunaga K, Schier AF, Resendez-Perez D, Affolter M, Otting G and Wutrich K (1994): Homeodomain-DNA recognition. Cell 78: 211-223.

Guazzi S, Lonigro R, Pintonello L, Boncinelli E, Di Lauro R and Mavilio F (1994): The thyroid transcription factor-1 gene is a candidate target for regulation by Hox proteins. EMBO J 13: 3339-3347

Hoppler S and Bienz M (1994): Specification of a single cell type by a *Drosophila* homeotic gene. Cell 76: 689-702

Hunt P, Gulisano M, Cook M , Sham M-H, Faiella A, Wilkinson D, Boncinelli E and Krumlauf R (1991): A distinct Hox code for the branchial region of the vertebrate head. Nature 353: 861-864.

Izpisua-Belmonte JC, De Robertis EM, Storey KG and Stern CD (1993): The homeobox gene *goosecoid* and the origin of organizer cells in the early chick blastoderm. Cell 74: 645-659

Jones B and McGinnis W (1993): The regulation of *empty spiracles* by *Abdominal-B* mediates an abdominal segment identity function. Genes Dev 7: 229-240

Kastury K, Druck T, Huebner K, Barletta C, Acampora D, Simeone A, Faiella A and Boncinelli E (1994): Chromosome locations of human *EMX* and *OTX* genes. Genomics 22: 41-45

Kessel M and Gruss P (1991): Homeotic transformations of murine vertebrae and concomitant alteration of Hox codes induced by retinoic acid. Cell 67: 89-104

Krumlauf R (1994): Hox genes in vertebrate development Cell 78: 191-201

Kuhlenbeck H (1973): "The Central Nervous System of Vertebrates. Vol. 3 Part II. Overall morphologic pattern". Basel: Karger.

Lawson KA, Meneses JJ and Pedersen RA (1991): Clonal analysis of epiblast fate during germ layer formation in the mouse embryo. Development 113: 891-911

Levine M and Hoey T (1988): Homeobox proteins as sequence specific transcription factors. Cell 55: 537-540

Luskin MB, Pearlman AL and Sanes JR (1988): Cell lineage in the cerebral cortex of the mouse studied in vivo and in vitro with a recombinant retrovirus. Neuron 1: 635-647

Martinez S, Wassef M and Alvarado-Mallard RM (1991): Induction of a mesencephalic phenotype in the 2-day-old chick prosencephalon is preceded by the early expression of the homeobox gene *en*. Neuron 6: 971-981

McGinnis W and Krumlauf R (1992): Homeobox genes and axial patterning. Cell 68: 283-302.

Pannese M, Polo C, Andreazzoli M, Vignali R, Kablar B, Barsacchi G and Boncinelli E (1995): The *Xenopus* homologue of *Otx2* is a maternal homeobox gene that demarcates and specifies anterior strctures in frog embryos. Development (in press).

Price M, Lemaistre M, Pischetola M, Di Lauro R and Duboule D (1991): A mouse gene related to *distal-less* shows a restricted expression in the developing forebrain. Nature 351: 748-751.

Price M, Lazzaro D, Pohl T, Mattei M-G, Rüther U, Olivo J-C, Duboule D and Di Lauro R (1992): Regional expression of the homeobox gene *Nkx-2.2* in the developing mammalian forebrain. Neuron 8: 241-255.

Porteus MH, Bulfone A, Ciaranello RD and Rubenstein JLR (1991): Isolation and characterization of a novel cDNA clone encoding a homeodomain that is developmentally regulated in the ventral forebrain. Neuron 7: 221-229.

Puelles L, Amat JA and Martinez-de-la-Torre M (1987): Segment-related, mosaic neurogenetic pattern in the forebrain and mesencephalon of early chick embryos: I. Topography of AchE-positive neuroblasts up to stage HH18. J Comp Neurol 266: 247-268.

Robinson GW, Wray S and Mahon, KA (1991): Spatially restricted expression of a member of a new family of murine *distal-less* homeobox genes in the developing forebrain. New Biologist 3: 1183-1194.

Sakai Y (1987): Neurulation in the mouse I. The ontogenesis of neural segments and the determination of topographical regions in a central nervous system. Anat Rec 218: 450-457

Salinas PC and Nusse R (1992): Regional expression of the *Wnt-3* gene in the developing mouse forebrain in relationship to diencephalic neuromeres. Mech Dev 39: 151-160.

Simeone A, Acampora D, Gulisano M, Stornaiuolo A and Boncinelli E (1992a): Nested expression domains of four homeobox genes in developing rostral brain. Nature 358: 687-690.

Simeone A, Gulisano M, Acampora D, Stornaiuolo A, Rambaldi M and Boncinelli E (1992b): Two vertebrate homeobox genes related to the *Drosophila empty spiracles* gene are expressed in the embryonic cerebral cortex. embo j 11: 2541-2550.

Simeone A, Acampora D, Mallamaci A, Stornaiuolo A, D'Apice M R, Nigro V and Boncinelli, E (1993): A vertebrate gene related to *orthodenticle* contains a homeodomain of the *bicoid* class and demarcates anterior neuroectoderm in the gastrulating mouse embryo. Embo J 12: 2735-2747

Simeone A, Acampora D, Pannese M, D'Esposito M, Stornaiuolo A, Gulisano M, Mallamaci A, Kastury K, Druck T, Huebner K and Boncinelli E (1993): Cloning and characterization of two new members of the vertebrate *dlx* gene family. Proc Natl Acad Sci USA 91: 2250-2254.

Tam PPL (1989): Regionalisation of the mouse embryonic ectoderm: allocation of prospective ectodermal tissues during gastrulation. Development 107: 55-67

Tessier-Lavigne M (1992): Axon guidance by molecular gradients. Curr Opinion Neurobiol 2: 60-65.

Walldorf U and Gehring WJ (1992): *empty spiracles*, a gap gene containing a homeobox involved in *Drosophila* head development. EMBO J 11: 2247-2259

Wieschaus E, Perrimon N and Finkelstein R (1992): *orthodenticle* activity is required for development of medial structures in the larval and adult epidermis of *Drosophila*. Development 115: 801-811

Wilkinson DG, Bhatt S, Cook M, Boncinelli E and Krumlauf R (1989): Segmental expression of *Hox-2* homeobox-containing genes in the developing mouse hindbrain. Nature 341: 405-409

POU DOMAIN TRANSCRIPTION FACTORS IN THE NEUROENDOCRINE SYSTEM

Bogi Andersen[1], Linda Erkman[1], Peng Li[2], Chijen R. Lin[1], Sheng-Cai Lin[1], Robert McEvilly[3], Marcus Schonemann[3], Eric Turner[1], Michael G. Rosenfeld[1,4]

[1]Eukaryotic Regulatory Biology Program, [2]Biomedical Sciences Graduate Program, [3]Biology Graduate Program, Department of Medicine, [4]Howard Hughes Medical Institute, University of California, San Diego, 9500 Gilman Drive, La Jolla, CA, 92093-0648, U.S.A.

INTRODUCTION

The precise molecular mechanisms by which distinct, mature neuronal phenotypes arise from a common primordium, in response to specific morphogens and signals, remains a fundamental question in neurobiology. We have discovered and characterized eight novel neuronally-expressed mammalian POU domain transcription factors. Based on the distinct spatial and temporal patterns of their expression, and their structural similarity to critical determining factors in other organ systems, we hypothesize that these POU domain factors exert essential roles in establishing specific neuronal and glial phenotypes and the patterns of connection between them. We have discovered the cognate DNA recognition elements for the three classes of neuronally-expressed POU domain factors for Brn-1, Brn-2, Tst-1, Brn-4 (Class III), and Brn-3.0, Brn-3.1, and Brn-3.2 (Class IV), revealing differential spacing and orientation between the core elements of the bipartite DNA binding motif. Genetic methods are being used to critically test our hypothesis concerning the roles of these POU domain factors in determining the generation and function of specific neuronal and glial phenotypes. Homologous recombination in embryonic stem cells will be used to generate mice null for these POU IV class genomic loci, alone and in combination. We plan to test the roles of these factors in differentiation and survival of sensory neurons in dorsal root ganglia, spinal cord, and retinal ganglion cells, and their connections. Furthermore, we plan to examine the role of the four Class III POU domain factors, Brn-1, Brn-2, Tst-1 and Brn-3 in establishing central neuronal and glial phenotypes by gene deletion.

Neural Cell Specification: Molecular Mechanisms and Neurotherapeutic Implications
Edited by Juurlink *et al.*, Plenum Press, New York, 1995

IDENTIFICATION OF THE POU DOMAIN FAMILY OF TRANSCRIPTION FACTORS

The mammalian nervous system consists of as many as 10^{12} neurons that are both biochemically and biophysically heterogeneous (McKay, 1989). These large numbers of neurons differ from one another in their signaling capabilities, neuronal and hormonal responsiveness, neurotransmitter use, expression of cell surface recognition molecules, as well as the number and shape of their neurites and the neuronal connections they make and receive. Understanding the molecular mechanisms underlying the appearance of specific neuronal phenotypes in the nervous system, therefore, is a major challenge for neuroscience. The study of neurogenesis in lower animals such as *Drosophila* and *C. elegans*, has been greatly facilitated by genetic approaches and led to the identification of a large number of genes involved in these processes (Baker et al., 1990; Chalfie and Au 1989). In mammals, the study of tissue-specific gene expression characteristic of particular cellular phenotypes has generally been most successfully approached by characterizing the cis-active elements that are necessary for tissue-specific gene expression, and subsequently isolating the tissue-specific transcription factors that bind to these elements (Mitchell and Tjian 1989; Johnson and McKnight 1989; Serfling 1989; Weintraub et al., 1991; Rosenfeld 1991). These two independent approaches in different species led to the documentation that hierarchies of transcriptional regulation underlie the establishment of cell phenotypes in all metazoans.

While the nuclear events underlying neuronal development and the accompanying diverse patterns of specific gene expression are only beginning to be clarified, it is likely that the precise temporal and spatial patterns characteristic of mammalian brain development reflect sequential activation of a complex array of regulatory factors similar to those presumed to account for establishing structural patterns in *Drosophila* (Gehring 1987; Scott and Carroll 1987). Morphogenesis in *Drosophila* is regulated, in part, by a family of genes that contain a 180 bp DNA sequence encoding a highly conserved 60 amino acid sequence referred to as the homeodomain. These homeobox genes were first identified by analysis of homeotic mutations (Scott et al., 1989), and are preferentially clustered within the *Drosophila* genome in loci controlling morphogenesis. Genes containing sequences homologous to the homeodomain have since been identified in mammals (Martin 1987; Holland and Hogan 1988), including the identification of the Hox gene clusters (Ingraham et al., 1988), and genetic experiments have established their critical functions in mammalian pattern formation (Wilkinson et al., 1987; Lufkin et al., 1991; Chisaka et al., 1992; LeMouellic et al., 1992; Ramirez-Solis et al., 1993; Gendron-Maquire et al., 1993; Rijli et al., 1993).

Our interest in defining the molecular mechanisms that underlie the appearance of neural phenotypes have led us to define and investigate the role of a distinct class of transcriptional regulators, referred to as the POU-domain factors. The simultaneous discovery of a pituitary-specific transcription factor (Pit-1) (McGinnis and Krumlauf 1992; Bodner et al., 1988), a B-cell specific transcription factor Oct-2 (Scheidereit et al., 1988; Clerc et al., 1988; Ko et al., 1988; Müller et al., 1988), the ubiquitous octamer binding protein, Oct-1, (Sturm et al., 1988), and the *C. elegans* unc-86 gene product (Finney et al. 1988) led to the discovery of a gene family that contains a novel DNA binding motif, referred to as the POU domain (Herr et al., 1988; Rosenfeld 1991). This domain is bipartite in nature, consisting of a highly conserved amino-terminal region of 60 amino acids, the POU-homeodomain, divergent from the *Drosophila* homeodomain, and an even more conserved 76-78 amino acid region, the POU-specific (POUs) domain separated by a less conserved linker region.

Detailed functional and genetic analyses of several POU domain factors have proven their roles as developmental regulators critical for determining cell phenotype. The mammalian anterior pituitary gland has presented a well-developed model for investigating the mechanisms by which five phenotypically distinct cell types arising from a common primordium appear during ontogeny in a stereotypical order (Voss and Rosenfeld 1992). One form of genetically transmitted dwarfism in mice, characterized as two allelic recessive mutations (Snell and Jackson), results in mice producing neither detectable growth hormone, prolactin, or thyroid stimulating hormone, nor mature somatotrope, lactotrope and thyrotrope cell types, is the consequence of mutations in the Pit-1 gene, rendering the Pit-1 protein impaired in binding to its cognate DNA recognition elements (Li et al., 1990). Pit-1 thus regulates proliferation and/or survival of these three pituitary cell types, and we have demonstrated that Pit-1 regulates the transcription of a specific receptor critical for somatotrope growth (Lin et al., 1993; Godfrey et al., 1993). Intriguingly, Pit-1 autoregulates the transcription of its encoding gene, based primarily on its actions on an enhancer localized >10 kb 5' of the transcription initiation site (Rhodes et al., 1993).

The second POU domain factor for which genetic evidence has proven its function as a determining factor is unc-86. Mutation of the unc-86 locus in *C. elegans* reveals that this POU domain factor is required for the commitment of several neuroblast lineages, and the specification of various neurons (Finney et al., 1988; Finney and Ruvkun 1990), causing cells to reiterate the lineage of their mothers and as a result, leads to over production of certain cell types and absence of others. Mutation in the unc-86 gene also affects the identity of many non-dividing cells, such as the mechanoreceptor and HSN neurons. The mechanoreceptor neurons are a subset of unc-86 expressing cells that also express a homeodomain protein, mec-3, an important protein in specifying the identity of these neurons. The unc-86 protein, required for the initial commitment of the mechanoreceptor neurons, appears to be directly involved in the transcriptional activation of the mec-3 gene (Way and Chalfie 1989; Way et al., 1991), while unc-86 and mec-3 act synergistically to bind as a heterodimer. In contrast to these functions in initial commitment, the unc-86 protein plays a terminal differentiation role in the maturation of HSN neurons, suggesting distinct roles in specifying different cellular phenotypes. Finally, based on a gene "knockout", Oct-2 is now established to exert important roles on terminal B cell development and survival (Corcoran et al., 1993).

We, therefore, initiated an experimental program to examine whether novel POU-domain proteins were expressed during establishment of the nervous system, potentially specifying neuronal phenotypes. We initially identified four new mammalian POU domain genes, referred to as Brain-1 (Brn-1), Brain-2 (Brn-2), Brain-3 (Brn-3), and Testes-1 (Tst-1/SCIP), that proved to be selectively expressed in the nervous system, except for detectable expression of Tst-1 in testes (He et al., 1989). These POU domain genes were shown to be expressed during ontogeny in all levels of the developing neural tube, with differential patterns of hybridization in the subependymal and subventricular layer, and to exhibit differential, overlapping patterns of restricted expression in the adult brain (He et al., 1989; Wegner et al., 1993). The identification of a POU domain protein in *Drosophila*, highly similar to Brn-3 and unc-86 (Treacy et al., 1991; Treacy et al., 1993), further suggested that this large gene family exerts conserved functions in all metazoans. Additional POU domain gene products have been identified in a variety of species such as *Drosophila*, *C. elegans*, and *X. laevis*, and we and others have identified numerous additional members of this family expressed in mammals (reviewed in Wegner et al.,

1993). The POU domain proteins are subdivided into six classes (I-VI) according to primary amino acid in several regions (Wegner et al., 1993; Andersen et al., 1993). Two classes, referred to as Class III and Class IV factors, encompass most of the neuronal POU domain factors, including Brn-1, Brn-2, Brn-4, and Tst-1/SCIP/Oct-6 (He et al., 1989; Monuki et al., 1989; He et al., 1991; Schöler 1991; Mathis et al., 1992; LeMoine and Young 1992), and Brn 3.0, Brn 3.1 and Brn 3.2 (Gerrero et al., 1993; Xiang et al., 1993; Turner et al., 1994), respectively. We identified additional POU domain factors selectively expressed in other organs, including skin and sperm.

While the POU domain factors that we discovered were the initial tissue-specific transcription factors identified in the mammalian forebrain, screening for *Drosophila* homologues has subsequently resulted in identification of other unique homeodomain transcription factors in forebrain (Evans 1988; Beato 1989; Porteus et al., 1991; Price et al., 1991; Robinson et al., 1991; Price et al., 1992; Simeone et al., 1992). By analogy with the functions of Pit-1, unc-86 and Oct-2, it is likely that neuronally-expressed POU domain proteins will prove to selectively control terminal differentiation events in specific neuronal cell types. The challenge now is to test this hypothesis by using genetic biochemical and cell biological approaches.

IDENTIFICATION OF NOVEL MEMBERS OF THE POU DOMAIN GENE FAMILY EXPRESSED IN THE BRAIN AND PERIPHERAL NERVOUS SYSTEM

Four new members of the POU domain gene family were initially identified using a degenerate oligonucleotide PCR technique, one of the initial reports of this approach (He et al., 1989). Using additional strategies, seven more family members were identified and characterized, including four expressed in the nervous system. Based on structural similarities, the known mammalian POU domain proteins appear to segregate into six classes (Fig. 1).

The POU Domain Family

We have determined the sequence of each of these factors (rat/Brn-1, Brn-2, Brn-3, Tst-1/SCIP/Oct-6, and Brn-4) and each has been characterized as a transcription factor. In situ hybridization analysis has revealed that each factor is expressed in distinct temporal and spatial patterns in the brain. All these genes are widely expressed at all levels in the neural tube (He et al., 1989). Tst-1 and Brn-4 are initially expressed in the subventricular zone, while Brn-1 and Brn-2 initially appear in mitotic subependymal cells. The time course of anatomic restriction in the developing neural tube is distinct for each factor, with the patterns for each gene product tending to reflect adult loci of expression. While Tst-1 is initially expressed in a single set of neurons in the PVH nucleus, it is progressively more widely expressed in ontogeny, and is expressed in layer V of the cortex (He et al., 1989). In addition to expression in a subset of neurons, Tst-1 is later expressed in myelinating Schwann cells (Monuki et al., 1989), and in central myelinating glia (LeMoine and Young 1992), with Tst-1 gene expression corresponding precisely to the onset of myelination in glial cells in vivo. Tst-1, although inhibitory to protein zero (P0) myelin gene expression by transient cotransfection analyses (LeMoine and Young 1992), can clearly function as a positive transcription factor (Wegner et al., 1993). Brn-4 transcripts, initially widely expressed, become restricted to only a few regions of adult forebrain, including supraoptic and paraventricular nuclei of

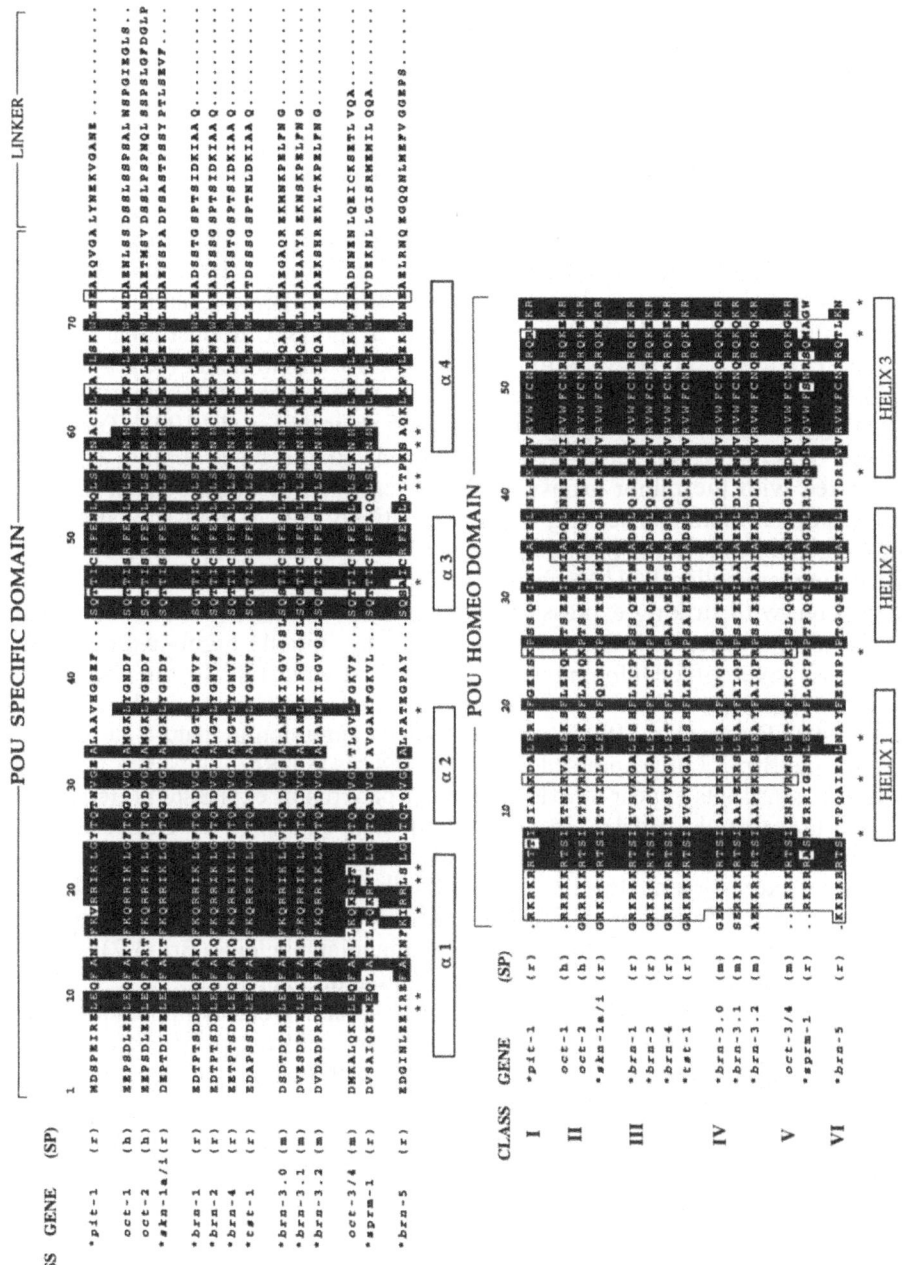

Figure 1. POU domains of the known members of the POU domain family. Optional activator on a potential neuronal target gene (Li et al., 1993).

hypothalamus (Mathis et al., 1992). Brn-1 and Brn-2 exhibit a virtually identical pattern of expression in the brain, with particularly high levels in the hypothalamus and isocortex (layers II-V). In the hypothalamus, the more abundant Brn-2 transcripts are present in the dorsomedial paracellular (mpd) region of the paraventricular nucleus (PVH), which synthesizes and secretes the corticotropin-releasing hormone (CRF) (Li et al., 1993). The ability of Brn-2 to bind with high affinity and activate (40-fold) the CRF promoter confirmed its function as a transcri-

The POU IV Family: BRN-3.0, BRN-3.1, BRN-3.2, I-POU

Cloning of Brn-3.0 (He et al., 1989) and the *Drosophila* homologue, I-POU (Treacy et al., 1991), revealed that these factors have clear homology to the unc-86 gene product that determines sensory lineages in *C. elegans*. Our further analyses has revealed that mammals express three highly related, but distinct genes; Brn-3.0 [421 amino acids], Brn-3.1 [338 amino acids] (Gerrero et al., 1993; and Brn-3.2 [411 amino acids] (Turner et al., 1994). Brn-3.0, Brn-3.1, Brn-3.2, and I-POU share a region of homology in the N-terminus outside of the POU domain, that we refer to as the "POU IV homology region", including 10 amino acids (RAEALAAVDI) entirely conserved from *Drosophila* to man (Gerrero et al., 1993), and that exhibits homology to the comparable region in unc-86 (Gerrero et al., 1993; Turner et al., 1994). I-POU, which lacks two amino acids in the basic amino acid cluster associated with the minor groove at the N-terminus of the POUHD, is a potent transcriptional inhibitor, serving as an alternatively spliced transactivator of Twin of I-POU that restores the two amino acids (Treacy et al., 1991; Treacy et al., 1992). The three mammalian POU-IV factors exhibit overlapping, but distinct, patterns of expression, with striking expression in developing retinal and dorsal root ganglia. By E15.5 in mouse, both Brn-3.0 and Brn-3.2 are initially highly expressed in the innermost layer of developing retinal cells that give rise to the ganglion cell population, and were absent from the neuroblastic layer (Gerrero et al., 1993; Xiang et al., 1993; Turner et al., 1994), with continued expression in mature ganglion cells. Brn-3.0 and Brn-3.1 are expressed in similar patterns in DRG. However, in the spinal cord, the transcripts appear to exhibit distinct patterns of expression (Gerrero et al., 1993; Turner et al., 1994). In contrast to their extensive co-expression in the peripheral sensory nervous system and retina, the developmental and adult patterns of expression of Brn-3.2 differ markedly from those of Brn-3.0 in the central nervous system. For example, near the anterior limit of expression, including the medial habenula (Brn-3.0) and the interpeduncular nucleus (Brn-3.2), the genes display exclusive regions of developmental and adult expression.

Further, Brn-3.0 is expressed in several B- and T-cell lymphocyte lines, and at very low levels in the anterior pituitary gland and in specific populations of splenic B cells (Gerrero et al., 1993). While Brn-3.0, Brn-3.1, and Brn-3.2 bind very weakly to classic octamer elements, they fail to transactivate these elements. In contrast, Brn-3.0, Brn-3.1, and Brn-3.2 can function in specific binding sites characterized as variants of GCAT (nnn) TAAT (Gerrero et al., 1993; Xiang et al., 1993; Li et al., 1993; Turner et al., 1994).

The POU VI Class: BRN-5.0.

We have identified rat cDNAs that encode the most divergent POU domain protein, referred to as Brn-5 (Wegner et al., 1993). Brn-5.0 is widely expressed, beginning on E12.5, with highest levels in the developing brain and spinal cord, with an enrichment in the inner granular layer (layer IV) of the neocortex, with expression localized over neurons (Wegner et al., 1993). Brn-5.0 binds and functions preferentially on a variant of an octamer binding site that is very similar to the optimal binding site for Oct-1.

SPACING AND ORIENTATION OF BIPARTITE DNA BINDING MOTIFS AS FUNCTIONAL DETERMINANTS FOR NEURONALLY-EXPRESSED POU DOMAIN TRANSCRIPTION FACTORS

Sequence comparison of the Brn-2 recognition elements, and random oligonucleotide selection, revealed that the Brn-2 binding site consists of two conserved core motifs separated by a variable spacer region CATnTAAT (n=0,2,3) consistent with the bipartite structure of the POU domain (Ingraham et al., 1990; Kristie and Sharp 1990; Verrijzer et al., 1990; Verrijzer et al., 1992). Brn-2 is able to recognize DNA elements with different spacing (0, 2 or 3) nucleotides, further reveals the remarkable flexibility of the Brn-2 POU domain, because altering the spacing from 0 to 3 represents about 10 Å linear distance and 108° rotation between the two core binding motifs. UV cross-linking of BrdU substituted DNA sites, and methylation interference assays demonstrated that the Brn-2 POUs domain contacted a core motif (CAT) in an inverted orientation compared to that of the ATG core motif in the Pit-1 binding element while the POUHD domain bound to the TAAT site (Li et al., 1993). Based on the co-crystal structures of l and phage 434 repressors, and phage 434 Cro protein (Harrison and Aggarwal 1990) and the existing model based on the NMR solution structure of the POUs domain of Oct-1 (Hochschild and Ptashne 1986; Assa-Munt et al., 1993), we adapted a strategy previously utilized to determine the direct amino acid and base pair contacts of l repressor (Hochschild and Ptashne 1986) by replacing a conserved residue glutamine44 with a much smaller amino acid alanine (Q44 -> A), indicating that the orientation of the Brn-2 POUs domain on binding to its preferred recognition motif is apparently opposite to that of the POUs domain of Oct-1 or Pit-1 on their response elements (Li et al., 1993).

In contrast, the class IV POU domain factor, Brn-3.0, was capable of specific, high affinity binding and function only on sites containing a spacing of 3 nucleotides between the core binding motifs. A detailed analysis of POU domain chimeras identified the precise residues regulating the spacing preference apparently resided in the POUHD. A marked (>75%) switch in Brn-2 sites involved an alteration of only the three amino acid residues distinct between Brn-2 and Brn-3 in the zinc finger in the POUHD N-terminus (R -> E in position 3 and RK -> KR in position 5,6). These three residues in the basic cluster of Brn-3 POUHD, together with the helix 2, transfer a complete switch in the spacing preference of Brn-2. These observations led us to propose a model for spacing preferences of the POU domain on its DNA binding sites, suggesting that minor groove, and perhaps phosphate backbone, contacts conferred by three amino acids in the Brn-3 N-terminal basic region of the POUHD, constrains the permitted major groove contacts by the POUs domain, restricting Brn-3 binding to DNA elements in which the core binding motifs are spaced by 3 nucleotides (Li et al., 1993).

Control of the JC Virus Life Cycle by Tst-1

We have linked a POU domain factor to cell-specific infection by a DNA virus that causes progressive multifocal leukoencephalopathy (PML), the human papovavirus JC (Wegner et al., 1993). Selective infection and cytolytic destruction of the myelin-producing oligodendrocyte of the central nervous system by JC virus results in severe demyelination (Major et al., 1992). The tissue tropism of JC virus is defined on the level of transcription and is conferred by the hypervariable non-coding region, that not only serves as origin of DNA replication but also as promoter for both the early regulatory and the late capsid genes.

Because oligodendrocytes are the physiological target of viral infection during PML and since both Schwann cells and oligodendrocytes are permissive for infection by JC virus in culture (Yogo et al., 1990), there is a striking correlation between expression of Tst-1 (Monuki et al., 1989; He et al., 1991) and susceptibility for JC virus. We have established site-specific DNA binding of Tst-1 to the JC viral regulatory region, and documented ability of Tst-1 to stimulate viral early and late promoter dependent on the ability of Tst-1 to bind one specific element. Tst-1 also stimulated (>30-fold) expression of the viral T antigen from its own promoter (Wegner et al., 1993). Tst-1 stimulates the expression of the viral regulatory genes, small and large tumor antigens that subvert cellular control mechanisms by altering the cellular transcription pattern, overriding control of cellular proliferation exerted by the retinoblastoma and p53 genes and stimulating quiescent cells to synthesize DNA. This positive effect of Tst-1 on various stages of the lytic life cycle of JC virus correlates well with its expression in those cells of the central nervous system that are the preferred target of JC viral infection, suggesting that Tst-1 is one of the factors that determines the specificity of JC virus for oligodendrocytes.

REFERENCES

Andersen B, Schonemann MD, Pearse II RV, Jenne K, Sugarman J, Sawchenko P and Rosenfeld MG (1993): Brn-5 is a divergent POU domain factor enriched in layer IV of the neocortex. J Biol Chem 268: 23390-23398

Assa-Munt N, Mortishire-Smith RJ, Aurora R, Herr W and Wright PE (1993): The solution structure of the Oct-1 POU-specific domain reveals a striking similarity to the bacteriophage lambda repressor DNA-binding domain. Cell 73: 193-205.

Baker NE, Mlodzik M and Rubin GM (1990): Spacing differentiation in the developing Drosophila eye: a fibrinogen-related lateral inhibitor encoded by scabrous. Science 250: 1370-7.

Beato M (1989): Gene regulation by steroid hormones. Cell 56: 335-44.

Bodner M, Castrillo JL, Ellisman M, and Karin M (1988): The pituitary-specific transcription factor GHF-1 is a homeobox-containing protein. Cell 55: 505-18.

Chalfie M and Au M (1989): Genetic control of differentiation of the Caenorhabditis elegans touch receptor neurons. Science 243:1027-33.

Chisaka O, Musci TS and Capecchi MR (1992): Developmental defects of the ear, cranial nerves and hindbrain resulting from targeted disruption of the mouse homeobox gene Hox-1.6. Nature 355: 516-20.

Clerc RG, Corcoran LM, Baltimore D and Sharp PA (1988): The B-cell-specific Oct-2 protein contains POU box- and homeo box-type domains. Genes and Dev 2: 1570-81.

Corcoran LM, Karvelas M, Nossal GJ, Ye ZS, Jacks T, and Baltimore D (1993): Oct-2, although not required for early B-cell development, is critical for later B-cell maturation and for postnatal survival. Genes and Dev 7: 570-82.

Evans RM (1988): The steroid and thyroid hormone receptor superfamily. Science 240: 889-95.

Finney M, Ruvkun G and Horvitz HR (1988): The C. elegans cell lineage and differentiation gene unc-86 encodes a protein with a homeodomain and extended similarity to transcription factors. Cell 55: 757-69.

Finney M and Ruvkun G (1990): The unc-86 gene product couples cell lineage and cell identity in C.elegans. Cell 63: 895-905.

Gehring WJ (1987): Homeo boxes in the study of development. Science 236: 1245-1252.

Gendron-Maquire M, Mallo M, Zhang M and Gridley T (1993): Hoxa-2 mutant mice exhibit homeotic transformation of skeletal elements derived from cranial neural crest. Cell 75:1317-1331.

Gerrero RM, McEvilly RJ, Turner E, Lin CR, O'Connell S, Jenne KJ, Hobbs MV and Rosenfeld MG (1993): Brn-3.0: A POU-domain protein expressed in the sensory, immune, and endocrine systems that functions on elements distinct from known octamer motifs. Proc Natl Acad Sci 90: 10841- 10845.

Godfrey P, Rahal JO, Beamer WG, Copeland NG, Jenkins NA and Mayo KE (1993): GHRH receptor of little mice contains a missense mutation in the extracellular domain that disrupts receptor function. Nature Genetics 4: 227-232.

Harrison SC and Aggarwal AK (1990): DNA recognition by proteins with the helix-turn-helix motif. Ann Rev Biochem 59: 933-69.

He X, Treacy MN, Simmons DM, Ingraham H, Swanson LW and Rosenfeld MG (1989): Expression of a large family of POU-domain regulatory genes in mammalian brain development. Nature 340: 35-42.

He X, Gerrero R, Simmons DM, Park RE, Lin CR, Swanson LW and Rosenfeld MG (1991): Tst-1, a member of the POU-domain gene family, binds the promoter of the gene encoding the cell surface adhesion molecule Po. Mol Cell Biol 3: 1739-1744.

Herr W, Sturm RA, Clerc RG, Corcoran LM, Baltimore D, Sharp PA, Ingraham HA, Rosenfeld MG, Finney M and Ruvkun G (1988): The POU domain: a large conserved region in the mammalian pit-1, oct-1, oct-2, and Caenorhabditis elegans unc-86 gene products. Genes Develop 2: 1513-6.

Holland PW and Hogan BL (1988): Expression of homeo box genes during mouse development: a review. Genes Develop 2: 773-82.

Hochschild A and Ptashne M (1986): Homologous interactions of lambda repressor and lambda Cro with the lambda operator. Cell 44: 925-33.

Ingraham HA, Chen RP, Mangalam HJ, Elsholtz HP, Flynn SE, Lin CR, Simmons DW, Sawanson L and Rosenfeld MG (1988): A tissue-specific transcription factor containing a homeodomain specifies a pituitary phenotype. Cell 55: 519-529.

Ingraham HA, Flynn SE, Albert VR, Kapiloff MS, Wilson L and Rosenfeld MG (1990): The POU-specific domain of the pituitary transcriptional factor, Pit-1, is essential for high DNA binding affinity, site specificity and cooperative interactions. Cell 61: 1021-1033.

Johnson PF and McKnight SL (1989): Eukaryotic transcriptional regulatory proteins. Ann Rev Biochem 58: 799-839.

Ko HS, Fast P, McBride W and Staudt LM (1988): A human protein specific for the immunoglobulin octamer DNA motif contains a functional homeobox domain. Cell 55: 135-44.

Kristie TM and Sharp PA (1990): Interactions of the Oct-1 POU subdomains with specific DNA sequences and with the HSV alpha-trans-activator protein. Genes Develop 4: 2383-96.

LeMoine CL and Young WS (1992): RHS2, a POU domain-containing gene, and its expression in developing and adult rat. Proc Natl Acad Sci USA 89: 3285-3289.

LeMouellic H, Lallemand Y and Brulet P (1992): Homeosis in the mouse induced by a null mutation in the Hox-3.1 gene. Cell 69: 251-264, 1992.

Li P, He X, Gerrero MR, Aggarwal A and Rosenfeld MG (1993): Spacing and orientation of bipartite DNA binding motifs as potential functional determinants for POU domain factors. Genes Develop 7: 2483-2496.

Li S, Crenshaw EB, Rawson EJ, Simmons DM, Swanson LW and Rosenfeld MG (1990): Dwarf locus mutants, which lack three pituitary cell types, result from mutations in the POU-domain gene, Pit-1. Nature 347: 528-533.

Lin SC, Lin CR, Gukovsky I, Lusis AJ, Sawchenko PE and Rosenfeld MG (1993): Molecular basis of the little mouse phenotype and implications for cell type-specific growth. Nature 364: 208-213.

Lufkin T, Dierich A, LeMeur M, Mark M and Chambon P (1991: Disruption of the Hox-1.6 homeobox gene results in defects in a region corresponding to its rostral domain of expression. Cell 66: 1105-19.

Major EO, Amemiya K, Tornatore CS, Houff SA and Berger JR (1992): Pathogenesis and molecular biology of progressive multifocal leukoencephalopathy, the JC virus-induced demyelinating disease of the human brain. Clin Microbiol Rev 5: 49-73.

Martin GR (1987): Nomenclature for homoeobox-containing genes. Nature 325: 21-2.

Mathis MJ, He X, Simmons DM, Swanson LW and Rosenfeld MG (1992): Brain 4: A novel mammalian POU-domain transcription factor exhibiting restricted brain-specific expression. EMBO J 11: 2551-2561.

McGinnis W and Krumlauf R (1992): Homeobox genes and axial patterning. Cell 68: 283-302.

McKay RD (1989): The origins of cellular diversity in the mammalian central nervous system. Cell 58: 815-21.

Mitchell PJ and Tjian R (1989): Transcriptional regulation in mammalian cells by sequence-specific DNA binding proteins. Science 245:371-8.

Monuki ES, Weinmaster G, Kuhn R and Lemke G (1989): SCIP: a glial POU domain gene regulated by cyclic AMP. Neuron 3: 783-93.

Müller MM, Ruppert S, Schaffner W and Matthias P (1988): A cloned octamer transcription factor stimulates transcription from lymphoid-specific promoters in non-B cells. Nature 336: 544-51.

Porteus MH, Bulfone A, Ciaranello RD and Rubenstein JL (1991): Isolation and characterization of a novel cDNA clone encoding a homeodomain that is developmentally regulated in the ventral forebrain. Neuron 7: 221-9.

Price M, Lemaistre M, Pischetola M, Di Lauro R and Duboule D (1991): A mouse gene related to Distal-less shows a restricted expression in the developing forebrain. Nature 351: 748-51.

Price M, Lazzaro D, Pohl T, Mattei MG, Ruther U, Olivo JC, Duboule D and Di Lauro R (1992): Regional expression of the homeobox gene Nkx-2.2 in the developing mammalian forebrain. Neuron 8: 241-55.

Ramirez-Solis R, Zheng H, Whiting J, Krumlauf R and Bradley A (1993): Hoxb-4 (Hox-2.6) mutant mice show homeotic transformation of a cervical vertebra and defects in the closure of the sternal rudiments. Cell 73: 279-94.

Rhodes SJ, Chen R, DiMattia GE, Scully KM, Kalla KA, Lin SC, Yu VC and Rosenfeld MG (1993): A tissue-specific enhancer confers Pit-1-dependent morphogen inducibility and autoregulation on the pit-1 gene. Genes Develop 7: 913-932.

Rijli FM, Mark M, Lakkaraju S, Dierich A, Dolle P and Chambon P (1993): A homeotic transformation is generated in the rostral bronchial region of the head by disruption of Hoxa-2 which acts as a selecter gene. Cell 75: 1333-49.

Robinson GW, Wray S and Mahon KA (1991): Spatially restricted expression of a member of a new family of murine distal-less homeobox genes in the developing forebrain. New Biologist 3: 1183-94.

Rosenfeld MG (1991): POU-domain transcription factors: pou-er-ful developmental regulators. Genes Develop 5: 897-907.

Scheidereit C, Cromlish JA, Gerster T, Kawakami K, Balmaceda CG, Currie RA and Roeder RG (1988): A human lymphoid-specific transcription factor that activates immunoglobulin genes is a homoeobox protein. Nature 336: 551-7.

Schöler HR (1991): Octamania: the POU factors in murine development. Trends Genetics 7: 323-9.

Scott MP and Carroll SB (1987): The segmentation and homeotic gene network in early *Drosophila* development. Cell 51: 689-98.

Scott MP, Tamkun JW and Hartzell GW 3d (1989): The structure and function of the homeodomain. Biochim Biophys Acta 989: 25-48.

Serfling E (1989): Autoregulation--a common property of eukaryotic transcription factors? Trends Genetics 5:131-3.

Simeone A, Acampora D, Gulisano M, Stornaiuolo A and Boncinelli E (1992): Nested expression domains of four homeobox genes in developing rostral brain. Nature 358: 687-90.

Sturm RA, Das G and Herr W (1988): The ubiquitous octamer-binding protein Oct-1 contains a POU domain with a homeo box subdomain. Genes and Dev 2: 1582-99.

Treacy MN, He X and Rosenfeld MG (1991): I-POU: a POU-domain protein that inhibits neuron-specific gene activation. Nature 350: 577-84.

Treacy MN, Neilson LI, Turner EE, He X and Rosenfeld MG (1992): Twin of I-POU: a two amino acid difference in the I-POU homeodomain distinguishes an activator from an inhibitor of transcription. Cell 68: 491-505.

Turner EE, Jenne KJ and Rosenfeld MG (1994): Brn-3.2: A Brn-3-related transcription factor with distinctive central nervous system expression and regulation by retinoic acid. Neuron 12: 1-20.

Verrijzer CP, Kal AJ and van der Vliet PC (1990): The oct-1 homeo domain contacts only part of the octamer sequence and full oct-1 DNA-binding activity requires the POU-specific domain. Genes Develop 4: 1964-74.

Verrijzer CP, Strating M, Mul YM and van der Vliet PC (1992): POU domain transcription factors from different subclasses stimulate adenovirus DNA replication. Nucleic Acids Res 20: 6369-75.

Voss JM and Rosenfeld MG (1992): Anterior pituitary development:Short tales from dwarf mice. Cell 70: 527-30.

Way JC and Chalfie M (1989): The mec-3 gene of Caenorhabditis elegans requires its own product for maintained expression and is expressed in three neuronal cell types. Genes Develop 3: 1823-33.

Way JC, Way L, Run JQ and Wang A (1991): The mec-3 gene contains cis-acting elements mediating positive and negative regulation in cells produced by assymmetric cell division in Caenorhabditis elegans. Genes Develop 5: 2199-2211.

Wegner M, Drolet DW and Rosenfeld MG (1993): Regulation of JC virus by the POU domain transcription factor Tst-1: Implications for progressive multifocal leukoencephalopathy. Proc Natl Acad Sci USA 90: 4743-4747.

Weintraub H, Davis R, Tapscott S, Thayer M, Krause M, Benezra R, Blackwell TK, Turner D, Rupp R and Hollenberg S (1991): The myoD gene family: nodal point during specification of the muscle cell lineage. Science 251: 761-6.

Wilkinson DG, Bailes JA and McMahon AP (1987): Expression of the proto-oncogene int-1 is restricted to specific neural cells in the developing mouse embryo. Cell 50: 79-88.

Xiang M, Zhou L, Peng YW, Eddy RL, Shows TB and Nathans J (1993): Brn-3b: a POU domain gene expressed in a subset of retinal ganglion cells. Neuron 11: 689-701.

Yogo Y, Kitamura T, Sugimoto C, Ueki T, Aso Y, Hara K and Taguchi F (1990): Isolation of a possible archetypal JC virus DNA sequence from nonimmuno-compromised individuals. J Vir 64: 3139-43.

MULTIPLE ROLES FOR PRONEURAL GENES IN DROSOPHILA NEUROGENESIS

Andrew P. Jarman[1] and Yuh Nung Jan

Howard Hughes Medical Institute and Departments of
Physiology & Biochemistry, University of California,
San Francisco, San Francisco, CA, 94143-0724, U.S.A.

[1]Present Address: Institute of Cell and Molecular
Biology, University of Edinburgh, Darwin Building,
King's Buildings, Edinburgh EH9 3JR, U.K.

INTRODUCTION

Important questions concerning early neurogenesis include: how do certain ectodermal cells decide to become neural precursors; and how do these cells choose to differentiate as one of many possible neural subtypes? Owing to its relative simplicity, and our detailed knowledge of its anatomy and ontogeny, the *Drosophila* peripheral nervous system (PNS) has been a good model system for identifying and characterizing genes involved in these fate decisions.

The *Drosophila* PNS comprises a stereotyped pattern of sensory neurons in and under the cuticle. There are four major classes of sensory element: external sense organs (such as the prominent mechanosensory bristles), chordotonal organs (stretch and vibration receptors), multiple dendritic neurons, and photoreceptors. Since there is little migration of neural precursors in *Drosophila*, the stereotyped pattern of the PNS results from the pattern in which neural precursors are selected from the ectoderm during development. Proneural genes are central to this selection process. These genes encode transcription factors of the helix-loop-helix (HLH) family. We briefly review the function of the proneural genes of the *achaete–scute* complex (AS-C), and describe our work on the new proneural gene, *atonal* (*ato*).

HOW THE AS-C PRONEURAL GENES WORK

Thanks to the efforts of many of researchers, neural precursor formation is best understood for external sense organs (reviewed in Ghysen and Dambly-Chaudière, 1989; Jan and Jan, 1990; Campuzano and Modolell, 1992). In this case,

Neural Cell Specification: Molecular Mechanisms and Neurotherapeutic Implications
Edited by Juurlink *et al.*, Plenum Press, New York, 1995

97

the selection of neural precursors from unpatterned ectoderm involves a cascade of events beginning with the activation of AS-C proneural genes, particularly the genes *achaete* (*ac*) and *scute* (*sc*) themselves. These genes are expressed transiently in well defined patches of ectodermal cells (proneural clusters), and endow these cells with competence to follow a neural fate (Fig. 1) (Romani et al., 1989; Cubas et al., 1991; Skeath and Carroll, 1991). The pattern of proneural clusters thus foreshadows the eventual pattern of neural precursors. *Ac/sc* activation is thought to be in direct response to the pattern information set up by segmentation, homeotic, and other positional information genes. In consequence, the AS-C genes have complex, modular regulatory regions, each module directing expression in one or a few proneural clusters (Ruiz-Gómez and Modolell, 1987; Skeath et al., 1992).

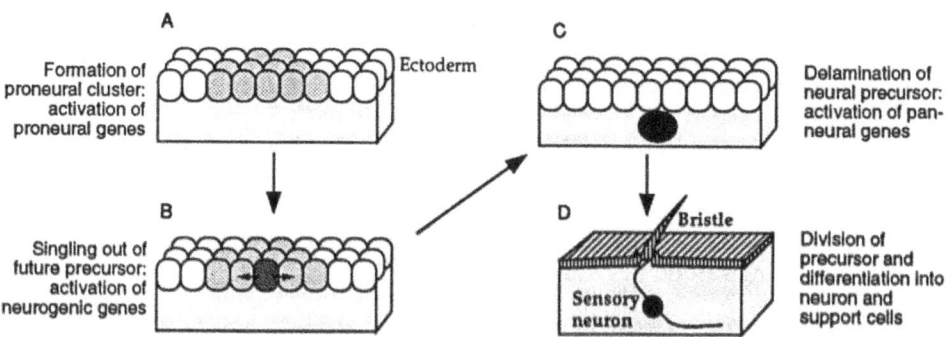

Figure 1. Schematic representation of early events in neurogenesis of the *Drosophila* PNS. (a) cross section of a proneural cluster of *ac/sc* expression in the ectodermal sheet; (b) singling out of future precursor by lateral inhibition; (c) restriction of *ac/sc* expression and delamination of neural precursor; (d) final external sense organ (bristle) after division and differentiation.

Within each proneural cluster, lateral inhibition then ensures that typically only one cell realizes its neural potential by restricting AS-C gene expression to a single cell; the remainder adopt the default epidermal fate. This process involves interplay of the proneural genes with the neurogenic genes (such as *Notch* and *Delta*), which encode an inhibitory cell communication pathway (Campos-Ortega, 1988; Artavanis-Tsakonis and Simpson, 1991; Ghysen et al., 1993). In essence, *ac/sc* products in each cell of the cluster activate this pathway, which in turn acts to inhibit *ac/sc* expression in neighbouring cells. From this state of mutual inhibition, a single cell eventually predominates and completely represses expression in the surrounding cells; expression thus becomes refined to that cell, the neural precursor (Skeath et al., 1992).

This restriction is the trigger for the delamination of the precursor from the ectoderm. Concomitantly, the *ac/sc* products activate a second group of genes, known as pan-neural or neural precursor genes, which seem to be required for determination and maintenance of neural precursor fate (Vaessin et al., 1991; Bier et al., 1992). Having achieved this, the proneural genes are rapidly shut off. The neural precursor then goes on to divide to give the four cells that will become the neuron and support cells of an external sense organ.

Molecularly, the AS-C products have DNA binding properties typical of bHLH proteins: they bind to a subset of E-boxes ($CAG^C/_GTG$) as heterodimers with the bHLH product of a second ubiquitously expressed gene, *daughterless* (Murre et al., 1989; Cabrera and Alonso, 1991). Such binding sites have been characterized in a few target genes so far (including the *ac* autoregulatory region [Van Doren et al., 1991], and the neural precursor gene, *asense* [Jarman et al., 1993b]).

So, the proneural step is the foundation of neural development, and AS-C action sets the pattern of the PNS. But this is true only for external sense organs (and the poorly characterized multiple dendritic neurons): mutants of the AS-C lack external sense organs (Fig. 2a, b), but still have chordotonal organs and photoreceptors (Dambly-Chaudière and Ghysen, 1987; Jiménez and Campos-Ortega, 1987). This raises the question of whether this model of neurogenesis applies to other neural elements or whether a different mechanism of neurogenesis pertains. Namely, do other proneural genes exist?

ATONAL: WIDESPREAD REQUIREMENT IN THE PNS FOR PRONEURAL GENES

Some evidence suggested other proneural genes should exist. In particular, mutations of the binding partner gene, *daughterless*, result in loss of chordotonal organs as well as external sense organs (Caudy et al., 1988). On this basis, we searched for other potential proneural genes with similarity to the AS-C. By PCR with degenerate primers based on conserved features of the bHLH domains of the AS-C genes, we were able to isolate a new gene, *atonal (ato)*, which indeed proved to be the proneural gene for chordotonal organs (Jarman et al., 1993a).

ato encodes an HLH protein that is similar to, but distinct from, the AS-C proteins. During development of both the larval and adult nervous system, *ato* is expressed in the patches of ectodermal cells from which chordotonal precursors arise (the proneural clusters) and in the precursors themselves (Jarman et al., 1993a). We have recently been able to isolate mutations of *ato*, which prove its proneural role in chordotonal organ formation (manuscript in preparation). Flies lacking *ato* function survive, but lack all adult chordotonal organs, including those in the leg, thorax, and abdomen, as well as the large array of chordotonal neurons in the antenna that comprise the fly's auditory apparatus (Fig. 2c, d). As would be expected for a proneural gene, we could show that this defect results from a failure of the chordotonal precursors to be selected in the mutant.

This leaves the question of whether photoreceptor formation also requires a proneural gene. To our surprise, *ato* mutant flies also have a severe eye defect. They lack all photoreceptors, both of the compound eye and the ocelli (Fig. 2e, f) (Jarman et al., 1994). Again, we could show that this defect is due to a failure in photoreceptor selection during development.

Therefore, we can now state that almost the entire PNS is formed by the action of two sets of proneural genes, the AS-C and *ato*. Curiously, a few sensory elements are still unaccounted for, such as a row of mechanosensory bristles on the wing margin, and the chemosensory organs of the antenna. Thus, it seems likely that there are still other proneural genes to find. This is also true for development of the CNS, which is only partly affected by the known proneural genes.

PRONEURAL GENES INFLUENCE NEURONAL DIVERSITY

Each major class of sense organ can be further subdivided on the basis of structural and functional variations. A useful view of *Drosophila* neurogenesis is that it is a process of progressive determination, in which cells undergo a hierarchical series of dichotomous fate choices (Ghysen and Dambly-Chaudière, 1989). Thus, in ectodermal cells proneural genes are concerned with choice of neural fate versus epidermal fate. In neural precursors, the homeobox gene, *cut*, directs external sense organ fate as opposed to chordotonal organ fate (Bodmer et al., 1987). In external sense organ precursors, the paired box gene, *poxn*, directs chemosensory fate as opposed to a mechanosensory one (Dambly-Chaudière et al., 1991). In mechanosensory organs, the homeobox gene, *BarH1*, determines a sensillum campaniformia structure rather than a sensillum trichodea (Higashijima et al., 1992).

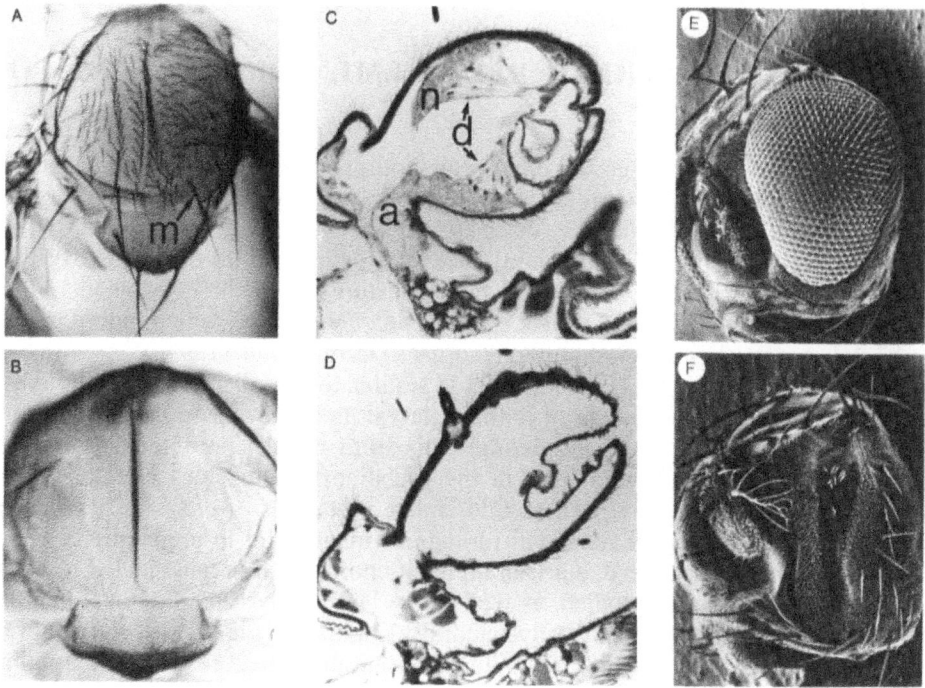

Figure 2. Proneural genes are required for most of the PNS. (A) Thorax of a wildtype fly showing external sense organs (bristles). (B) Fly mutant for *ac* and *sc* (*sc10-1*). Almost all bristles are missing. (C) Section of the second antennal segment of wildtype fly showing some of the hundreds of chordotonal neurons (arrows) of Johnston's Organ, the fly's ear. (D) Section from *ato* mutant, no chordotonal organs remain. (E) Wildtype compound eye. Each of the ~800 ommatidia contain a cluster of eight photoreceptors. (F) *ato* mutant. Only pigment cells are left.

Given this, we may ask whether proneural genes simply act at one level in this hierarchy (the neural-epidermal decision) to provide 'generic' neural precursors as raw material to be diversified later, or do they influence these subsequent choice steps concerned with neural subtype identity. Experimental evidence for the latter comes from misexpression studies. Generalized expression of any AS-C gene (in transgenic flies under the control of a heat-shock-inducible promoter) results in ectopic sense organs (Rodríguez et al., 1990; Brand et al., 1993), but these are exclusively external sense organs (Jarman et al., 1993a). Identical misexpression of *ato* also yields ectopic sense organs, but in this case they are largely (although not exclusively) chordotonal organs (Jarman et al., 1993a).

Thus, proneural genes do influence at least the first, major neuronal subtype fate choice (external sense organ vs. chordotonal organ). Presumably, they do this through differential regulation of *cut*, although it is not yet known how this is achieved molecularly (limited in vitro studies have not revealed great differences in DNA properties between the proneural proteins, even though there are clear sequence differences in their DNA-binding domains).

Subsequent subtype choices, however, probably depend not on the particular proneural gene that is expressed, but more on the position in which a neural precursor finds itself. For instance, generalized expression of any AS-C gene gives a remarkably similar pattern of ectopic external sense organs, which includes many different subtypes depending on the position and time at which the ectopic precursors are induced (Rodríguez et al., 1990; Brand et al., 1993). Also, *ato*'s function must presumably be modified by positional influences in the eye to form photoreceptors rather than chordotonal organs. One possible influence is the eye-specific expression of the zinc-finger protein from the *glass* gene. As may be expected if this were the case, photoreceptors in *glass* mutants still begin general neural development, but never express photoreceptor-specific markers (Moses et al., 1989). While these later fate choices are not under direct proneural control, they are clearly still influenced insofar as the proneural genes set the positions of the precursors.

DOES *ATONAL* CONTROL NEURAL CLUSTERING?

Given that proneural genes do influence neuronal subtype identity, it seems surprising that chordotonal organs and photoreceptors should require the same proneural gene, *ato*. Structurally, chordotonal organs have more in common with external sense organs than with photoreceptors (McIver, 1985), except in one interesting respect. Chordotonal organs and photoreceptors both exist as organized, closely apposed clusters of neurons. Chordotonal organ arrays can range from a few to many hundreds of aligned units, each requiring its own precursor. Similarly, in the eye, each ommatidium is an organized unit of eight photoreceptors (R1–8), each deriving from a separate precursor (Ready, 1989). These sensory neurons may be likened to bosons. External sense organs, on the other hand, are like mesons: they are generally solitary features of the cuticle, separated from other neurons by an expanse of epidermis, a situation that is guaranteed by the mechanism of lateral inhibition (compare the cluster of chordotonal organs with the evenly spaced bristles in Fig. 2).

If clustering is what links these chordotonal organs and photoreceptors, this implies a shared mechanistic basis that may be under the influence of *ato*. A clue for this comes from requirement for *ato* in the eye. Within each ommatidium, mosaic analysis has revealed that although *ato* is required for all eight photoreceptors, it is only directly required for one of these, R8 (Jarman et al., 1994), which is the first one of the cluster to appear during development (Tomlinson and Ready, 1987). This result demonstrates that photoreceptor clustering is due to two mechanisms of neurogenesis: R8 is selected by a proneural route requiring *ato*; R1-7 are then determined by local inductive signals emanating from R8. This substantiates previous findings that cell signalling events are important in photoreceptor determination (see Ready, 1989; Banerjee and Zipursky, 1990; Basler and Hafen, 1991; Rubin, 1991) and that R8 may be the 'founding' photoreceptor of the ommatidium (Tomlinson and Ready, 1987; Karpilow et al., 1989).

Does this two step neural clustering process hold for chordotonal organs also? Analysis of *ato* expression during embryonic chordotonal precursor formation suggests that it does, i.e. there appear to be founding and recruited chordotonal precursors (unpublished). We are currently testing by mosaic analysis whether, as in the eye, *ato* is only directly required for the founding precursors. At present it seems very likely that *ato* generally activates a clustering process in its precursors, which leads to the recruitment of further neural precursors by a homoiogenetic induction mechanism (Gurdon et al., 1994), and that it differs from the AS-C in its ability to do this (Fig. 3).

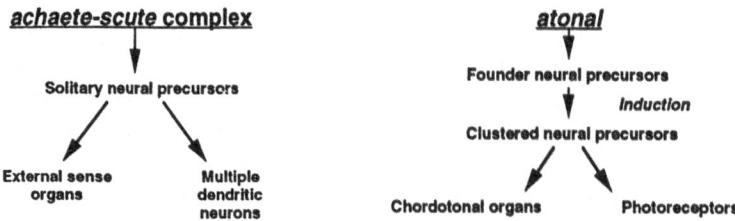

Figure 3. Proneural gene function in the *Drosophila* PNS.

IMPLICATIONS FOR VERTEBRATE NEUROGENESIS

In *Drosophila*, the precursors of sensory neurons of diverse structure and function are determined in a similar process requiring proneural genes. This makes it all the more likely that the genetics of neurogenesis may be conserved in vertebrates even if there are apparently large differences in structural and ontological details. Indeed, murine AS-C homologues have been known for some time (*Mash-1* and *-2*: Johnson et al., 1990), and there are vertebrate homologues of many other fly genes involved in early neurogenesis (a partial list includes *daughterless, hairy, Enhancer(split), Notch, prospero, cut*). Mouse knockouts show that *Mash-1* plays an important role in the development of a subset of PNS neurons (Guillemot et al., 1993). It is not yet clear what this role is: *Mash-1* may not be expressed early enough for a 'classical' proneural role in the epidermal-neural decision (Lo et al., 1991), although another homologue, *Xash-3*, has an early expression pattern that suggests it may fill this role (Zimmerman et al., 1993; Turner and Weintraub, 1994). Instead, the function of *Mash-1* may be in neural precursor survival, as suggested for the late-expressing AS-C gene *asense* (Brand et al., 1993; Jarman et al., 1993b). Given our recent findings that proneural genes endow their precursors with different neural subtype potentials, it may also be that the major role of *Mash-1* is in subtype restriction within the neural crest. This possibility implies that other vertebrate proneural genes may remain to be discovered, each directing the development of different neural subtypes. It seems almost inevitable that there must be vertebrate homologues of *ato*. But yet other proneural genes may be quite divergent in sequence (such as seen between *ato* and the AS-C), requiring very low stringency conditions (and some luck!) for their detection and isolation.

REFERENCES

Artavanis-Tsakonas S and Simpson P (1991): Choosing a cell fate: a view from the *Notch* locus. Trends Genet 7:403-408.

Banerjee U and Zipursky SL (1990): The role of cell-cell interaction in the development of the *Drosophila* visual system. Neuron 4:177-187.

Basler K and Hafen E (1991): Specification of cell fate in the developing eye of *Drosophila*. Bioessays 13:621-631.

Bier E, Vaessin H, Younger-Shepherd S, Jan LY and Jan YN (1992): *deadpan*, an essential pan-neural gene in *Drosophila* encodes a helix-loop-helix protein similar to the *hairy* gene product. Genes Dev 6:2137-2151.

Bodmer R, Barbel S, Shepherd S, Jack JW, Jan LY and Jan YN (1987): Transformation of sensory organs by mutations of the *cut* locus of *D. melanogaster*. Cell 51:293-307.

Brand M, Jarman AP, Jan LY and Jan YN (1993): *asense* is a *Drosophila* neural precursor gene and is capable of initiating sense organ development. Development 119:1-17.

Cabrera CV and Alonso MC (1991): Transcriptional activation by heterodimers of the *achaete–scute* and *daughterless* gene products of *Drosophila*. EMBO J 10:965-973.

Campos-Ortega JA (1988): Cellular interactions during early neurogenesis in *Drosophila melanogaster*. Trends Neurosci 11:400-405.

Campuzano S and Modolell J (1992): Patterning of the *Drosophila* nervous system: the *achaete-scute* gene complex. Trends Genet 8:202-208.

Caudy M, Grell EH, Dambly-Chaudière C, Ghysen A, Jan LY and Jan YN (1988): The maternal sex determination gene *daughterless* has a zygotic activity necessary for the formation of peripheral neurons in *Drosophila*. Genes dev 2:843-852.

Cubas P, de Celis J-F, Campuzano S and Modolell J (1991): Proneural clusters of *achaete-scute* expression and the generation of sensory organs in the *Drosophila* wing disc. Genes Dev 5:996-1008.

Dambly-Chaudière C and Ghysen A (1987): Independent subpatterns of sense organs require independent genes of the *achaete-scute* complex in *Drosophila* larvae. Genes Dev 1:297-306.

Dambly-Chaudière C, Jamet E, Burri M, Bopp D, Basler K, Hafen E, Dumont N, Spielmann P, Ghysen A and Noll M (1991): The paired box gene *pox neuro*: a determinant of poly-innervated sense organs in *Drosophila*. Cell 69:159-172.

Ghysen A and Dambly-Chaudière C (1989): Genesis of the *Drosophila* peripheral nervous system. Trends Genet 5:251-255.

Ghysen A, Dambly-Chaudière C, Jan LY and Jan YN (1993): Cell interactions and gene interactions in peripheral neurogenesis. Genes Dev 7:723-733.

Guillomot F, Lo L-C, Johnson JE, Auerbach A, Anderson DJ and Joyner AL (1993): Mammalian *achaete-scute* homolog 1 is required for the early development of olfactory and autonomic neurons. Cell 75:463-476.

Gurdon JB, Lemaire P and Kato K (1993): Community effects and related phenomena in development. Cell 75:831–834.

Higashijima S, Michiue T, Emori Y and Saigo K (1992): Subtype determination of *Drosophila* embryonic external sensory organs by redundant homeobox genes *BarH1* and *BarH2*. Genes Dev 6:1005-1018.

Jan YN and Jan LY (1990): Genes required for specifying cell fates in *Drosophila* embryonic sensory nervous system. Trends Neurosci 13:493–498.

Jarman AP, Grau Y, Jan LY and Jan YN (1993a): *atonal* is a proneural gene that directs chordotonal organ formation in the *Drosophila* peripheral nervous system. Cell 73:1307–1321.

Jarman AP, Brand M, Jan LY and Jan YN (1993b): The regulation and function of the helix-loop-helix gene, *asense*, in *Drosophila* neural precursors. Development 119:19–29.

Jarman AP, Grell E, Ackerman L, Jan LY and Jan YN (1994): *atonal* is the proneural gene for *Drosophila* photoreceptors. Nature 369:398-400.

Jiménez F and Campos-Ortega JA (1987): Genes in the subdivision 1b of the *Drosophila melanogaster* X-chromosome and their influence on neural development. J

Neurogenet 4:179.

Johnson JE, Birren SJ and Anderson DJ (1990): Two rat homologues of *Drosophila achaete-scute* specifically expressed in neuronal precursors. Nature 346:858–861.

Karpilow J, Kolodkin A, Bork T and Venkatesh T (1989): Neuronal development in the *Drosophila* compound eye: *rap* gene function is required in photoreceptor cell R8 for ommatidial pattern formation. Genes Dev 3:1834–1844.

Lo L-C, Johnson JE, Wuenschell CW, Saito T and Anderson DJ (1991): Mammalian *achaete-scute* homolog 1 is transiently expressed by spatially-restricted subsets of early neuroepithelial and neural crest cells. Genes Dev 5:1524–1537.

McIver S (1985): Mechanoreception. In Gilbert LI and Kerkut DA (eds): "Comprehensive Insect Physiology, Biochemisty and Pharmacology," Vol 6. New York/London: Pergamon Press.

Moses K, Ellis MC and Rubin GM (1989): The *glass* gene encodes a zinc-finger protein required by *Drosophila* photoreceptor cells. Nature 340:531–536.

Murre C, Schonleber McCaw P, Vaessin H, Caudy M, Jan LY, Jan YN, Cabrera CV, Buskin JN, Hauschka SD, Lassar AB, Weintraub H and Baltimore D (1989): Interactions between heterologous helix-loop-helix protein generate complexes that bind specifically to a common DNA sequence. Cell 58:537–544.

Ready DF (1989): A multifaceted approach to neural development. Trends Neurosci 12:102–110.

Rodríguez I, Hernandez R, Modolell J and Ruiz-Gómez M (1990): Competence to develop sensory organs is temporally and spatially regulated in *Drosophila* imaginal primordia. EMBO J 9:3583–3592.

Romani S, Campuzano S, Macagno E and Modolell J (1989): Expression of *achaete* and *scute* genes in *Drosophila* imaginal discs and their function in sensory organ development. Genes Dev 3:997–1007.

Rubin GM (1991): Signal transduction and the fate of the R7 photoreceptor in *Drosophila*. Trends Genet 7:372–377.

Ruiz-Gómez M and Modolell J (1987): Deletion analysis of the *achaete-scute* locus of *Drosophila melanoaster*. Genes Dev 1:1238–1246.

Skeath JB and Carroll SB (1991): Regulation of *achaete-scute* gene expression and sensory organ formation in the *Drosophila* wing. Genes Dev 5:984–995.

Skeath JB, Panganiban G, Selegue J and Carroll SB (1992): Gene regulation in two dimensions: the proneural *achaete* and *scute* genes are controlled by combinations of axis-patterning genes through a common intergenic control region. Genes Dev 6:2606–2619.

Tomlinson A and Ready DF (1987): Neuronal differentiation in the *Drosophila* ommatidium. Dev Biol 120:366–376.

Turner DL and Weintraub H (1994): Expression of achaete-scute homolog 3 in *Xenopus* embryos converts ectodermal cells to a neural fate. Genes Dev 8:1434-1447.

Vaessin H, Grell E, Wolff E, Bier E, Jan LY and Jan YN (1991): *prospero* is expressed in neuronal precursors and encodes a nuclear protein that is involved in the control of axonal outgrowth in *Drosophila*. Cell 67:941–953.

Van Doren M, Ellis HM and Posakony JW (1991): The *Drosophila extramacrochaetae* protein antagonizes sequence-specific DNA binding by *daughterless/achaete-scute* protein complexes. Development 113:245–255.

Zimmerman K, Shih J, Bars J, Cellazo A and Anderson DJ (1993): XASH-3, a novel *Xenopus achaete-scute* homolog, provides an early marker of planar neural induction and position along the mediolateral axis of the neural plate. Development 119:221-232.

GENETIC ANALYSIS OF NEURONAL MIGRATION IN THE NEMATODE *Caenorhabditis elegans*

Gian Garriga

Department of Molecular and Cell Biology, 401 Barker Hall, University of California, Berkeley, CA 94720-3204, U.S.A.

INTRODUCTION

During metazoan development the migrations of neurons and their growth cones play a major role in generating the final pattern and connectivity of the adult nervous system. Although many migration pathways have been described, the mechanisms regulating these cellular movements are poorly understood. In order to understand how cell migration is regulated, there are several key questions that must be addressed. How do cells decide to migrate? What defines the pathways taken by migrating cells? How do migrating cells know that they have reached their final destinations? One approach to elucidating the mechanisms controlling cell migration is to define the molecules involved by identifying mutations that perturb cell migration. The nematode *Caenorhabditis elegans* is particularly useful for the study of cell migration because of its anatomical simplicity. Nomarski optics can be used to observe cells as they migrate in living animals (Sulston and Horvitz, 1977; Sulston et al., 1983). Moreover, potential cell interactions involved in both cell migration (Thomas et al., 1990) and axonal guidance (Durbin, 1987; Walthall and Chalfie, 1988; Li and Chalfie, 1990; Garriga et al., 1993a) are easily tested. Our research is directed at understanding how a specific pair of motor neurons, the HSNs, migrate.

THE EGG-LAYING SYSTEM

The egg-laying system of the *C. elegans* hermaphrodite is shown in Fig. 1. Eggs are stored in the uterus and laid through the vulva, a passageway connecting the uterus to the environment. The vulval and uterine muscle cells, collectively termed the egg-laying muscles, contract synchronously pulling open the vulva and squeezing the uterus to expel eggs. The egg-laying muscles are directly innervated by the two bilaterally symmetric HSN and six ventrally positioned VC motor neurons. Laser microsurgery has shown that the vulval cells, the egg-laying

Neural Cell Specification: Molecular Mechanisms and Neurotherapeutic Implications
Edited by Juurlink *et al.*, Plenum Press, New York, 1995

105

muscles and the HSNs are necessary for egg laying; killing any of these cells results in an egg-laying defective (Egl) phenotype. Egl hermaphrodites bloat with late stage embryos and are easily detected using a dissecting microscope.

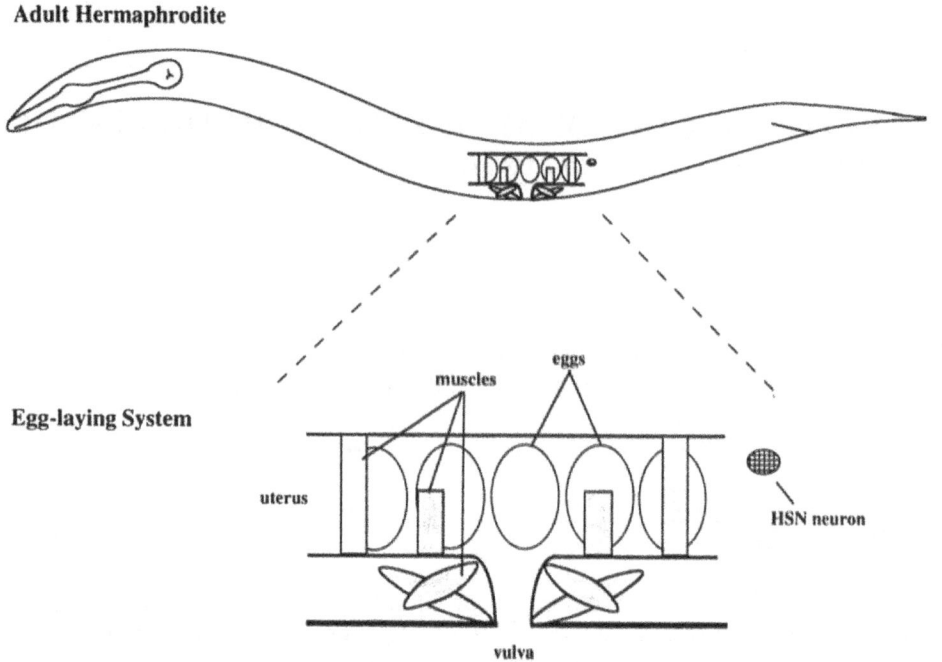

Figure 1. Schematic representation of the adult *C. elegans* hermaphrodite and the components of the egg-laying system. Eight of the sixteen egg-laying muscles and one of the two HSNs are shown. The VC neurons and the HSN axon are not shown.

The HSNs are born in the tail and migrate embryonically to positions flanking the gonad primordium (Sulston et al., 1983; Fig. 2). The HSNs are well suited for the study of cell migration for several reasons. First, the HSNs are necessary for normal egg laying; migrations of the HSNs appear to be required to position these cells so that they can function properly. Second, the HSN migrations can be followed *in vivo* using Nomarski optics. Third, HSN development can be assessed using morphological and immunocytochemical criteria. HSN development can be monitored by Nomarski microscopy to follow changes in size and shape of the HSN associated with HSN differentiation. In addition, expression of the UNC-86 protein (Finney and Ruvkun, 1990) and the neurotransmitter serotonin (Desai et al., 1988) can be used to monitor HSN differentiation.

HSN MIGRATION MUTANTS

Mutant screens

Most of the HSN abnormal migration (Ham) mutants were identified using two approaches: (1) genetic screens designed to identify genes required for HSN stimulation of egg laying; and (2) scoring a set of previously isolated uncoordinated (Unc) mutants for HSN positioning defects (Trent et al., 1983; Desai et al., 1988; Desai and Horvitz, 1989). The HSN cell bodies of these mutants either fail to migrate or migrate partially.

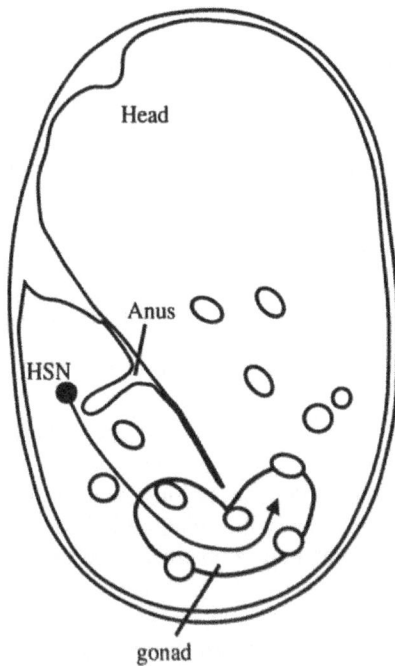

Figure 2. Schematic representation of a lateral view of an embryo at 400 minutes after fertilization, at the time of the HSN birth. Only one of the two HSNs is shown and only a few of the other cells (indicated by circles and ovals) present at this time are shown. The arrow indicates the route of HSN migration.

The screens described above, however, are unlikely to identify all genes that function in HSN migration. For example, mutations that also affect viability or fertility would have been missed. To circumvent this problem, we have conducted a preliminary screen for mutants with displaced HSNs that allows recovery of mutations in essential genes (Fig. 3). So far, we have identified at least 15 recessive mutations, seven of which also lead to a lethal or sterile phenotype (G Garriga, D Parry and P Baum, unpublished results). Most of these mutations are only partially characterized and are not included in the description of the Ham genes below.

Ham genes

Mutations in 19 genes cause a Ham phenotype. The phenotypic analysis of the mutants has led us to place these genes into three classes based on their additional defects in HSN development and other cell migrations (Table 1). Class I mutations affect multiple aspects of HSN development. Class II mutations affect the migrations of other cell types. Class III mutations affect the development of other non-migrating cell types. While the classification of these genes is tentative at this time since the null phenotypes for most of the genes have not been defined, it is still useful for thinking about the potential functions of these genes.

Mutations in class I genes disrupt HSN differentiation. For example, the HSNs of *egl-5* mutants fail to migrate normally, fail to express the neurotransmitter serotonin, and extend axons abnormally (Desai et al., 1988). The defective differentiation of the HSNs in class I mutants is not a consequence of misplaced HSN cell bodies since most Ham mutants do not display general defects in HSN development. Hence, aberrant cell fate determination can affect cell migration.

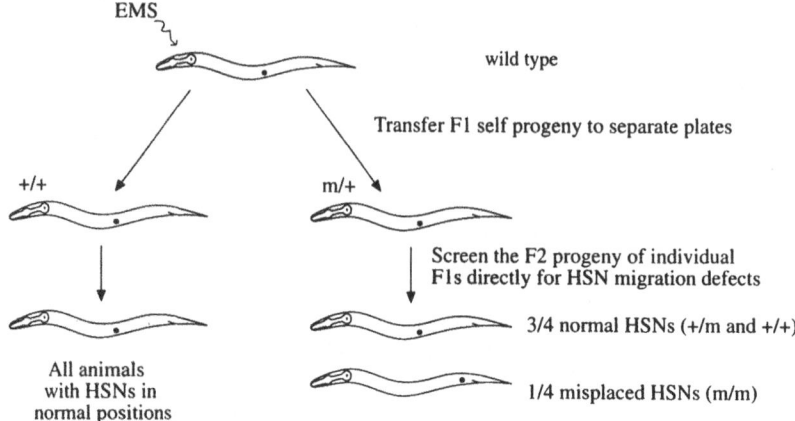

Figure 3. Strategy for isolating HSN migration mutants. This approach uses Nomarski optics to screen directly for mutants with misplaced HSN. If a mutation also leads to a lethal or sterile phenotype, the mutation can be recovered from F2 heterozygous siblings.

Mutations in class II genes perturb multiple cell migrations. The number of cell migrations affected by these mutations varies dramatically with different genes. For example, mutations in *mig-1* disrupt only HSN and QL neuroblast migrations, whereas mutations in *mig-2* disrupt the migrations of many cell types and the outgrowth of axons (Hedgecock et al., 1987; Desai et al., 1988). The genes in this class may encode molecules that function generally in cell migration; for example, components of the extracellular matrix or signal transduction molecules.

Mutations in class III genes disrupt HSN migration specifically. Although these mutations do affect the development of other cell types, they do not affect other aspects of HSN development or other cell migrations. One explanation for the specific effects on HSN migration is that these mutations are weak alleles of genes that function more generally in either HSN cell fate determination or cell migration. Assuming that this is not the case, these genes may define some aspect of migration that is unique to the HSNs; for example, a pathway used by the HSNs but not other migratory cells. Alternatively, these genes could specifically regulate HSN migration by controlling genes that function generally in cell migration.

Transcription factors in the regulation of HSN migration

Molecular analysis of two genes functioning in HSN migration, *egl-5* and *egl-43*, suggests that they encode transcription factors. The class I gene *egl-5* encodes a homeodomain containing protein that acts cell autonomously in the development of the HSNs and additional cells positioned in the tail (Chisholm, 1991; Wang et al., 1993). Thus, EGL-5 may act in the HSNs to transcriptionally regulate other genes involved in HSN differentiation. The class III gene *egl-43* encodes two proteins with six and three zinc-finger motifs that are similar to six of the ten zinc fingers of the mouse oncoprotein Evi-1 (Garriga et al., 1993b). The *Evi-1* gene encodes a 145 kD nuclear protein that binds DNA in a sequence-specific manner (Matsugi et al., 1990; Perkins et al., 1991; Delwel et al., 1993), consistent with a role in transcriptional regulation. Antibodies raised to the EGL-43 proteins show that they are transiently expressed in the HSNs during their migrations (C Guenther and G Garriga, unpublished results). These observations suggest that EGL-43 functions in the HSNs to transcriptionally regulate genes that act more directly in HSN migration. As additional genes required for HSN migration are cloned, we will be able to test if EGL-5 and EGL-43 regulate these genes directly.

Table 1. Genes required for HSN migration

Class I: HSN cell fate	Class II: general cell migration	Class III: HSN migration specific
egl-5	*egl-20*	*egl-18*
ham-1	*epi-1*	*egl-27*
ham-3	*mig-1*	*egl-43*
ham(gm34)	*mig-2*	*ham-2*
lin-32	*mig-10*	
	mig(gm2)	
	mig(gm38)	
	unc-34	
	unc-71	
	unc-73	

egl <u>e</u>gg-<u>l</u>aying defective   *epi* <u>epi</u>thelialization defective
ham <u>H</u>SN <u>a</u>bnormal <u>m</u>igration   *mig* <u>mig</u>ration defective
unc <u>unc</u>oordinated

ACKNOWLEDGEMENTS

This work was supported by grants from the National Institutes of Heath and the McKnight Endowment Fund for Neuroscience.

REFERENCES

Chisholm A (1991): Control of cell fate in the tail region of *C. elegans* by the gene *egl-5*. Development 111: 921-932.

Delwel R, Funabiki T, Kreider BL, Morishita K and Ihle JN (1993): Four of the seven zinc fingers of the *Evi-1* myeloid-transforming gene are required for sequence-specific binding to GA(C/T)AAGA(T/C)AAGATAA. Mol Cell Biol 13: 4291-4300.

Desai C, Garriga G, McIntire SL and Horvitz HR (1988): A genetic pathway for the development of the *Caenorhabditis elegans* HSN motor neurons. Nature 336: 638-646.

Desai C and Horvitz HR (1989): *Caenorhabditis elegans* mutants defective in the functioning of the motor neurons responsible for egg laying. Genetics 121: 703-721.

Durbin RM (1987): 'Studies in the Development and Organization of the Nervous System of *Caenorhabditis elegans*'. Ph.D. dissertation, King's College, Cambridge, England.

Finney M and Ruvkun G (1990): The *unc-86* gene product couples cell lineage and cell identity in *C. elegans*. Cell 63: 895-905.

Garriga G, Desai C and Horvitz HR (1993a): Cell interactions control the direction of outgrowth, branching and fasciculation of the HSN axons of *Caenorhabditis elegans*. Development 117: 1071-1087.

Garriga G, Guenther C and Horvitz HR (1993b): Migrations of the *Caenorhabditis elegans* HSNs are regulated by egl-43, a gene encoding two zinc finger proteins. Genes & Dev 7: 2097-2109.

Hedgecock EM, Culotti JG, Hall DH and Stern BD (1987): Genetics of cell and axon migrations in *Caenorhabditis elegans*. Development 100: 365-382.

Li C and Chalfie M (1990): Organogenesis in *C. elegans*: Positioning of neurons and muscles in the egg-laying system. Neuron 4: 681-695.

Matsugi T, Morishita, K and Ihle JN (1990): Identification, nuclear localization, and DNA-binding activity of the zinc finger protein encoded by the *Evi-1* myeloid transforming gene. Mol Cell Biol 10: 1259-1264.

Perkins AS, Fishel R, Jenkins NA and Copeland NG (1991): *Evi-1*, a murine zinc finger proto-oncogene, encodes a sequence-specific DNA-binding protein. Mol Cell Biol 11: 2665-2674.

Sulston JE, Schierenberg E, White JG and Thomson JN (1983): The embryonic cell lineage of the nematode *Caenorhabditis elegans*. Dev Biol 100: 64-119.

Sulston JE and Horvitz HR (1977): Post-embryonic cell lineages of the nematode *Caenorhabditis elegans*. Dev Biol 56: 110--156.

Thomas JH, Stern MJ and Horvitz HR (1990): Cell interactions coordinate the development of the *C. elegans* egg-laying system. Cell 62: 1041-1052.

Trent C, Tsung N, Horvitz HR (1983): Egg-laying defective mutants of the nematode *Caenorhabditis elegans*. Genetics 104: 619-647.

Walthall WW and Chalfie M (1988): Cell-cell interactions in the guidance of late-developing neurons in *Caenorhabditis elegans*. Science 239: 643-645.

Wang BB, Muller-Immergluck MM, Austin J, Robinson NT, Chisholm A and Kenyon C (1993): A homeotic gene cluster patterns the anteroposterior body axis of *C. elegans*. Cell 74: 29-42.

INDUCTION AND DIFFERENTIATION OF MOTOR NEURONS

S.L. Pfaff[1], T. Yamada[2], T. Edlund[3] and T.M. Jessell[1]

[1]Howard Hughes Medical Institute, Department of Biochemistry and Molecular Biophysics, Center for Neurobiology and Behavior, Columbia University, New York, NY 10032, U.S.A.; [2]Centre for Molecular Biology and Biotechnology, University of Queensland, Brisbane, AUSTRALIA; [3]Department of Microbiology, Umeå University, Umeå, SWEDEN.

INTRODUCTION

During the early development of the vertebrate nervous system, distinct cell types appear at specific locations, establishing a primitive pattern within the neural tube. In the caudal region of the neural tube that gives rise to the spinal cord, the first differentiated cell types are found at distinct dorsoventral positions. Floor plate cells occupy the ventral midline of the neural tube, and motor neurons appear in a ventrolateral position. Cells in the dorsal neural tube give rise to sensory relay neurons, to neural crest cells, and, at the dorsal midline to roof plate cells. Thus, the embryonic spinal cord contains two midline cell groups and several neuronal cell types that are distributed in a bilaterally symmetric manner with respect to the midline (Fig. 1A).

Embryonic manipulations in chick embryos have provided evidence that the pattern of cell types generated in the ventral half of the neural tube is critically dependent on signals provided by axial mesodermal cells of the notochord. The best characterized influence of the notochord on the differentiation and patterning of the neural tube is the induction of floor plate cells at its ventral midline. Notochord grafts induce the differentiation of floor plate cells at ectopic locations in the neural tube, and removal of the notochord leads to the absence of floor plate cells in the spinal cord (van Straaten et al., 1988; Placzek et al., 1990b, 1993; Yamada et al., 1991). Strikingly, once induced, the floor plate can recruit additional neural plate cells to a floor plate fate by a homeogenetic induction process (Yamada et al., 1991; Hatta et al., 1991; Placzek et al., 1993).

At later stages in spinal cord development, the floor plate provides both long-range and local guidance cues that promote the growth of axons to and across the ventral midline of the spinal cord. First, the floor plate secretes a

Neural Cell Specification: Molecular Mechanisms and Neurotherapeutic Implications
Edited by Juurlink *et al.*, Plenum Press, New York, 1995

111

diffusible chemoattractant, Netrin-1, which can orient the growth of axons of commissural neurons in vitro (Tessier-Lavigne et al., 1988; Placzek et al., 1990a; Serafini et al., 1994; Kennedy et al., 1994) and may account for the homing of these

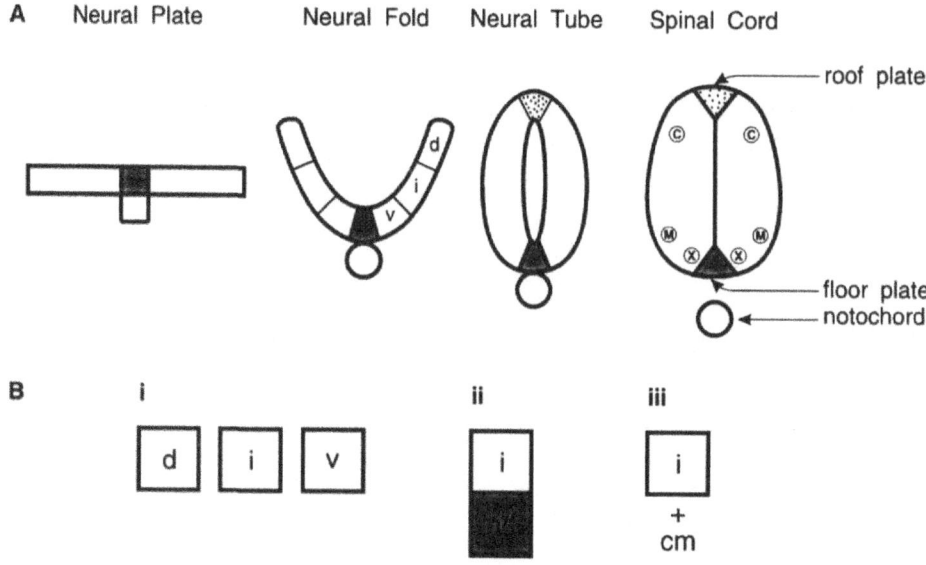

Figure 1. Establishment of the dorso-ventral axis in the developing spinal cord. (A) The four diagrams show successive stages in the development of the neural tube and spinal cord. The neural plate consists initially of a simple columnar epithelium. Cells at the midline of the neural plate are contacted directly by axial mesoderm cells of the notochord. During neurulation, the neural plate buckles at its midline to form the neural folds. The neural tube is formed when the dorsal tips of the neural folds fuse. Cells in the region of fusion form a specialized group of dorsal midline cells, the roof plate, and cells at the ventral midline of the neural tube differentiate into the floor plate. After neural tube closure neuroepithelial cells continue to proliferate and eventually differentiate into defined classes of neurons at different dorso-ventral positions within the spinal cord. For example, commissural (C) and other classes of dorsal neurons differentiate near to the roof plate, and motor (M) neurons differentiate ventrally near to the floor plate, which by this time is no longer in contact with the notochord. Other classes of neurons (X) differentiate in the region of the neural epithelium that intervenes between the floor plate and motor neurons. (B) In vitro induction assay using the dorsal (d), intermediate (i), and ventral (v) regions isolated at the neural fold stage shown in A. Diagram i: to determine the degree of commitment of cells in the neural plate, dorsal (d), intermediate (i), and ventral neural plate (v) explants were grown separately in collagen gels. Diagrams ii and iii: for induction assays, intermediate neural plate explants (i) were grown in contact with notochord (n) or floor plate (f) explants (diagram ii) or in the presence of notochord- or floor plate-conditioned medium (cm) (diagram iii). For details see Yamada et al., (1993).

axons to the floor plate in vivo (Placzek et al., 1990b; Bovolenta and Dodd, 1991; Yaginuma and Oppenheim, 1991). Second, the floor plate appears to control the change in trajectory of commissural axons from the transverse to the longitudinal plane that occurs immediately after axons have crossed the ventral midline (Holley and Silver, 1987; Dodd et al., 1988; Bovolenta and Dodd, 1990). In support of this, genetic mutations in mice and zebrafish that result in the absence of the floor plate during embryonic development lead to errors in the pathfinding of commissural axons at the midline of the spinal cord (Bernhardt et al., 1992; Bovolenta and Dodd, 1991). Third, the floor plate may promote the fasciculation of commissural axons after they cross the midline of the spinal cord (Holley and Silver, 1987) by regulating the expression of glycoproteins of the immunoglobulin superfamily (Dodd et al., 1988; Schachner et al., 1990; Furley et al., 1990). Thus, the floor plate appears to have a role in both the early patterning of the neural tube and in aspects of axonal pathfinding.

Intercellular signaling molecules and transcription factors that participate in floor plate differentiation have recently been identified. A vertebrate homolog of the Drosophila gene hedgehog, vhh-1/shh, encodes a putative secreted protein and is expressed by the notochord and the floor plate at the time that these cell groups exhibit their inductive activities (Riddle et al., 1993; Krauss et al., 1993; Echelard et al., 1993; Roelink et al., 1994). The samecell groups also express three members of the winged helix (HNF-3/fork head) family of DNA-binding transcription factors (Lai et al., 1991; Weigel and Jäckle, 1990), Pintallavis (Ruiz i Altaba and Jessell, 1992) and HNF-3β (Sasaki and Hogan, 1993; Monaghan et al., 1993; Ruiz i Altaba et al., 1993b; Ang et al., 1993) and HNF-3α (Bolce et al., 1993; Monaghan et al., 1993).

Widespread expression of vhh-1/shh leads to the ectopic expression of floor plate markers in the neural tube in vivo (Echelard et al., 1993; Roelink et al., 1994) and vhh-1 expression in COS cells induces floor plate differentiation in neural plate explants in vitro (Roelink et al., 1994). Similarly, transgenic mice that express HNF-3β, or the related gene Pintallavis, throughout the midbrain express floor plate markers ectopically (Ruiz i Altaba et al., 1993a; Sasaki and Hogan, 1994). Mouse embryos with targeted disruption of the HNF-3β gene display a perturbation in node development, lack a notochord and consequently exhibit a loss of floor plate cells (Weinstein et al., 1994; Ang and Rossant, 1994). Taken together, these results provide strong evidence that vhh-1/shh and HNF-3β are key components in a signaling pathway that controls floor plate differentiation.

By contrast, the cellular and molecular events that control the development of motor neurons remain less well defined. Motor neurons mediate the central control of movement through the connections formed by motor axons with skeletal muscles in the periphery. In chick embryos, spinal motor neurons are derived from progenitor cells in the neural tube that also give rise to other classes of neurons and to glial cells (Leber et al., 1990). Studies in zebrafish and avian embryos have provided evidence that the acquisition of general motor neuron properties precedes the acquisition of specific motor neuron identities (Eisen, 1994; Landmesser, 1992; Matise and Lance-Jones, 1992). The generation of discrete subclasses of motor neurons determines the pathfinding choices of motor neurons and the peripheral targets that they innervate (Eisen, 1994; Landmesser, 1992). Thus, the differentiation of specific motor neuron subclasses appears to be a sequential but still poorly understood process.

To begin to define the cellular interactions and the nature of signals that regulate cell fate in the neural tube, we have studied motor neuron differentiation in chick embryos using grafting and extirpation methods and have developed assays in which the differentiation of motor neurons from precursor cells in the neural plate can be examined in vitro. The results described below provide evidence that the differentiation of motor neurons from progenitor cells begins at the neural plate stage and is dependent on a diffusible factor that is secreted from both notochord and floor plate cells (see also Yamada et al., 1991; 1993).

MOLECULAR MARKERS FOR MOTOR NEURONS

In order to examine the differentiation of motor neurons at a molecular level we have used as markers, several genes that have recently been isolated and shown to be expressed in motor neurons (Table 1). To assess motor neuron differentiation in the studies described below we have used three markers: the LIM homeodomain protein Islet-1 (Ericson et al., 1992) (Fig. 2), the SC1 immunoglobulin-like glycoprotein (Tanaka and Obata, 1984) and choline acetyltransferase (ChAT) which is the rate-limiting enzyme in the synthesis of acetycholine, the neurotransmitter used by motor neurons. In vitro, motor neurons can be distinguished from other spinal neurons and from dorsal root ganglion neurons by their coordinate expression of Islet-1, SC1, and ChAT (Table 1).

Table 1. Genes expressed in embryonic motor neurons

Gene	Function	Reference
Islet-1	transcription factor	Karlsson et al., 1990
Islet-2	transcription factor	Tsuchida et al., 1994
Lim-1	transcription factor	Tsuchida et al., 1994
Lim-3	transcription factor	Tsuchida et al., 1994
Gsh-4	transcription factor	Li et al., 1994
COUP-TFII	transcription factor	Lutz et al., 1994
ChAT	acetylcholine biosynthesis	Ishii et al., 1990
GGF/ARIA	acetylcholine receptor induction	Falls et al.,1993
Agrin	acetycholine receptor clustering	Tsim et al., 1992
SC1	cell adhesion, neurite extension	Tanaka et al. 1991
TAG-1	cell adhesion, neurite extension	Furley et al., 1990

MOTOR NEURON DIFFERENTIATION IN VIVO DEPENDS ON SIGNALS FROM THE NOTOCHORD AND FLOOR PLATE

In the embryonic spinal cord, motor neurons initially occupy a position close to, but not immediately adjacent to, the floor plate (Fig. 3). The ventral location of motor neurons raised the possibility, that like floor plate cells, motor neuron differentiation is also dependent on inductive signals from the notochord. To test this possibility, the distribution of motor neurons has been examined in the spinal cord of embryos in which a segment of notochord had been grafted at different dorsoventral positions. Placing an extra notochord next to the neural tube results in a marked alteration in the pattern of cell differentiation (Fig. 3) in the spinal cord (Yamada et al., 1991). The region of spinal cord immediately adjacent to notochord grafts differentiates into floor plate and motor neurons are found in the dorsal region of the spinal cord at a distance from the ectopic notochord graft. Floor plate grafts adjacent to the neural tube also induce ectopic dorsal motor neurons, providing evidence that floor plate cells mimic the inductive properties of the notochord.

To examine whether the notochord and floor plate are required for motor neuron differentiation in the neural tube during normal development, a segment of notochord underlying the ventral region of the caudal neural tube was removed and the spinal cord of operated embryos examined after 48-96h (Yamada et al.,

1991). A short distance into the notochord-free region, the expression of floor plate markers disappears although distinct lateral motor columns are still present. Beyond this point, the normal bilateral column of motor neurons merges to form a single expanded column at the ventral midline of the neural tube. Thus, an arc of motor neurons differentiates around the end of the floor plate. At greater distances
from the end of the floor plate (Fig. 3), motor neurons are absent (Yamada et al., 1991; Placzek et al., 1991). These and other results are consistent with a model in

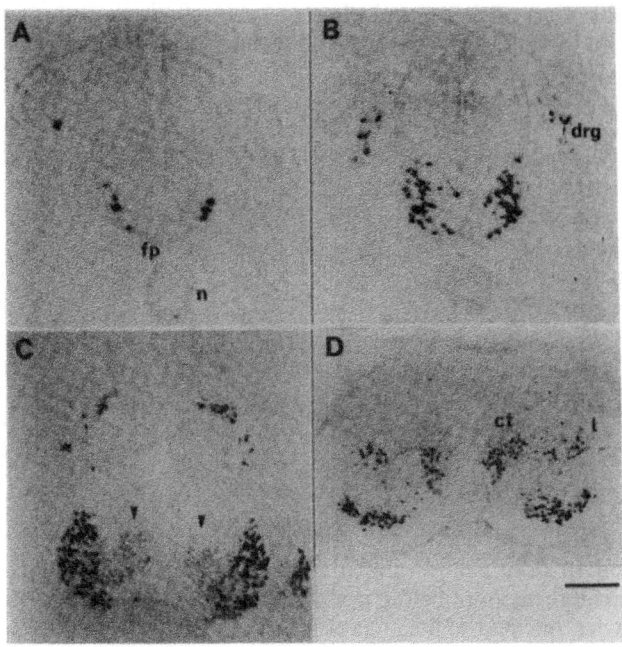

Figure 2. Expression of Islet-1 during embryonic chick spinal cord development. (A) Cross sections of the upper thoracic spinal cord of a stage 15 to 16 embryo. Islet-1 protein was detected with antibody to Islet-1 and HRP-conjugated secondary antibody. Islet-1 is expressed in a small number of cells in the ventral part of the spinal cord lateral to the floor plate (fp) and notochord (n). (B) By stages 17 to 18, there is a marked increase in the number of Islet-1-positive cells in the ventral spinal cord. Note that cells close to the midline of the spinal cord express Islet-1. Islet-1 is also expressed in cells of the dorsal root ganglion (drg). (C) Section through thoracic spinal cord at stage 24, showing Islet-1 expression in cells in the ventrolateral spinal cord. The medial-most groups of cells (arrowheads) appear to exhibit lower amounts of Islet-1 immunoreactivity at this stage. A small number of cells in the dorsal region of the spinal cord also express Islet-1. (D) Section through thoracic spinal cord at stage 29, showing Islet-1 expression in the cells in the lateral motor column and in presumptive visceral motor neurons that are beginning to form the Column of Terni (ct). A lateral group (L) of Islet-1 positive cells is also present. Scale bar: (A) 60 mm; (B and C) 70 mm; (D) 90 mm. Reprinted with permission from: Ericson J, Thor S, Edlund T, Jessell TM and Yamada T (1992): Early stages of motor neuron differentiation revealed by expression of homeobox gene Islet-1. Science 256: 1555-1560. Copyright 1992 American Association for the Advancement of Science

which motor neuron differentiation depends on signals from the notochord and floor plate and that motor neurons differentiate at a constant position with respect to the floor plate.

SIGNALS THAT TRIGGER MOTOR NEURON DIFFERENTIATION BEGIN PRIOR TO NEURAL TUBE CLOSURE

To begin to address the mechanisms of motor neuron differentiation, we have examined whether motor neuron differentiation can be elicited from neural progenitor cells in vitro. We first monitored cell differentiation in explants isolated from the caudal region of the neural plate of stage 10 chick embryos. Segments of the neural plate were divided along the future dorsoventral axis into ventral, intermediate, dorsal and midline regions (Fig. 1B). Each region was placed in a collagen gel and maintained in vitro for up to 72 hr.

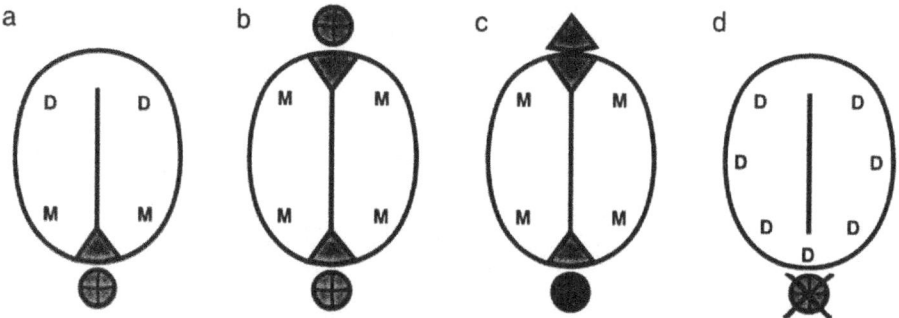

Figure 3. Diagram summarizing the results obtained from experiments in chick embryos in which a notochord or floor plate is grafted to the dorsal midline of the neural tube and in which the notochord is removed before neural tube closure. (A) Neural organization of the spinal cord showing the ventral location of motor neurons (M) and the dorsal location of sensory relay neurons (D). B. Dorsal grafts of a notochord result in the induction of a floor plate at the dorsal midline and in the induction of ectopic dorsal motor neurons. (C) Dorsal grafts of a floor plate also result in the induction of a floor plate at the dorsal midline, and in the induction of ectopic dorsal motor neurons. (D) Removal of the notochord results in the elimination of the floor plate and motor neurons and the expression of dorsal cell types (D) in the ventral region of the spinal cord. For details see Yamada et al. (1991) and Placzek et al. (1991).

Ventral neural plate explants do not express Islet-1, SC1, or ChAT mRNA at the time of isolation, but when maintained in vitro for 48 hr, are found to contain Islet-1 cells and to express ChAT mRNA (Fig. 4). These results show that cells derived from prospective ventral regions of the neural plate can adopt motor neuron fates in vitro and suggest that the specification of motor neurons begins at the neural plate stage. In contrast, intermediate neural plate explants contain few Islet-1 cells and do not express detectable levels of ChAT mRNA when grown alone for up to 72 hr in vitro (Fig. 4). Although motor neurons are not detected, many neurons differentiate over the 48-72 hr period that intermediate neural plate explants are maintained in vitro (see Fig. 5C).

INDUCTION OF MOTOR NEURON DIFFERENTIATION IN NEURAL PLATE EXPLANTS

The lack of expression of motor neuron markers in intermediate neural plate explants grown alone in vitro provided the basis for assays to assess the nature of signals that direct the fates of neural plate cells to differentiate as motor neurons (Yamada et al., 1993). To determine whether signals from the notochord and floor plate can induce motor neurons in vitro, intermediate neural plate explants are placed in contact with segments of stage 10-20 notochord or floor plate and grown in vitro for 48 hr (Fig. 5). The notochord and floor plate induce a 50-100 fold increase in the number of Islet-1 cells over controls assayed at 48 hr. The maximum number of Islet-1 cells induced by the notochord or floor plate ($\sim$ 400 cells per explant) represents 25%-30% of the total number of cells present in intermediate neural plate explants after 48 hr.

Motor Neuron Induction By Diffusible Factors

Notochord grafting experiments have shown that motor neurons appear at a distance from inducing tissue (Yamada et al., 1991). This observation raised the possibility that motor neuron induction may not be dependent on contact with inducing tissues as is the case with floor plate induction (Placzek et al., 1993). To examine this, intermediate neural plate explants have been grown alone in vitro for 48 hr in the presence of medium conditioned by the notochord or floor plate (Fig. 5M-O). Notochord- and floor plate-conditioned medium produces a concentration-dependent induction of Islet-1 cells (not shown, Yamada et al., 1993). Medium conditioned by an equivalent mass of dorsal spinal cord tissue isolated from stages 16-26 does not result in a significant increase in the number of Islet-1[+] cells (not shown; Yamada et al., 1993). Moreover, the number of Islet-1 cells induced by concentrated floor plate-conditioned medium is similar to that induced by contact with a single notochord explant (Fig. 5). These results indicate that diffusible factors from the notochord can induce motor neurons.

Involvement Of Vertebrate Hedgehog Genes In Motor Neuron Differentiation

Studies on the vertebrate homolog of the *Drosophila* hedgehog gene vhh-1 (also known as sonic hedgehog, Kraus et al., 1993; Echelard et al., 1993; Riddle et al., 1993) have shown that the vhh-1 protein can induce motor neurons as well as floor plate cells in neural plate explants in vitro (Roelink et al., 1994). Moreover, vhh-1 is expressed in the notochord and floor plate at the time that these cell groups possess motor neuron inducing activities. These experiments, however, have not resolved whether motor neuron induction by vhh-1 is direct and thus do not establish whether vhh-1 could represent the diffusible motor neuron inducing activity present in notochord-conditioned medium. Since vhh-1 can induce floor plate differentiation (Roelink et al., 1993; Echelard et al., 1993; Kraus et al., 1993), the induced floor plate could, in turn, secrete a motor neuron-inducing factor distinct from vhh-1.

CONTRIBUTION OF THE NOTOCHORD AND FLOOR PLATE TO MOTOR NEURON INDUCTION

Taken together the results of in vivo grafting experiments and in vitro assays have provided evidence that the notochord and floor plate can induce motor neuron differentiation. The present in vitro assays suggest that the induction of motor neurons in vivo is initiated directly by the notochord (Fig. 6). Four related findings support this idea. First, notochord-conditioned medium can induce motor neurons under conditions in which floor plate differentiation does not occur, providing direct evidence that motor neuron differentiation does not require prior induction of a floor plate. Second, notochord grafting studies in chick embryos have provided supportive, although indirect, evidence that the notochord is able to induce motor neurons directly (van Straaten et al., 1985; Yamada et al., 1991). Third, the studies described above (see also Yamada et al., 1993) and those of Placzek et al. (1993) show that floor plate and motor neuron induction in vitro requires a similar period of exposure to inducing factors. Thus, by the time that floor plate cells acquire motor neuron-inducing activity, many cells in the neural plate are likely to have committed to a motor neuron fate. Fourth, cells in ventral neural plate explants give rise to motor neurons, whereas the ventral midline region of the neural plate do not acquire floor plate properties when placed in vitro (M. Placzek and J. Dodd, unpublished data). Thus, the commitment of neural plate cells to a motor neuron fate may indeed occur before commitment to a floor plate fate. Finally, the presence of both primary and secondary motor neurons in zebrafish cyclops embryos lacking a floor plate supports the idea that signals from the notochord are sufficient to induce motor neuron differentiation in vivo (Hatta, 1992).

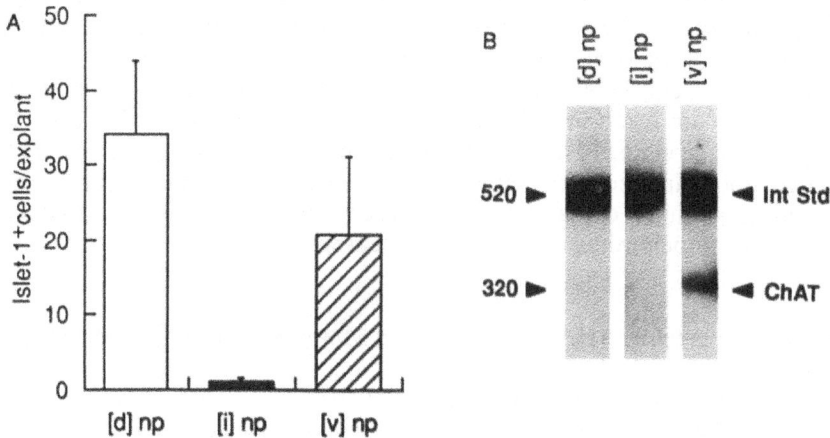

Figure 4. Early specification of motor neuron fate. (A) Quantitation of Islet-1$^+$ cells present in dorsal (d), intermediate (i), and ventral (v) neural plate (np) explants grown in isolation for 48 hr. Each column shows the mean $\pm$ SEM of 9-16 explants. (B) Detection of mRNA encoding ChAT by PCR analysis in dorsal (d), intermediate (i), and ventral (v) neural plate (np) explants grown in vitro for 72 hr. Upper band indicates internal standard (Int Std). Lower band shows endogenous ChAT mRNA levels (ChAT). Note that cells in dorsal (d) explants expressed Islet-1, but did not express ChAT mRNA suggesting that these cells are not motor neurons. From Yamada et al., (1993).

SPECIFICATION OF CELL FATE ALONG THE DORSOVENTRAL AXIS OF THE NEURAL TUBE BEGINS IN THE NEURAL PLATE

The pattern of cell differentiation within the caudal neural tube becomes evident after neural tube closure with the appearance of distinct cell types at different dorsoventral positions (Fig. 1). The differentiation of motor neurons from appropriate regions of the neural plate in vitro indicates that the signals that trigger the differentiation of this class of neurons has begun at the neural plate stage (Fig. 6). There are few post mitotic cells within the neural plate (Sechrist and Bronner-Fraser, 1992); thus, the restriction in fate of neural plate cells may be acquired prior to their final cell division. Since the cell cycle time in the chick neural plate is ~ 8 hr (Langman et al., 1966), several cell divisions appear to occur between the time that neural plate cells acquire the potential to give rise to motor neurons and the onset of expression of definitive motor neuron markers such as Islet-1.

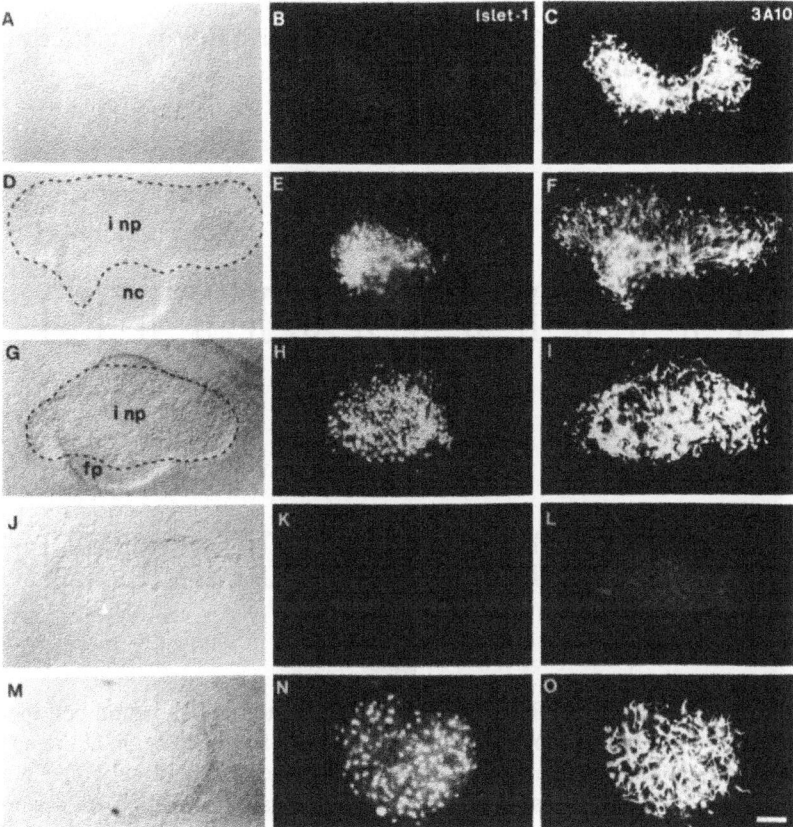

Figure 5. Induction of Islet-1 expression in neural plate explants. Intermediate neural plate explants were placed in collagen gels, grown for 48 hr in vitro, and labeled with rabbit anti-Islet-1 antibodies (middle panels) and MAb 3A10 (right hand panels). Nomarski images of explants are shown in left-hand panels. (A-C) Cells in an intermediate neural plate explant grown alone in vitro do not express Islet-1 (B), but numerous 3A10[+] neuronal cell bodies and axons are present (C). (D-F) Numerous Islet-1[+] cells are present in an intermediate neural plate (i np) explant grown in contact with notochord (nc). There are no obvious changes in the density of 3A10[+] cells and axons (F). (G-I) Numerous Islet-1[+] cells are detected in an intermediate neural plate (i np) explant grown in contact with floor plate (fp), with no obvious change in 3A10[+] cells (I) compared with controls. (J-L) A floor plate explant isolated from stage 17 chick spinal cord and grown alone in vitro does not contain Islet-1[+] (K) nor 3A10[+] (L) cells. (M-O) Many Islet-1[+] cells are present in intermediate neural plate explants grown for 48 hr in the presence of floor plate-conditioned medium. All micrographs show representative explants from 10-60 similar experiments. Scale bar: (A-C), 60 mm; (D-L) 50 mm, (M-O) 60 mm. From Yamada et al., (1993).

Although our findings indicate that motor neuron differentiation begins within the neural plate, retroviral lineage analysis of cells in the chick neural tube has shown that the commitment of an individual cell to a motor neuron fate need not occur until its penultimate or final cell division and that clonally related cells give rise to other cell types (Leber et al., 1990). Inductive signals from the notochord and floor plate may therefore act on dividing neural plate cells to confer them with the potential to generate motor neurons, although the selection of motor neuron fate within the clonal progeny may occur later and may be influenced by other factors. The proliferation of committed motor neuron progenitors may also be influenced by other factors. Indeed, the neurotrophin NT-3 has been shown to increase the number of motor neurons generated in cultures of dissociated neural tube cells that have already been exposed to inductive signals from the notochord (Averbuch-Heller et al., 1994).

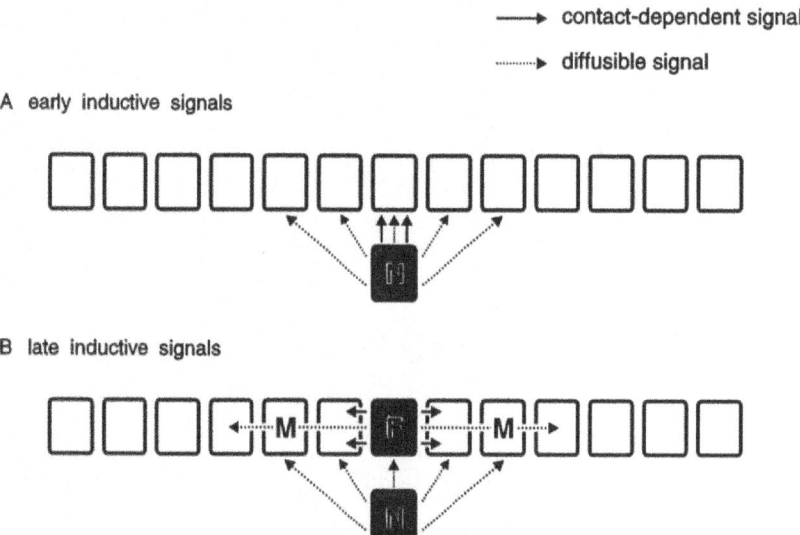

Figure 6. Model of the early inductive interactions that establish neural cell identity and pattern in the ventral neural tube. Inductive interactions appear to begin at the neural plate stage, although the eventual pattern of cell types becomes apparent only after neural tube closure. (A) Early inductive signals. Cells in the neural plate, which are assumed to be equivalent, respond to contact-dependent (solid line) and diffusible (broken line) signals from the notochord (N). Cells immediately overlying the neural plate that are destined to give rise to the floor plate probably receive both contact-dependent and diffusible signals, whereas cells located in more lateral regions of the neural plate that give rise to motor neurons and other ventral neurons are exposed only to diffusible signals. Thus, the floor plate-inducing signal appears to dominate over motor neuron-inducing signals at the midline of the neural plate. (B) Early inductive signals have specified the first group of floor plate cells (F), motor neurons (M) that are shown in a position that corresponds to their eventual dorsoventral location after neural tube closure. The initial group of floor plate cells acquires the inductive properties of the early notochord (N) and provides both contact-dependent signals that recruit additional floor plate cells and diffusible signals that induce later differentiating motor neurons and other ventral cell types. By this time, the notochord has lost its contact-dependent inductive ability but retains its diffusible signaling properties, which may act in conjunction with signals from the floor plate. This model is based on the results presented here and those of Placzek et al. (1993) and Yamada et al. (1993).

GENERATION OF MOTOR NEURON SUBCLASSES

The experimental studies described above have considered motor neurons as a homogenous population, although functional and physiological evidence shows that motor neurons are diverse and can be divided into several distinct subtypes. These motor neuron subtypes can be identified based on their columnar position within the spinal cord, and the axonal pathways along which they project in the periphery (Landmesser, 1992; Eisen et al., 1994). Studies in zebrafish embryos have suggested that neuroepithelial cells first acquire a general motor neuron identity and later acquire specific identities that determines the guidance cues they will respond to and the peripheral target they will innervate (Eisen, 1991, 1994). Therefore, motor neuron development may be initiated by a generic inductive process that is refined to generate unique populations of motor neurons that innervate specific targets.

The molecular mechanisms that control motor neuron diversification and that result in their unique identities remain unknown. Analysis of genes involved in cell fate decisions in diverse organisms has identified a class of transcription factors that contain a cysteine-rich repeat, termed a LIM domain, first recognized in three homeobox genes, Lin 11, Islet-1, and Mec 3. Several LIM-homeobox genes have now been identified in vertebrates and invertebrates The combinatorial expression of four of these genes Islet-1, Islet-2, Lim-1 and Lim-3 defines functionally distinct groups of motor neurons (Tsuchida et al., 1994). The molecular nature of these factors raises the possibility that they specify the identities of subtypes of motor neurons and control the response of motor axons to guidance cues in their peripheral environment. In addition, these genes may provide molecular markers with which to examine the cellular interactions that control motor neuron subtype diversification.

ACKNOWLEDGEMENTS

We thank V. Leon and I. Schieren for help in preparing the manuscript. This work was supported by grants from the Swedish Medical Research Council (T.E.), from the American Paralysis Association and NIH grant NS 33245 (T.M.J.). T.M.J. is an Investigator of the Howard Hughes Medical Institute.

REFERENCES

Ang S-L and Rossant J (1993): Anterior mesendoderm induces engrailed genes in explant cultures. Development 118: 139-149.

Ang S-L and Rossant J (1994): HNF-3β is essential for node and notochord formation in mouse development. Cell 78: 561-574.

Averbuch-Heller L, Pruginin M, Kehane N, Tsoulfas P, Parada L, Rosenthal A, Kalcheim C (1994): Neurotrophin 3 stimulates the differentiation of motoneurons from avian neural tube progenitor cells. Proc Natl Acad Sci 91: 3247-3251.

Bernhardt RR, Nguyen N and Kuwada JY (1992): Growth cone guidance by floor plate cells in the spinal cord of zebrafish embryos. Neuron 8: 869.

Bolce ME, Hemmati-Brivanlou A and Harland RM (1993): XFKH2, a Xenopus HNF-3a homologue, exhibits both activin inducible and autonomous phases of expression in early embryos. Dev Biol 160: 413-423.

Bovolenta P and Dodd J (1990): Guidance of commissural growth cones at the floor plate in the embryonic rat spinal cord. Development 109: 435-447.

Bovolenta P and Dodd J (1991): Perturbation of neuronal differentiation and axon guidance in the spinal cord of mouse embryos lacking a floor plate: analysis of Danforth's short-tail mutation. Development 113: 625-639.

Dodd J, Morton SB, Karagogeos D, Yamamoto M and Jessell TM (1988): Spatial regulation of axonal glycoprotein expression on subsets of embryonic spinal neurons. Neuron 1: 105-116.

Echelard Y, Epstein DJ, St-Jacques B, Shen L, Mohler J, McMahon JA and McMahon AP (1993): Sonic hedgehog, a member of a family of putative signaling molecules, is implicated in the regulation of CNS polarity. Cell 75: 1417-1430.

Eisen J (1991): Determination of primary motoneuron identity in developing zebrafish embryos. Science 252: 569-572.

Eisen JS (1994): Development of motoneuronal phenotype. Annu Rev Neurosci 17: 1-30.

Ericson J, Thor S, Edlund T, Jessell TM and Yamada T (1992): Early stages of motor neuron differentiation revealed by expression of homeobox gene Islet-1. Science 256: 1555-1560.

Falls DL, Rosen KM, Confas G, Lane WS, Fischbach GD (1993): ARIA, a protein that stimulates acetylcholine receptor synthesis is a member of the Neu ligand family. Cell 72: 801-815.

Furley AJ, Morton SB, Manalo D, Karagogeos D, Dodd J and Jessell TM (1990): The axonal glycoprotein TAG-1 is an immunoglobulin superfamily member with neurite outgrowth-promoting activity. Cell 61: 157-170.

Hatta K, Kimmel CB, Ho RK and Walker C (1991): The cyclops mutation blocks specification of the floor plate of the zebrafish central nervous system. Nature 350: 339-341.

Hatta, K. (1992) Role of the floor plate in axonal patterning in the zebrafish CNS. Neuron 9: 629-642.

Holley J and Silver J (1987): Growth pattern of pioneering chick spinal cord axons. Dev Biol 123: 375-388.

Ishii K, Oda Y, Ichikawa T and Deguchi T (1990): Complementary DNAs for choline acetyltransferase from spinal cord of rat and mouse nucleotide sequences, expression in mammalian cells, and in situ hybridization. Mol Brain Res 7: 151-159.

Karlsson O, Thor S, Norbert T, Ohlsson H and Edlund T (1990): Insulin gene enhancer binding protein Isl-1 is a member of a novel class of proteins containing both a homeo and a Cys-His domain. Nature 344: 879-882.

Kennedy TE, Seragini T, de la Torre JR, Tessier-Lavigne M (1994): Netrins are diffusible chemotrophic factors for commissural axons in the embryonic spinal cord. Cell 78: 425-435.

Krauss S, Concordet JP and Ingham PW (1993): A functionally conserved homolog of the Drosophila segment polarity gene hedgehog is expressed in tissues with polarizing activity in zebrafish embryos. Cell 75: 1431-1444.

Lai E, Prezioso VR, Tao W, Chen WS and Darnell JE, Jr (1991): Hepatocyte nuclear factor 3α belongs to a gene family that is homologous to the Drosophila homeotic gene fork head. Genes Dev 5: 416-427.

Landmesser LT (1992): Growth cone guidance in the avian limb: a search for cellular and molecular mechanisms. In Letourneau PC, Kater SB and Macagno ER (eds): "The Nerve Growth Cone," New York: Raven Press. pp 373-85.

Langman J, Guerrant RL and Freeman BG (1966): Behavior of neuroepithelial cells during closure of the neural tube. J Comp Neurol 127: 399.

Leber SM, Breedlove SM and Sanes JR (1990): Lineage arrangement and death of clonally related motoneurons in chick spinal cord. J Neurosci 10: 2451-2462.

Li H, Witte DP, Branford WW, Aronow BJ, Weinstein M, Kaur, S, Wert S, Singh G, Schreiner CM, Whitsett JA, Scott WJ and Potter SS (1994): Gsh-4 encodes a LIM type homeodomain, is expressed in the developing central nervous system and is required for early postnatal survival. EMBO J 13: 2876-2885.

Lutz B, Kuratani S, Cooney AJ, Wawersik S, Tsai SY, Eichele G, Tsai M (1994): Developmental regulation of the orphan receptor coup-TFII gene in spinal motor neurons. Development 120: 25-36.

Matise MP and Lance-Jones C (1992): The timing of motoneuron commitment in the developing chick spinal cord. Soc Neurosci Abstr 18: 1111.

Monaghan AP, Kaestner KH, Grau E and Schütz G (1993): Postimplantation expression patterns indicate a role for the mouse forkhead/HNF-3 α, β, and γ genes in determination of the definitive endoderm, chordamesoderm and neuroectoderm. Development 119: 567-578.

Placzek M, Tessier-Lavigne M, Jessell TM and Dodd J (1990a): Orientation of commissural axons in vitro in response to a floor plate derived chemoattractant. Development 110: 19-30.

Placzek M, Tessier-Lavigne M, Yamada T, Jessell TM and Dodd J (1990b): Mesodermal control of neural cell identity: floor plate induction by the notochord. Science 250: 985-988.

Placzek M, Yamada T, Tessier-Lavigne M, Jessell TM and Dodd J (1991): Control of dorso-ventral pattern in vertebrate neural development induction and polarizing properties of the floor plate. Development 113(Suppl. 2): 105-122.

Placzek M, Jessell TM and Dodd J (1993): Induction of floor plate differentiation by contact-dependent, homeogenetic signals. Development 117: 205-218.

Riddle R, Johnson RL, Laufer E and Tabin C (1993): Sonic hedgehog mediates the polarizing activity of the ZPA. Cell 75: 1401-1416.

Roelink H, Augsburger A, Heemskerk J, Korzh V, Norlin S, Ruiz i Altaba A, Tanabe Y, Placzek M, Edlund T, Jessell TM and Dodd J. (1994): Floor plate and motor neuron induction by vhh-1, a vertebrate homolog of hedgehog expressed by the notochord. Cell 76: 761-775.

Ruiz i Altaba A and Jessell TM (1992): Pintallavis, a gene expressed in the organizer and midline cells of frog embryos: involvement in the development of the neural axis. Development 116: 81-93.

Ruiz i Altaba A, Cox C, Jessell TM and Klar A (1993a): Ectopic neural expression of a floor plate marker in frog embryos injected with the midline transcription factor Pintallavis. Proc Natl Acad Sci 90: 8268-8272.

Ruiz i Altaba A, Prezioso VR, Darnell JE and Jessell TM (1993b): Sequential expression of HNF-3β and HNF-3α by embryonic organizing centers: the dorsal lip/node, notochord and floor plate. Mech Dev 44: 91-108.

Sasaki H and Hogan BLM (1994): HNF-3β as a regulator of floor plate development. Cell 76: 103-115.

Schachner M, Antonicek H, Fahrig T, Faissner A, Fischer G, Kunemund V, Martini R, Meyer A, Persohn E, Pollerberg E, Probstmeier R, Sadoul K, Sadoul R, Seiheimer B and Thor G (1990): Families of neural cell adhesion molecules. In Edelman GM, Cunningham BA and Thiery JP (eds): "Morphoregulatory Molecules," New York: John Wiley & Sons, pp 443-468.

Sechrist J and Bronner-Fraser M (1991): Birth and differentiation of reticular neurons in the chick hindbrain: ontogeny of the first neuronal population. Neuron 7: 947-963.

Serafini T, Kennedy TE, Galko MJ, Mirzayan C, Jessell TM and Tessier-Lavigne M (1994): The netrins define a family of axon outgrowth-promoting proteins homologous to C. elegans UNC-6. Cell 78: 409-424.

Tanaka H, and Obata K (1984): Developmental changes in unique cell surface antigens of chick embryo spinal motor neurons and ganglion cells. Dev Biol 106: 26-37.

Tanaka H, Matsui T, Agata A, Tomura M, Kubota I, McFarland KC, Kohr B, Lee A, Phillips HS and Shelton DL (1991): Molecular cloning and expression of a novel adhesion molecule SC1. Neuron 7: 535-545.

Tessier-Lavigne M, Placzek M, Lumsden AGS, Dodd J and Jessell TM (1988): Chemotropic guidance of developing axons in the mammalian central nervous system. Nature 336: 775-778.

Tsim KWK, Ruegg MA, Eschar G, Kroger S, and McMahon YJ (1992): cDNA that encodes active agrin. Neuron 8: 677-689.

Tsuchida T, Ensini M, Morton SB, Baldassare M, Edlund T, Jessell TM and Pfaff SL (1994): Topographic organization of embryonic motor neurons defined by expression of LIM homeobox genes. Cell 79: 957-970.

van Straaten HMW, Hekking JWM, Wiertz-Hoessels EL, Thors F and Drukker J (1988): Effect of the notochord on the differentiation of a floor plate area in the neural tube of the chick embryo. Anat Embryol 177: 317-324.

van Straaten HMW, Thors F, Wierz-Hoessels EL, Hekking JWM and Drukker J (1985): Effect of a notochordal implant on the early morphogenesis of the neural tube and neuroblasts: histometrical and histological results. Dev Biol 110: 247-254.

Weigel D and Jäckle H (1990): The fork head domain: a novel DNA-binding motif of eukaryotic transcription factors? Cell 63: 466-456.

Weinstein DC, Ruiz i Altaba A, Chen WS, Hoodless P, Prezioso VR, Jessell TM and Darnell JE, Jr (1994): The winged-helix transcription factor HNF-3β is required for notochord development in the mouse embryo. Cell 78: 575-588.

Yaginuma H and Oppenheim RW (1991): An experimental analysis of in vivo guidance cues used by axons of spinal interneurons in the chick embryo: evidence for chemotropism and related guidance mechanisms. J Neurosci 11: 2598-2613.

Yamada T, Placzek M, Tanaka H, Dodd J and Jessell TM (1991): Control of cell pattern in the developing nervous system: polarizing activity of the floor plate and notochord. Cell 64: 635-647.

Yamada T, Pfaff SL, Edlund T and Jessell TM (1993): Control of cell pattern in the neural tube: motor neuron induction by diffusible factors from notochord and floor plate. Cell 73: 673-686.

NEURAL CELL DIFFERENTIATION

NEURONAL DEVELOPMENT IN THE RAT SYMPATHOADRENAL LINEAGE

S. J. Birren[1], J. M. Verdi[2], and D. J. Anderson[2]

[1]Department of Biology, Center for Complex Systems,
Brandeis University, Waltham, MA 02254, U.S.A.
[2]Howard Hughes Medical Institute, California Institute
of Biology, Pasadena, CA 91125, U.S.A.

INTRODUCTION

The generation of cellular diversity in the mammalian nervous system requires that multipotential precursor cells become restricted to their final cell fate and undergo appropriate differentiation. This developmental process is a function of position, environment, and intrinsic factors. In the peripheral nervous system neural precursor cell populations have been identified and characterized, permitting the experimental investigation of these developmental parameters.

The peripheral nervous system is derived from a transient embryonic structure, the neural crest, that arises along the dorsal aspect of the neural tube following neurulation (Le Douarin, 1982). The cells of the neural crest migrate throughout the embryo and give rise to such diverse cell types as melanocytes, the bones of the face and skull, neuroendocrine cells such as adrenal chromaffin cells, and the neurons and glia of the sympathetic nervous system (Le Douarin et al., 1993). Experiments in which neural crest cells have been marked and followed through their migratory pathway have demonstrated that individual neural crest cells can develop into several different cell types (Bronner-Fraser and Fraser, 1988; 1989). Indeed, transplantation studies in which the axial position of neural crest cells was altered have indicated that many neural crest cells are multipotent and have the capacity to develop into any of a wide array of neural crest derivatives (Le Douarin, 1980). These experiments suggest that environmental factors play a critical developmental role in cell fate decisions in peripheral lineages.

The sympathoadrenal lineage is one peripheral nervous system sublineage in which the actions of environmental factors have been implicated in differentiation and survival (Patterson, 1990; Anderson, 1993). Developmentally restricted precursor cells with the potential to give rise to either sympathetic neurons or adrenal chromaffin cells have been isolated and characterized (Anderson and Axel, 1986; Carnahan and Patterson, 1991). Primary cultures of these cells have been used to examine the choice of cell fate in this sublineage and

Neural Cell Specification: Molecular Mechanisms and Neurotherapeutic Implications
Edited by Juurlink *et al.*, Plenum Press, New York, 1995

127

have resulted in at least a partial understanding of the factors, their receptors and the timing of their actions and expression. This understanding has been increased by the ability to immuno-purify the bipotential precursor cells and study the effects of various factors in the absence of contaminating cell populations. The MAH cell line, a *v-myc* immortalized cell line established from immuno-purified precursor cells, has also provided insight into neuronal development in the sympathetic lineage (Birren and Anderson, 1990).

The development of the precursor cells into chromaffin cells is controlled by the actions of glucocorticoid hormones (GC) present in the adrenal microenvironment (Unsicker et al., 1978; Doupe et al., 1985; Anderson and Axel, 1986; Seidl and Unsicker, 1989b). The actions of CG are required at two distinct phases of chromaffin cell development, as early inhibitors of neuronal differentiation, and later, as inducers of phenylethanolamine-N-methyltransferase (PNMT) (Michelsohn and Anderson, 1992). This late induction of PNMT in chromaffin cells reflects a gain in competence of the cells to express PNMT in response to GC.

The development of precursor cells into sympathetic neurons is a more complex problem that appears to require the interaction of several different factors. In a possible model for sympathetic neuron development, mature chromaffin cells are able to transdifferentiate into sympathetic neurons under the influence of nerve growth factor (NGF) (Doupe et al., 1985; Seidl and Unsicker, 1989a). This, combined with the observation that sympathetic neurons are dependent upon NGF for survival both in vivo (Levi-Montalcini and Booker, 1960) and in culture (Levi-Montalcini and Angeletti, 1963), suggested the possibility that NGF would be an important differentiation factor for neuronal development in the sympathoadrenal lineage. More recent work, however, has indicated that sympathetic precursor cells are unable to respond to NGF (Coughlin and Collins, 1985; Anderson and Axel, 1986; Ernsberger et al., 1989), and that there is not a local source of NGF in the developing sympathetic ganglia (Davies et al., 1987; Korsching and Thoenen, 1988). This raises the question of the mechanism by which NGF-non-responsive precursor cells initiate neuronal differentiation, gain NGF-responsiveness, and eventually, as mature neurons, become dependent upon NGF for survival. In addition to studying these questions in primary cultures of sympathetic neuroblasts, many of these issues have been addressed using MAH cells as a model system for sympathetic neuron development.

NEURONAL DIFFERENTIATION AND ACQUISITION OF TROPHIC RESPONSIVENESS

Fibroblast Growth Factor Is A Neuronal Differentiation Agent For Sympathetic Precursor Cells

Fibroblast growth factor (FGF) was originally identified as a potent mitogen for a variety of non-neuronal cell types (Gospodarowicz, 1975). FGF also acts upon a number neuronal precursor cells and neurons, with effects that include proliferation (Gensburger, 1987), neurite outgrowth (Hatten et al., 1988) and survival (Morrison, 1986; Walicke, 1986). In the peripheral nervous system, FGF supports the survival of chick ciliary parasympathetic neurons (Unsicker, 1987). PC12 cells respond to FGF by extending neurites, in a manner similar to their response to NGF. Finally, when neonatal rat chromaffin cells are treated with FGF, they proliferate, undergo neuronal differentiation, and acquire a dependence upon NGF for survival (Claude et al., 1988; Stemple et al., 1988). These results suggest a role for FGF in the differentiation of sympathetic neuroblasts and a possible link between exposure to FGF and the later acquisition of NGF-dependence.

MAH cells show both a proliferative and a differentiation response to FGF, but FGF is not a long term survival factor for the resulting MAH neurons (Birren and Anderson, 1990). The efficiency of initiation of neuronal differentiation in response to FGF is very high; virtually all of the MAH cells in the culture extend neurites in response to FGF. FGF also induces the up-regulation of neuron-specific gene expression. Following this initial differentiation however, the cells undergo a crisis in which all of the differentiated MAH cells die. This raised the question of whether in addition to causing neuronal differentiation, FGF led to NGF-dependence. In fact, when NGF as well as FGF was included in the culture medium, a very small number of cells underwent further morphological development resulting in very long, interconnecting neurites, enlarged cell bodies, and large, prominent nucleoli. This morphology is typical of mature sympathetic neurons in culture. In addition, like mature sympathetic neurons, the MAH neurons resulting from treatment with FGF and NGF were dependent upon NGF for survival. When a blocking antibody to NGF was included in the culture medium, the MAH neurons died within about 24 hours.

The result that FGF is an efficient differentiation factor for MAH cells has also been borne out in primary cultures of embryonic sympathetic precursor cells (Birren and Anderson, unpublished). Immuno-isolated sympathetic precursor cells responded to FGF by extending neurites, although in the primary cells this effect is a potentiation of process outgrowth seen in the absence of any added factor. It is not clear whether this phenomenon reflects a default differentiation pathway for these cells, or whether the precursor cells had been previously exposed to differentiation factors before being removed from the animal. In either case, these data, taken together with the observation that FGF is expressed in the developing sympathetic ganglia (Kalcheim and Neufeld, 1990), implicate FGF as an early differentiation agent for sympathetic neuroblasts.

Acquisition Of NGF-Responsiveness In MAH Cells

While the evidence for FGF as a sympathetic differentiation factor is strong, the question of whether FGF is responsible for the acquisition of NGF-responsiveness and NGF-dependence is not as compelling. A small percentage of FGF treated MAH cells go on to become dependent upon NGF for survival, but the vast majority (over 99%) of the cells are not rescued. It was therefore likely that other factors were required for the acquisition of NGF-responsiveness by MAH cells. MAH cells do not respond to NGF because they lack the receptors to mediate a functional response. Two receptors for NGF are known, a low affinity receptor, p75 (Johnson et al., 1986; Radeke et al., 1987), and trk, a receptor tyrosine kinase (Kaplan et al., 1991; Klein et al., 1991), and the first member of the neurotrophin receptor family to be identified and cloned. Trk is able to confer NGF-responsiveness to a variety of non-neuronal cell types (Cordon-Cardo et al., 1991). The role of p75, and the interaction between the p75 and trk proteins is still not clear. Neither p75 nor trk are constitutively expressed in MAH cells. In order to understand how MAH cells acquire the ability to respond to NGF, therefore, it was necessary to identify conditions that induced the expression of functional NGF receptors.

A variety of growth and neurotrophic conditions, many of which have either differentiation or survival effects in the sympathetic lineage, were examined for their ability to induce trk expression in MAH cells (Birren et al., 1992). These included neurotrophins such as neurotrophin 3 (NT-3) and brain-derived neurotrophic factor (BDNF), FGF, membrane depolarization, and retinoic acid. Of the factors tested, only membrane depolarization, achieved by including 40 mM KCl in the culture medium, was able to induce trk expression. Increased levels of trk expression were detected within 24 hours of treatment with KCl. Induction of trk expression by depolarization was sufficient to mediate a biological response to NGF in a significant percentage of MAH cells; following treatment with KCl, MAH cells responded to NGF by assays of process outgrowth and cell survival. Depolarization also resulted in an increased efficiency of NGF-responsive neuron formation from MAH cells (see below).

While it is tempting to speculate on a role for depolarization through synaptic activity in the acquisition of trophic response, there is as yet no direct evidence that depolarization results in the induction of trk mRNA in freshly isolated sympathetic neuroblasts (Wyatt and Davies, unpublished, Verdi and Anderson, unpublished). Depolarization does not lead to trk expression in sensory neuronal precursor cells (Wyatt and Davies, 1993). One possibility is the induction of trk is dependent not upon synaptic transmission, but on a cell-autonomous mechanism, for example the cessation of division by sympathetic neuroblasts. KCl acts as an anti-mitotic agent on MAH cells (Birren, Verdi and Anderson, unpublished), and if mitotic arrest were a requirement for expression of trk mRNA, KCl might lead to the induction of trk through this mechanism.

Modulation Of NGF-Response In MAH Cells By CNTF And Depolarization

Depolarization of MAH cells with KCl induces the expression of trk mRNA and leads to NGF-responsiveness in MAH cells. But the effect of depolarization on other components of the factor response pathway in these cells is not known. The question of whether the expression of trk alone was sufficient to mediate a biological response to NGF in MAH cells was addressed by introducing a trk expression construct into the cells (Verdi et al., 1994b). Trk-expressing MAH cells showed both proliferative and neurite outgrowth responses to NGF that were not seen in MAH cells transfected with vector alone. These responses to NGF were similar to those seen in KCl-treated MAH cells. Trk-expressing MAH cells also were able to give rise to post-mitotic neurons with a mature sympathetic neuron morphology. The efficiency of this neuronal maturation was much higher in trk-expressing MAH cells treated with NGF (8%), than in parental MAH cells grown in NGF + FGF (0.25%) suggesting that insufficient induction of trk is an explanation for inefficient neuronal differentiation. But even at this very enhanced level of neuron formation, the vast majority of MAH cells failed to develop into mature post-mitotic neurons. These data suggest that additional factors might collaborate with NGF to increase the neuronal differentiation of these cells.

Membrane depolarization and ciliary neurotrophic factor (CNTF) were both able to collaborate with FGF to enhance neuronal differentiation in MAH cell cultures (Birren et al., 1992; Ip et al., 1994). The neurons generated in these cultures could subsequently be supported by NGF alone. The CNTF/FGF effect did not lead to a significant increase in levels of trkA expression, but to changes in the phosphorylation patterns of downstream signal transduction molecules. When the effects of CNTF and membrane depolarization were examined in trk-expressing MAH cells in the absence of FGF, each of these factors independently increased the efficiency with which post-mitotic neurons were generated in the presence of NGF (Verdi et al., 1994b). These effects on neuronal efficiency were additive, when both CNTF and KCl were included in the culture medium, the neuronal efficiency reached 70%. Thus, both CNTF and depolarization act in the absence of FGF treatment, suggesting that they act to enhance signaling through trk. Interestingly, there was no significant increase in trk expression in the trk-transfected MAH cells following depolarization or CNTF treatment, suggesting a downstream effect on the signal transduction pathway. These results suggest that the low efficiency of MAH neuron formation seen following treatment of the cells with FGF and NGF reflects additional requirements for collaborating factors for neuronal maturation in this lineage. What these factors are in vivo, and how they interact with the trk signaling cascade remains to be determined.

Modulatory Role Of p75 In MAH Cell Development

The role of p75 in generating a biological response to NGF has been a topic of much debate. NGF binds to p75 with low affinity, and is not sufficient to mediate a biological response to NGF (Johnson et al., 1986; Hempstead et al., 1989). Whether or not p75 is required in addition to trk for an NGF response in neuronal cells has been a controversial issue (Hempstead et al., 1991; Jing et al., 1992). Current models suggest that a direct interaction of the trk and p75 proteins is unlikely, but maintain the possibility that p75 may influence the trk signal transduction pathway in some way (Meakin and Shooter, 1992). In MAH cells expressing trk, however, p75 is not required for a response to NGF. MAH cells treated for 5 days with KCl show a clear response to NGF, yet do not express detectable levels of p75 by an immunostaining assay on a fluorescence-activated cell sorter (Birren et al., 1992). Although it is clear that p75 is not necessary for NGF response through the trk receptor, recent evidence has emerged that implicates p75 as yet another modulator of NGF-responsiveness in MAH cells (Verdi et al., 1994a).

NGF-responsiveness was examined in MAH cells expressing trk, or both trk and p75. In cells expressing trk and p75, the level of expression of p75 mRNA was approximately 20 fold greater than the level of trk, a ratio in the range of that found on PC12 cells. The autophosphorylation of trk was examined in cells treated with NGF for 5 minutes by Western blotting with anti-trk and anti-phosphotyrosine antibodies. p75 expressing MAH cells showed an 8-fold increase in tyrosine phosphorylation on trk compared to similar cells in which p75 was not expressed. This increase in phosphorylation was not seen when a mutant NGF protein, able to bind to trk, but not to p75 (Ibáñez et al., 1992), was used to stimulate the MAH cells. This result demonstrated that the increase in phosphorylation was a function of NGF binding to the p75 receptor. Enhanced phosphorylation was also observed in p75-expressing MAH cells on other targets of the trk tyrosine kinase, phospholipase C-γ and the SNT protein (Verdi et al., 1994a). Thus, the potentiation of tyrosine phosphorylation by p75 affected several components of the trk signal transduction pathway and raised the question of whether the expression of p75 would alter the biological response of trk-expressing MAH cells to NGF.

While early responses to NGF appeared no different in cells with or without p75, several different assays of later development showed that the expression of p75 in trk-expressing MAH cells had a modulating effect on the biological response of the cells to NGF (Verdi et al., 1994a). The effect of NGF on cell number was determined for MAH cells expressing trk or trk and p75. Although no differences in cell number were observed after 24 hours, by three days significantly more cells were present in the trk plus p75 cultures. This result was not seen when the mutant NGF protein was used in the assay. The expression of p75 in trk-containing MAH cells also increased the rate at which the cells exited the cell cycle as they differentiated into mature neurons under the influence of NGF. MAH cells expressing trk and p75 underwent mitotic arrest 2-3 days earlier than MAH cells expressing trk alone. The p75-expressing cells also achieved their final mature, post-mitotic morphology earlier than cells lacking p75. These results indicate that although p75 may not be required for a biological response to NGF, the effects of NGF are enhanced by its interaction with p75 as well as trk. These effects of p75 are more pronounced late in the developmental pathway of MAH cells. Interestingly, during the development of normal sympathetic neuroblasts, trk mRNA expression precedes the expression of p75 (Verdi and Anderson, unpublished).

A SWITCH IN NEUROTROPHIN RESPONSIVENESS LEADS TO NGF-DEPENDENCE

While NGF is required for the maturation and survival of sympathetic neurons, and both trk and p75 are involved in the NGF signal transduction pathway, NGF does not support the survival of early sympathetic precursor cells (Coughlin and Collins, 1985; Anderson and Axel, 1986; Ernsberger et al., 1989; Birren and Anderson, 1990). FGF, a neuronal differentiation agent for sympathetic precursor cells (Birren and Anderson, 1990), is also not a survival agent for these cells. What factors then, if any, are responsible for the survival of early sympathetic precursor cells. It has been proposed that all cells in an organism require trophic factors to maintain their survival (Raff, 1992), but early indications were that neural precursor cells exhibited trophic factor-independent behavior (Ernsberger et al., 1989; Vogel and Davies, 1991). These studies were performed in mixed populations of cells and in serum containing medium, and trophic influences may have been supplied from these sources. Recent studies of the development of sympathetic neuroblasts suggest that trophic factors are required early in development to support the survival of precursor cells before they become responsive to NGF (Birren et al., 1993; Dechant et al., 1993; DiCicco-Bloom, 1993).

The expression of neurotrophin receptors in developing rat sympathetic ganglia was examined by in situ hybridization (Birren et al., 1993) or Northern analysis (DiCicco-Bloom, 1993). At early embryonic stages (E14.5 -E15.5), there was no detectable expression of trk in sympathetic neuroblasts, consistent with the observation that early sympathetic precursor cells do not respond to NGF. E14.5 sympathetic ganglia and E15.5 neuroblasts did show expression of trkC, the receptor tyrosine kinase for NT-3, suggesting a possible role for this neurotrophin during sympathetic development. Interestingly, expression of NT-3 has also been reported in the developing sympathetic ganglia (Schecterson and Bothwell, 1992), further arguing for a role for this neurotrophin. In mice lacking the NT-3 gene, there is a 50% loss of cell number in neonatal sympathetic ganglia (Farinas et al., 1994). Taken together, these data strongly support a role for NT-3 at early developmental stages in the sympathetic lineage. The expression of the BDNF receptor, trkB was also examined by in situ hybridization, but although trkB was detected in the region of the developing sympathetic ganglia, it did not appear to be expressed in the developing neuroblasts.

Although sympathetic precursor cells do not respond to NGF, sympathetic neurons are both responsive to and dependent upon NGF. A developmental time course of trk expression in the rat sympathetic ganglia showed that although trk expression was undetectable at E14.5, by E17.5 there were significant levels of trk mRNA (Birren et al., 1993). Interestingly, the pattern of trkC expression showed reciprocal, but overlapping regulation; trkC mRNA was decreasing at E17.5, and by E19.5, when trk expression was high, trkC mRNA was not detectable. These results suggest a model in which NT-3 acts as an interim survival factor for developing sympathetic neurons during the developmental period before the cells can respond to NGF. As the cells begin to express trk mRNA, and presumably gain the capacity to respond to NGF, the expression of trkC and the ability to be supported by NT-3 is down-regulated. This model is attractive because it offers an explanation for both the early survival of sympathetic precursor cells, and for the developmental switch from NGF-responsiveness to NGF-dependence. This switch to NGF-dependence is necessary if, as has been proposed, the final complement of sympathetic neurons is determined by competition for target-derived NGF (Cowan et al., 1984). Two predictions of this neurotrophin switch model have been tested, first, that NT-3 should act as a survival factor for early sympathetic precursor cells (Birren et al., 1993; Dechant et al., 1993; DiCicco-Bloom, 1993), and second that there should be a developmental switch in the responsiveness of sympathetic neuroblasts from NT-3 to NGF.

Rat sympathetic neuroblasts from mid-gestational embryos were specifically labeled by dissecting developing sympathetic ganglia and incubating with the B2 antibody. These neuronal precursor cells were then isolated by fluorescence-activated cell sorting and cultured in the absence of contaminating cell populations. When cultured in serum-free medium, the majority of these sympathetic neuroblasts, immuno-isolated from E14.5 or E15.5 embryos died within 24 hours of plating. These cells could be rescued by the addition of NT-3, but not NGF, to the culture medium, demonstrating that the expression of trkC observed in the sympathetic ganglia at this developmental stage resulted in a functional response to NT-3. Although these embryonic precursor cells were still mitotic, several lines of evidence indicated that the predominant effect of NT-3 on these cells was as a survival agent. Proliferation assays indicated that both FGF, a mitogen for sympathetic precursor cells, and NT-3 acted to increase the percentage of cells incorporating BrdU during the last 8 hours of a 24 hour culture period. In that same 24 hour period, FGF did not lead to an increase in cell recovery, indicating that few cells actually completed the mitotic cycle during that period. Cultures treated with NT-3 during that same period, however, showed a significant increase in cell recovery, suggesting that NT-3 prevented the death of the neuroblasts. Further evidence of NT-3 as a survival factor for the B2$^+$ cells came from a single cell tracking experiment in which cells were plated at clonal density and followed over a 72 hour culture period. The number of colonies counted at the end of this period was significantly higher in the NT-3 treated cultures than the control cultures (40% versus 18% survival). If NT-3 had been acting solely as a mitogen, the prediction from this experiment would be that an equal number of colonies should be alive at the end of the assay period, but that the colonies should contain more cells in the NT-3 condition. The result that was obtained, that there was an increase in the number of colonies recovered in NT-3, was further proof that NT-3 was a survival factor for sympathetic neuroblasts.

Responsiveness to neurotrophins changed as the sympathetic neuroblasts developed. Immuno-isolated precursor cells from different developmental stages (E14.5 though postnatal day 1) were cultured with NGF or NT3, and the percentage cell recovery was measured after 24 hours. At E14.5 the cells showed a response to NT-3, but not to NGF. The cells showed clear NGF-responsiveness by E17.5 and this response continued to increase at E19.5. In contrast, the cells showed a decrease in NT-3-responsiveness at E19.5, and in sympathetic neurons isolated from neonatal superior cervical ganglia, the NT-3 response was small compared to the NGF response (Birren et al., 1993). It was somewhat surprising to see a residual NT-3 response in the E19.5 and neonatal sympathetic neurons, as in situ hybridization did not detect expression of trkC mRNA by E19.5. While this activity could be due to residual receptor protein on the cell surface, the high concentrations of NT-3 used in these experiments might have triggered a signal through another tyrosine kinase receptor expressed by these cells, such as trkA.

During neuronal development, sympathetic precursor cells undergo a developmental switch in trophic dependence from an early reliance upon NT-3, a factor expressed in the local area of the sympathetic ganglia, to NGF, a target derived factor. Control of cell number during development may take place therefore, not just at the level of target derived trophic factors, but earlier in development, before the axons of developing neuroblasts make contact with the target. Changes in response to neurotrophins during neuronal development have been reported for other neuronal populations both in the peripheral (Buchman and Davies, 1993) and central (Segal et al., 1992) nervous systems, suggesting that these types of switches in trophic responsiveness may be a common developmental motif.

CONCLUDING REMARKS

The central tenet of the neurotrophic theory is that the final number of neurons generated in a ganglion is a function of competition for target-derived trophic factor. This provides a mechanism for matching the number of neurons to the size of the target. This model requires, first, that the neurotrophic factor being competed for not be produced in the local region of the developing ganglia, and second, that the developing neurons not be supported by other factors in their environment. The development of neurons in the sympathetic lineage, the system in which this theory was first formulated, provides an opportunity not just to further our understanding of trophic competition, but to put it into the context of the entire neuronal developmental pathway. In the developing sympathetic ganglia, NGF is produced by the targets of sympathetic innervation, but is not found in the region of the ganglia themselves (Davies et al., 1987; Korsching and Thoenen, 1988). However, other locally-derived survival factors, such as NT-3, have now been identified. The developmental decrease in trkC expression and NT-3-responsiveness observed in sympathetic neuroblasts, coupled with an increase in trk expression and responsiveness to NGF, provides a developmental mechanism for the acquisition of NGF-dependence in these cells. As additional factors are identified that support the survival of these early sympathetic neuroblasts, the prediction would be that during the period when the neurons are making connections to their targets that there would be a decrease either in the responsiveness to, or the availability of these factors.

Once the sympathetic ganglia have attained their final complement of neurons, dependence upon a single target-derived neurotrophin is no longer necessary, and in fact may not be desirable. The period of trophic competition might therefore be thought of as a developmental window, before and after which sympathetic neurons can be supported by a variety of factors. This model would be consistent with the observation that mature sympathetic neurons become less dependent upon NGF over time (Koike et al., 1989). Recent evidence demonstrated that mature sympathetic neurons develop the ability to be supported by CNTF and LIF late in development (Kotzbauer et al., 1994), further suggesting that additional factors may support the survival of adult sympathetic neurons.

The loss of complete dependence upon NGF of mature sympathetic neurons can be placed in a developmental context in which not only the number and type of survival factors changes over time, but different developmental progression factors also act at specific developmental times (Fig. 1). These include the actions of FGF as an early differentiation factor, and additional factors, (depolarization and CNTF are two candidates), for the induction and modulation of NGF-responsiveness. While the actions of several of these factors have been characterized, the identity of others and the mechanisms of their actions remain to be investigated. It is the complex interplay between the expression, timing, and responsiveness to such factors that results in the generation of the sympathetic nervous system.

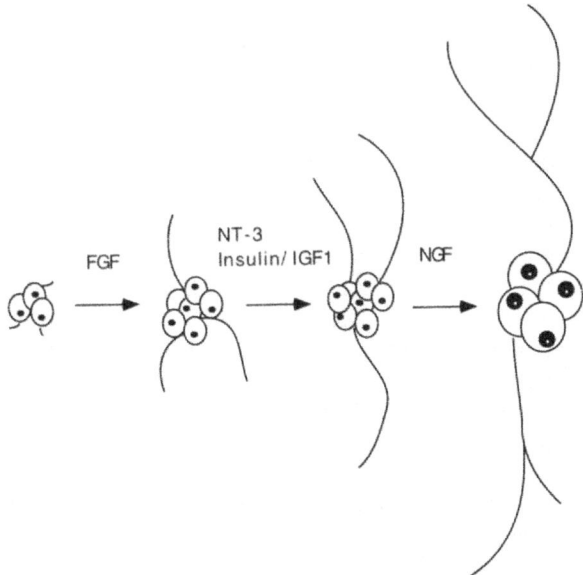

Figure 1. Model for the development of sympathetic neurons. Neuronal differentiation is initiated or potentiated by FGF. Early survival factors such as NT-3 permit the survival and proliferation of neuroblasts. During this period the cells acquire NGF-responsiveness, and this responsiveness is modulated by additional factors. As the cells lose the ability to be supported by early survival factors, they become dependent upon target-derived NGF and cells are lost through programmed cell death.

ACKNOWEDGEMENTS

This work was supported by NIH grant NS23476 and a PEW Faculty Fellowship in Neuroscience to D. J. A. D. J. A. is an Associate Investigator of the Howard Hughes Medical Institute. J. M. V. is supported by an NIH NRSA.

REFERENCES

Anderson DJ (1993): Molecular control of cell fate in the neural crest: the sympathoadrenal lineage. Annu Rev Neurosci 16: 129-158.

Anderson DJ and Axel R (1986): A bipotential neuroendocrine precursor whose choice of cell fate is determined by NGF and glucocorticoids. Cell 47: 1079-1090.

Birren SJ and Anderson DJ. (1990): A v-myc-immortalized sympathoadrenal progenitor cell line in which neuronal differentiation is initiated by FGF but not NGF. Neuron 4: 189-201.

Birren SJ, Lo LC and Anderson DJ (1993): Sympathetic neurons undergo a developmental switch in trophic dependence. Development 119: 597-610.

Birren SJ, Verdi J and Anderson DJ (1992): Membrane depolarization induces p140^{trk} and NGF-responsiveness, but not p75^{LNGFR}, in MAH cells. Science 257: 395-397.

Bronner-Fraser M and Fraser S (1988): Cell lineage analysis shows multipotentiality of some avian neural crest cells. Nature 335: 161-164.

Bronner-Fraser M and Fraser S (1989): Developmental potential of avian trunk neural crest cells in situ. Neuron 3: 755-766.

Buchman VL and Davies AM (1993): Different neurotrophins are expressed and act in a developmental sequence to promote the survival of embryonic sensory neurons. Development 118: 989-1001.

Carnahan JF and Patterson PH (1991): Isolation of the progenitor cells of the sympathoadrenal lineage from embryonic sympathetic ganglia with the SA monoclonal antibodies. J Neurosci 11: 3520-3530.

Claude P, Parada IM, Gordon KA, D'Amore PA and Wagner JA (1988): Acidic fibroblast growth factor stimulates adrenal chromaffin cells to proliferate and to extend neurites, but is not a long-term survival factor. Neuron 1: 783-790.

Cordon-Cardo C, Tapley P, Jing S, Nanduri V, O'Rourke E, Lamballe F, Kovary K, Klein R, Jones KR, Reichardt LF and Barbacid M (1991): The *trk* tyrosine protein kinase mediates the mitogenic properties of nerve growth factor and neurotrophin-3. Cell 66: 173-183.

Coughlin MD and Collins MD (1985): Nerve growth factor-independent development of embryonic mouse sympathetic neurons in dissociated cell culture. Dev Biol 110: 392-401.

Cowan WM, Fawcett JW, O'Leary DDM and Stanfield BB (1984): Regressive events in neurogenesis. Science 225: 1258-1265.

Davies AM, Bandtlow C, Heumann R, Korsching S, Rohrer H and Thoenen H (1987): The site and timing of nerve growth factor (NGF) synthesis in developing skin in relation to its innervation by sensory neurons and their expression of NGF receptors. Nature 326: 353-358.

Dechant G, Rodríguez-Tébar A, Kolbeck R and Barde Y-A (1993): Specific high-affinity receptors for Neurotrophin-3 on sympathetic neurons. J Neurosci 13: 2610-2616.

DiCicco-Bloom E, Friedman WJ and Black IB (1993): NT-3 stimulates sympathetic neuroblast proliferation by promoting precursor survival. Neuron 11: 1101-1111.

Doupe AJ, Landis SC and Patterson PH (1985): Environmental influences in the development of neural crest derivatives: glucocorticoids, growth factors and chromaffin cell plasticity. J Neurosci 5: 2119-2142.

Ernsberger U, Edgar D and Rohrer H (1989): The survival of early chick sympathetic neurons in vitro is dependent on a suitable substrate but independent of NGF. Dev Biol 135: 250-262.

Fariñas I, Jones KR, Backus C, Wang X-Y and Reichardt LF (1994): Severe sensory and sympathetic deficits in mice lacking neurotrophin-3. Nature 369: 658-661.

Gensburger C, Labourdette G and Sensenbrenner M (1987): Brain basic fibroblast growth factor stimulates the proliferation of rat neuronal precursor cells in vitro. FEBS Lett 217: 1-5.

Gospodarowicz D (1975): Purification of fibroblast growth factor from bovine pituitary. J Biol Chem 250: 2515-2520.

Hatten ME, Lynch M, Rydel RE, Sanchez J, Joseph-Slverstein J, Moscatelli D and Rifkin DB (1988).: In vitro neurite extension by granule neurons is dependent upon astroglial-derived fibroblast growth factor. Dev Biol 125: 280-289.

Hempstead BL, Martin-Zanca D, Kaplan DR, Parada LF and Chao MV (1991): High-affinity NGF binding requires coexpression of the *trk* proto-oncogene and the low-affinity NGF receptor. Nature 350: 678-683.

Hempstead BL, Schleifer LS and Chao MV (1989): Expression of functional nerve growth factor receptors after gene transfer. Science 243: 373-375.

Ibáñez CF, Ebendal T, Barbany G, Murray-Rust J, Blundell TL and Persson H. (1992): Disruption of the low affinity receptor-binding site in NGF allows neuronal survival and differentiation by binding to the *trk* gene product. Cell 69: 329-341.

Ip NY, Boulton TG, Li Y, Verdi JM, Birren SJ, Anderson DJ and Yancopoulos GD (1994): CNTF, FGF and NGF collaborate to drive the terminal differentiation of MAH cells into post-mitotic neurons. Neuron 13: 443-455.

Jing S, Tapley P and Barbacid M (1992). Nerve Growth Factor mediates signal transduction through trk homodimer receptors. Neuron 9: 1067-1079.

Johnson D, Lanahan A, Buck CR, Sehgal A, Morgan C, Mercer E, Bothwell M and Chao MV (1986): Expression and structure of the human NGF receptor. Cell 47: 545-554.

Kalcheim C and Neufeld G (1990): Expression of basic fibroblast growth factor in the nervous system of early avian embryos. Development 109: 203-215.

Kaplan DR, Hempstead BL, Martin-Zanca D, Chao MV and Parada LF (1991): The *trk* proto-oncogene product: a signal transducing receptor for nerve growth factor. Science 252: 554-558.

Klein R, Jing S, Nanduri V, O'Rourke E and Barbacid M (1991): The *trk* proto-oncogene encodes a receptor for nerve growth factor. Cell 65: 189-197.

Koike T, Martin DP and Johnson EM Jr. (1989): Role of Ca^{2+} channels in the ability of membrane depolarization to prevent neuronal death induced by trophic-factor deprivation: Evidence that levels of internal Ca^{2+} determine nerve growth factor dependence of sympathetic ganglion cells. Proc Natl Acad Sci USA 86: 6421-6425.

Korsching S and Thoenen H (1988): Developmental changes of nerve growth factor levels in sympathetic ganglia and their target organs. Dev Biol 126: 40-46.

Kotzbauer PT, Lampe A, Estus S, Milbrandt J and Johnson EM (1994): Postnatal development of survival responsiveness in rat sympathetic neurons to leukemia inhibitory factor and ciliary neurotrophic factor. Neuron 12: 763-773.

Le Douarin NM (1980): The ontogeny of the neural crest in avian embryo chimeras. Nature 286: 663-669.

Le Douarin NM (1982): "The Neural Crest". Cambridge University Press. Cambridge.

Le Douarin NM, Ziller C and Couly GF (1993): Patterning of neural crest derivatives in the avian embryo: *in vivo* and *in vitro* studies. Develop Biol 159: 24-49.

Levi-Montalcini R and Angeletti PU (1963): Essential role of the nerve growth factor in the survival and maintenance of dissociated sensory and sympathetic embryonic nerve cells *in vitro*. Develop Biol 7: 653-659.

Levi-Montalcini R and Booker B (1960): Destruction of the sympathetic ganglia in mammals by an antiserum to a nerve growth protein. Proc Natl Acad Sci USA 46: 384-391.

Meakin SO and Shooter EM (1992): The nerve growth factor family of receptors. Trends Neurosci 15: 323-331.

Michelsohn A and Anderson D J (1992): Changes in competence determine the timing of two sequential glucocorticoid effects on sympathoadrenal progenitors. Neuron 8: 589-604.

Morrison RS, Sharma A, de Vellis J and Bradshow RA (1986): Basic fibroblast growth factor supports the survival of cerebral cortical neurons in primary culture. Proc Natl Acad Sci USA 83: 7537-7541.

Patterson PH (1990): Control of cell fate in a vertebrate neurogenic lineage. Cell 62: 1035-1038.

Radeke MJ, Misko TP, Hsu C, Herzenberg LA and Shooter EM (1987): Gene transfer and molecular cloning of the rat nerve growth factor receptor. Nature 325: 593-597.

Raff MC (1992): Social controls on cell survival and cell death. Nature 356: 397-400.

Schecterson LC and Bothwell M (1992): Novel roles for neurotrophins are suggested by BDNF and NT-3 mRNA expression in developing neurons. Neuron 9: 449-463.

Segal RA, Takahashi H and McKay RDG (1992): Changes in neurotrophin responsiveness during the development of cerebellar granule neurons. Neuron 9: 1041-1052.

Seidl K and Unsicker K (1989a): Survival and neuritic growth of sympathoadrenal (chromaffin) precursor cells in vitro. Int J Devl Neuroscience 7: 465-473.

Seidl K and Unsicker K (1989b): The determination of the adrenal medullary cell fate during embryogenesis. Devel Biol 136: 481-490.

Stemple DL, Mahanthappa NK and Anderson DJ (1988): Basic FGF induces neuronal differentiation, cell division, and NGF dependence in chromaffin cells: a sequence of events in sympathetic development. Neuron 1: 517-525.

Unsicker K, Reichert-Preibsch H, Schmidt R, Pettmann B, Labourdette G and Sensenbrenner M (1987): Astroglial and fibroblast growth factors have neurotrophic functions for cultured peripheral and central nervous system neurons. Proc Natl Acad Sci USA 84: 5459-5463.

Unsicker K, Drisch B, Otten J and Thoenen H (1978): Nerve growth factor-induced fiber outgrowth from isolated rat adrenal chromaffin cells: impairment by glucocorticoids. Proc Natl Acad Sci USA 75: 3498-3502.

Verdi JM, Birren SJ, Ibáñez CF, Persson H, Kaplan DR, Benedetti M, Chao MV and Anderson DJ (1994a): p75^{LNGFR} regulates trk signal transduction and NGF-induced neuronal differentiation in MAH cells. Neuron 12: 733-745.

Verdi JM, Yancopoulos GD, Ip NY and Anderson DJ (1994b): Expression of p140trk is sufficient, but not limiting for NGF-induced differentiation of MAH cells to post-mitotic neurons. Proc Natl Acad Sci USA 91: 3949-3953.

Vogel KS and Davies AM (1991): The duration of neurotrophic factor independence in early sensory neurons is matched to the time course of target field innervation. Neuron 7: 819-830.

Walicke P, Cowan MW, Ueno N, Baird A and Guillemin R. (1986): Fibroblast growth factor promotes survival of dissociated hippocampal neurons and enhances neurite extension. Proc Natl Acad Sci USA 83: 3012-3016.

Wyatt S and Davies AM (1993): Regulation of expression of mRNAs encoding the nerve growth factor receptors p75 and trkA in developing sensory neurons. Development 119: 635-647.

SPECIFICATION OF CELL FATE IN THE VERTEBRATE RETINA

C.P. Austin and C.L. Cepko

Howard Hughes Medical Institute and Department of
Genetics, Harvard Medical School, 200 Longwood
Avenue, Boston, MA 02115, U.S.A.

INTRODUCTION

The strategies and mechanisms that govern commitment and differentiation in the vertebrate central nervous system (CNS) remain largely unknown. The retina is an attractive part of the CNS in which to investigate these questions. It is made up of a small number of cell types organized into laminae; each major cell type has a relatively well-defined morphology and can be identified using antibodies. Progenitor cells in retinal explants divide and produce progeny that can commit and differentiate in vitro, in the absence of signals from other CNS locations. Studies of cell lineage, proliferation, and differentiation have begun to lend insight into the developmental mechanisms which bring about retinal cell specification.

MANY RETINAL PROGENITORS ARE MULTIPOTENT

An initial question concerned whether dividing progenitors in the early retina were dedicated to producing only a single cell type. This issue was addressed in our laboratory and others using retroviral vectors (Turner and Cepko, 1987; Turner et al., 1990) or tracers such as horseradish peroxidase or fluorescent dextran (Holt et al., 1988; Wetts and Fraser, 1988). The conclusion of each of these studies was similar, that many retinal precursors are multipotent, and many remain so throughout development. After retroviral infection of the mouse retina at embryonic time points, clone size and cell type composition varied widely, indicating variable mitotic and differentiative behavior of progenitors. Clones contained from 1-234 cells, and from 1-6 cell types in various combinations (Turner et al., 1990). After infection of rat retina at postnatal time points, clones of up to 3 cell types were observed, in keeping with the cell types being born at this time; 2 cell clones could comprise two different cell types. This argued that retinal progenitors may remain multipotent up to their last cell division (Turner and Cepko, 1987).

Neural Cell Specification: Molecular Mechanisms and Neurotherapeutic Implications
Edited by Juurlink *et al.*, Plenum Press, New York, 1995

The variability in the numbers and cell types produced by single infected progenitors in these experiments suggested that cell fate specification in the retina was not solely lineage-based, and led to a search for environmental factors that might influence the number or types of cells that a progenitor might produce.

CONTROL OF PROLIFERATION OF RETINAL PROGENITORS

At the time the optic cup is formed as an outgrowth of the developing diencephalon, all progenitors are mitotic. Growth of the retina is therefore initially exponential, but as cells gradually withdraw from the cell cycle and differentiate, growth slows (Dutting et al., 1983). The factors that determine when during development a cell withdraws from the cell cycle, and whether these factors simultaneously influence cell type choice, has been the subject of recent study.

The point at which a progenitor stops proliferating and differentiates may reflect two independent decisions concerning mitosis and differentiation, or may be two aspects of the same decision. It may be that the progenitor is competent to respond to determinative signals only once it has made a decision to leave, or has already left, the cell cycle; in this case, the cell's fate would be determined only at or after the terminal cell division. Alternatively, it may be that the determinative influences simultaneously signal the progenitor to withdraw from the cell cycle. These factors may be intrinsic or extrinsic to the cell. Watanabe and Raff (1990) showed that rat retinal progenitors from embryonic day 15 were intrinsically more mitotically active than progenitors from postnatal day 1; their mitotic behavior was not affected by mixing with an excess of progenitors from a different age. This suggested that the proliferative potential of progenitors changes during development, perhaps due to changes in their response to mitogens in the environment. Anchan et al. (1991) found that the peptide growth factors EGF and TGFα were mitogens for cultures of rat retinal cells. Lillien and Cepko (1992) demonstrated that responsiveness of retinal cultures to TGFα, as well as aFGF and bFGF, changes over the course of retinal neurogenesis. The mitotic response to bFGF was 100-fold greater in cultures initiated early in development compared to late, and the response to TGFα was 100-fold greater late in development than early. In addition, TGFα and EGF blocked production and/or differentiation of rods, whereas it had no effect on the generation of other cell types born at the same time (Lillien and Cepko, 1992). Whether these growth factors act via a single mechanism to both induce proliferation and block differentiation is not clear. However, other factors must be involved, because cells ceased cycling after a few days in explant culture despite the continued presence of the mitogens. Negative regulators of mitosis, as well as intrinsic factors, may contribute to the limitation of cell division during development.

CONTROL OF CELL TYPE SPECIFICATION

The choice of cell fate, like the decision of when to exit the cell cycle, could be influenced by factors intrinsic to the progenitor, factors in its environment, or both. The intrinsic pathway may lead the cell into what has been termed a "default" state; that is, the cell type that is produced if no environmental signals are present, or have been eliminated. Such a pathway may be followed only rarely in retinal development, as there are probably environmental signals present throughout development. The one possible exception is the generation of ganglion cells, as they are the first-born cell type in every vertebrate species examined (reviewed in Altshuler et al., 1991), and thus arise from a uniform undifferentiated neuroepithelium which may lack differentiative signals.

The cell type that represents the "default" state during retinal development has been the subject of some investigation, using dissociated cultures to minimize cell-cell interactions. Reh and Kljavin (1989) dissociated rat retinal cells at E14, the peak of ganglion cell genesis, and found that 50% differentiated as ganglion cells after 4 days in sparse monolayer culture. This suggested that the default state of early rat retinal progenitors is ganglion cell differentiation. Adler and Hatlee (1989) dissociated chick retinal cells at E6, during ganglion cell, horizontal, and early photoreceptor genesis; and E8, during photoreceptor, bipolar, and Muller glia genesis (Prada et al., 1991). They found that of the cells that differentiated during 4 days in culture, the majority were photoreceptors after dissociation at E6, but "multipolar neurons" after dissociation at E8. These authors concluded that photoreceptor differentiation appears to be an early default pathway for chick retinal progenitors. These results may be reconciled if the default state of the progenitor changes during development; intrinsic and extrinsic cues may be envisioned to determine what the default state is and how many of the progenitors would differentiate as this default cell type.

A combinatorial model for the generation of cell identity in the retina has been discussed by both Reh and Kljavin (1989) and Cepko (1993). Such a scheme is similar to that proposed for the generation of cellular diversity in the Drosophila retina (Cagan, 1993). Retinal progenitors are envisioned to change in competence as development proceeds, and the "default" cell type seen in dissociated culture might correlate with this competence. Positively and negatively acting factors in the environment would then act on the progenitors to determine how many of the progenitors differentiate as the default cell type, and how many stay mitotic or differentiate into another cell type. Both progenitor competence and default state as well as the environmental differentiation signals are envisioned to change over time as each cell type is produced during retinal histogenesis.

The regulation of the hypothesized environmental differentiative signals has been the subject of recent investigation. Watanabe and Raff (1990) made aggregates of retinal cells of different developmental ages, and found that at each developmental age cells differentially influenced the production of rods in the other. E15 cells cultured in the presence of a 50-fold excess of P1 cells formed rods at a much higher rate than if the E15 cells were cultured alone, but at a much lower rate than P1 cells cultured alone. These results suggested that the E15 progenitors were limited in their competence to produce rods, but that rod differentiation could be affected by environmental factors. In addition, P1 cells cultured in the presence of a 50-fold excess of E15 cells made rods at a rate appreciably lower than P1 cells cultured alone. This argued for presence of an inhibitor of rod development, or the lack of a rod inducer, at E15.

Altshuler and Cepko (1992) used a dissociated culture system to minimize cell-cell interactions and identify factors required for rod differentiation. It was found that rat P0 progenitors cultured at a high density in collagen gels formed rods at the same rate as they do in vivo (25-30% of all cells after 7 days in culture). However, decreasing the density at which the cells were cultured by four-fold decreased the generation of rods in vitro to less than 1% in the same time interval. Culturing low-density cells in the presence either of cells at high density, or conditioned medium made from retinal explants, resulted in low-density cells producing rods in virtually the same numbers as the high density cells did. These results suggested the presence of a diffusible inducer of rod production. Co-cultures of P0 cells with cells from retinas of different ages showed that this inducing activity was only present during the period of rod generation in vivo.

Watanabe and Raff (1992) have obtained similar results using a reaggregate culture system, and have shown that the diffusible activity induces rod differentiation, but not amacrine cell differentiation.

Identification of some of the diffusible activities implicated in these studies has been accomplished using biochemical and candidate molecule approaches. Fractionation of conditioned medium demonstrated the presence of a high molecular weight inhibitory activity and a low molecular weight stimulatory activity (Altshuler et al., 1993). Further characterization revealed part of the stimulatory activity to be attributable to taurine, an amino acid previously known to be present in high levels in the retina and required for normal CNS development. Furthermore, incubation of retinal explants with a taurine antagonist, amino-methyl-sulphonic acid, decreased rod production (Altshuler et al., 1993). The identity of the remaining stimulatory activity, and the inhibitory activity present in conditioned medium are currently under investigation.

CONCLUSIONS

The basic principles underlying cell fate specification in the vertebrate retina are beginning to be understood, and some of the factors involved in the implementation of these principles are being identified. Cell lineage studies have shown that most retinal precursors are multipotent throughout development. Studies of progenitor proliferation have identified EGF, TGFα, aFGF, and bFGF as mitogens for retinal progenitors, and the latter three factors influence rod development as well. In vitro studies of rod generation have shown that progenitors change in their intrinsic competence to generate rods during development, and that environmental signals, of which taurine is one, change in their level of activity during development as well. Recent studies of retinal ganglion cells have demonstrated that the dissociated culture technique is useful for identifying basic principles and factors operative in the development of other retinal cell types (C.P. Austin and C.L.C., unpublished). These in vitro approaches hold promise in identifying further mitotic and differentiative factors that can then be tested in vivo for activity. Through this strategy, the mechanisms that regulate progenitor cell specification in this area of the vertebrate CNS may be determined.

ACKNOWLEDGEMENTS

Supported by grants K11EY00321 (C.P.A.) and RO1EY09676 (C.L.C.) from the National Eye Institute.

REFERENCES

Adler R and Hatlee M (1989): Plasticity and differentiation of embryonic retinal cells after terminal mitosis. Science 243: 391-393.

Altshuler D, Turner DL and Cepko CL (1991): Specification of cell type in the vertebrate retina. In Lam DM-K and Shatz CJ (eds): "Development of the Visual System" Cambridge: MIT Press, pp. 37-58.

Altshuler D and Cepko C (1992): A temporally regulated, diffusable activity is required for rod photoreceptor development in vitro. Development 114: 947-957.

Altshuler D, Lo Turco JJ, Rush J and Cepko C (1993): Taurine promotes the differentiation of a vertebrate retinal cell type in vitro. Development 119: 1317-1328.

Anchan RM, Reh TA, Angello J, Balliet A, and Walker M (1991): EGF and TGF-α stimulate retinal neuroepithelial cell proliferation in vitro. Neuron 6: 923-936.

Cagan R (1993): Cell fate specification in the developing Drosophila retina. Development (Suppl): 19-28.

Cepko C (1993): Retinal cell fate determination. Prog Retina Res 12: 1-12.

Dutting D, Gierer A and Hansmann G (1983): Self-renewal of stem cells and differentiation of nerve cells in the developing chick retina. Dev Brain Res 10: 21-32.

Holt CE, Bertsch TW, Ellis HM and Harris WA (1988): Cellular determination in the Xenopus retina is independent of lineage and birth date. Neuron 1: 15-26.

Lillien L and Cepko C (1992): Control of proliferation in the retina: temporal changes in responsiveness to FGF and TGFα. Development 115: 253-266.

Reh TA and Kljavin IJ (1989): Age of differentiation determines rat retinal germinal cell phenotype: Induction of differentiation by dissociation. J Neurosci 9: 4179-4189.

Turner DL and Cepko CL (1987): A common progenitor for neurons and glia persists in rat retina late in development. Nature 328: 131-136.

Turner DL, Snyder EY and Cepko CL (1990): Lineage-independent determination of cell type in the embryonic mouse retina. Neuron 4: 833-845.

Watanabe T and Raff MC (1990): Rod photoreceptor development in vitro: intrinsic properties of proliferating neuroepithelial cells change as development proceeds in the rat retina. Neuron 4: 461-467.

Watanabe T and Raff MC (1992): Diffusible rod-promoting signals in the developing rod retina. Development 114: 899-906.

Wetts R and Fraser SE (1988): Multipotent precursors can give rise to all major cell types in the frog retina. Science 239: 1142-1145.

NEUROTROPHINS AND TRK RECEPTORS IN HIPPOCAMPAL DEVELOPMENT

Diana Collazo and Ron McKay

LMB/NINDS, Building 36, Room 3D02, NIH, Bethesda, MD 20892, U.S.A.

The mammalian nervous system is composed of a large number of differentiated neuronal and glial cell types, which develop from precursor cells during the early development of the brain. Many aspects of the transition of proliferating precursors into differentiated cell types remain elusive because of the complexity of the brain. One way to reduce this complexity is to use in vitro systems to study these developmental steps in a simplified context. This chapter briefly reviews the role of neurotrophins in hippocampal development. The reason for studying the hippocampus is primarily because of its involvement in learning and memory processes as described below. The neurotrophin gene family was chosen because of the widespread distribution of the neurotrophins and their high affinity receptors in the developing and adult hippocampus. This distribution not only suggested previously unidentified hippocampal targets, but it also raised the possibility of autocrine or paracrine as well as target-derived mechanisms of trophic support.

GENERAL FEATURES OF CENTRAL NERVOUS SYSTEM DEVELOPMENT

The mature nervous system is composed of a complex array of neuronal and glial cells. This complex pattern arises from an apparently homogeneous monolayer of cells called the neuroepithelium or neuroectoderm. At the gastrula stage, the mesoderm induces the overlying ectoderm to become the neuroectoderm (Spemann, 1938). After a series of changes in cell shape, the neuroepithelium rolls up into a tube known as the neural tube. A group of cells known as the neural crest detach from the apical neural tube and migrate away to form the sensory and autonomic ganglia of the peripheral nervous system (PNS). Following neural tube closure, a second inductive event occurs from a second tube of cells of mesodermal origin, the notochord, which underlies the neural tube. A signal from

Neural Cell Specification: Molecular Mechanisms and Neurotherapeutic Implications
Edited by Juurlink *et al.*, Plenum Press, New York, 1995

the notochord induces cells in the neural tube to form the floor plate. Floor plate cells secrete a diffusible factor that has chemotropic effects on a group of spinal cord axons, the commissural axons (Hirano et al., 1991; Tessier-Lavigne et al., 1988). During this time, the central nervous system (CNS) becomes divided rostrocaudally into forebrain, midbrain, hindbrain and spinal cord. Following these steps, there is an ordered sequence of proliferation and fate determination events in the germinal layer of the neuroepithelium, the ventricular zone, that produce the major cell types in the CNS: astrocytes, oligodendrocytes and neurons.

The production of differentiated cells from the neuroepithelium depends on complex interactions between intrinsic and extrinsic determinants (reviewed in (Hayes and McKay, 1992; 1993). Intrinsic determinants can be defined as cell autonomous; meaning that once a particular genetic program is established either by lineage or environment, cells do not depend on extracellular signals for their fate choice. Extrinsic determinants involve environmental cues such as components of the extracellular matrix, cell adhesion molecules, and soluble peptide growth factors (Lander, 1987; Walicke, 1988). These signals are transduced by receptors on the surface of precursor cells, regulating their proliferation and fate determination. The frequently used term, commitment, refers to a point in which a fate of a cell no longer depends on extrinsic signals, but it is determined by an intrinsic cellular program.

Fate mapping studies support the view that extrinsic signals control the production of differentiated neurons from multipotential stem cells (Galileo et al., 1990; Gray et al., 1988; Holt et al., 1988; Price and Thurlow, 1988; Renfranz et al., 1991; Temple, 1989; Turner and Cepko, 1987; Turner et al., 1990; Wetts and Fraser, 1988). These extrinsic signals are precisely regulated, spatially and temporally, giving rise to successive waves of postmitotic neurons. Transplantation studies show that the commitment of neuronal precursors to specific laminar fate in the visual cortex is determined by the environment, and occurs close to the last cell cycle, in the G2 or M phase, prior to the migration of the cells away from the ventricular zone (McConnell, 1988; McConnell and Kaznowski, 1991). After these events, young postmitotic neurons migrate to their appropriate layers in the brain, extending axons and dendrites. Initially, neurons are produced in excess; the final number of neurons is determined after the arrival of axons at their targets during a period known as naturally occuring cell death (Barde, 1989). During this time, the surviving neurons are those that compete successfully for retrogradely transported trophic factors supplied in limited quantities by the target fields. If excess neurotrophic factors are experimentally introduced to the developing PNS, neurons that ordinarily would have died are rescued (Hamburger et al., 1981). Conversely, if the concentration of factors is reduced (for example by using blocking antibodies), then the neurons that would have survived die (Rohrer et al., 1988). Trophic factors also play a role in the maintenance of the adult nervous system by modulating neuronal properties such as neurotransmitter and peptidergic phenotype, ion channel expression, and neuronal architecture (Purves, 1988).

HIPPOCAMPAL MORPHOLOGY, DEVELOPMENT AND FUNCTION

Hippocampal Morphology And Connectivity

The potential involvement of the hippocampus in learning and memory has made this brain region one of the most extensively characterized areas of the mammalian brain. The study of hippocampal development is important because early developmental events such as fate choice, migration and synapse formation among others are critical in the establishment of a functional hippocampus. The hippocampal formation is composed of a series of areas or subfields located at the medial edge of the cortical mantle. The morphology and physiology of the hippocampus have been described in detail by various authors (Lorente de Nó, 1934; Ramón y Cajal, 1990; Witter, 1989). A distinct feature of the hippocampal formation is the lamellar distribution of its structures in which each lamella is oriented more or less perpendicular to the longitudinal axis of the hippocampus. In mammals, the hippocampus consists of two C-shaped, interlocking principal neuronal subfields: the pyramidal cell layer of Ammon's horn and the granular cell layer of the dentate gyrus. Based on distinct morphological criteria in Golgi preparations, several authors have subdivided the Ammon's horn into discrete regions. A widely used nomenclature was introduced by Lorente de Nó (1934) which divides the Ammon's horn into four areas, CA (cornu Ammonis) 1 to CA4. Subsequent studies have defined further these field borders on the basis of synaptic connectivity and neurochemical markers (Bayer, 1980; Blackstad, 1956; Swanson and Cowan, 1977). The dentate gyrus consists of a densely packed cell layer of granule cells. This region encloses the hilar region which contains the polymorphic or infragranular layer of the dentate gyrus, and the proximal part of the pyramidal cell layer of CA3. The granule cell layer can be subdivided in relation to its location to the CA3 pyramidal cells into a suprapyramidal and an infrapyramidal blade, which merge at the crest of the dentate gyrus. The architecture of the principal neuronal subfields is very similar, consisting of a single layer of neurons in which the apical dendrites project to cell-poor zones: the stratum moleculare in the dentate gyrus and the subiculum, the stratum lacunosum-moleculare and stratum radiatum in the Ammon's horn. Interneurons are found in all hippocampal subfields forming very important inhibitory connections with the excitatory pyramidal and granular cells (Brodal, 1978; Witter, 1989).

An important feature of the hippocampal connectivity is its trisynaptic fiber pathway containing a sequence of almost completely unidirectional connections from the dentate gyrus to the subiculum, via CA3 and CA1 (Witter et al., 1989). The main excitatory afferent projections to the hippocampus are the perforant path, which projects from the entorhinal cortex to the granule cells. Then, the dentate granule axons, known as the mossy fiber tract, project to CA3. Finally, the Schaffer collateral-commisural system connects ipsilateral and contralateral CA3 to CA1 cells. Excitation leaves the hippocampus via pyramidal axons that form a bundle of fibers known as the fimbria. These fibers project to the thalamic nucleus, the preoptic region, the lateral hypothalamus and the septum, among other brain regions (Brodal, 1978).

In addition to the major entorhinal afferents, other hippocampal inputs are made by the septal nucleus, the hypothalamus, anterior thalamic nucleus, and the brainstem. Some of these inputs have been carefully studied because of their potential clinical roles. In particular, the cholinergic fibers from the septal nuclei, which project to all hippocampal subfields, have been shown to undergo profound neurodegeneration in conditions such as senile dementia of the Alzheimer type (SDAT). The known trophic effects of the neurotrophins on septal cholinergic neurons in vitro and in vivo have led to the hypothesis that these growth factors could counteract the cholinergic deficits of SDAT (Bartus et al., 1982; Hefti, 1986).

Hippocampal Development

Neurogenesis in the hippocampus begins during embryonic life and continues postnatally. Among the earliest born hippocampal neurons are the inhibitory, GABAergic cells, which become postmitotic during embryonic days 13 through 17 (E13-17) (Amaral and Kurz, 1985; Rozenberg et al., 1989). Detailed studies on the neurogenesis, migration and settling of the pyramidal and granular cells have divided the hippocampal neuroepithelium into three components: The Ammonic neuroepithelium, the primary dentate neuroepithelium, and the fimbrial glioepithelium (Altman and Bayer, 1990a; 1990b; 1990c). These components show distinct locations, morphologies, and times of origin. The Ammonic neuroepithelium, which is located adjacent to the subiculum, shows high level of proliferative activity between E16 and E19. This structure gives rise to the neurons in the stratum radiatum and stratum oriens (E16-17), and the pyramidal neurons (E17-20). The primary dentate epithelium located around a ventricular indentation, the dentate notch, contains the granule cell precursors. These precursors continue to proliferate after leaving the neuroepithelium to form several germinal matrices: the primary, secondary and tertiary dentate matrices. These matrices show distinct patterns of migration and timecourse of proliferative activities, which account for the perinatal and postnatal production of granule cells (E17 onward) (Altman and Das, 1965; Bayer, 1980). The tertiary dentate matrix (P3-10) is postulated to produce the late postnatal granule cell population. Finally, the fimbrial glioepithelium produces the glial cells of the fimbria, and it is present by E16 at the tip of the hippocampal rudiment.

Retroviral marking experiments of the embryonic hippocampal formation have shown clones of pyramidal neurons which transgress the boundaries between subfields CA1-4, suggesting that the subdivision of hippocampal subfields does not result from a strict program present in the ventricular zone as previously proposed in the neocortex (Grove et al., 1992; Rakic, 1988). These results are also consistent with a model in which the same hippocampal precursor may respond to its environment to generate neurons with distinct morphological and functional properties. Finally, the migration of neuronal clones in the hippocampus from the ventricular zone showed a restricted radial array, distinct from the widespread neuronal migration seen in other cortical areas (Austin and Cepko, 1990; Grove, et al., 1992).

Hippocampal Function

Long term potentiation (LTP) in the CA1 region of the hippocampus is the best understood form of activity-dependent changes in synaptic transmission in the mammalian brain. LTP is a long lasting increase in the efficiency of synaptic transmission in response to high frequency stimulation. This synaptic potentiation has been observed to last several days in vivo (Bliss and Collingridge, 1993). Although originally identified in the dentate gyrus, LTP has been found in all excitatory pathways in the hippocampus, as well as in other brain regions, and it has been proposed to underlie certain forms of memory (Bliss and Lomo, 1973; Doyere and Laroche, 1992; Morris et al., 1990). There are two principal categories of synaptic potentiation divided on the basis of whether or not induction is blocked by antagonists of a subtype of the ionotropic, glutamate receptors known as the N-methyl-D-aspartate (NMDA) receptors. NMDA-mediated LTP is initiated postsynaptically, but a strengthening of the presynaptic cell during LTP was revealed by quantal analysis [the study of the changes in the probability of neurotransmitter released by the presynaptic cell] (Bekkers and Stevens, 1990; Malinow and Tsien, 1990). A subsequent study showed that LTP results in an increase in the amplitude of the postsynaptic miniature currents, which probably results from enhanced postsynaptic sensitivity (Manabe et al., 1992). These findings support the view that the induction and maintenance of LTP may require an interplay between pre- and post-synaptic components.

The inductive phase of LTP is characterized by three basic properties: cooperativity, associativity and input-specificity (reviewed in Bliss and Collingridge, 1993). These properties are elegantly explained by the properties of the NMDA receptor, in particular the voltage-dependent block of the NMDA receptor channel by magnesium ions [Mg^{++}] (Ascher and Nowak, 1988). The activation of this receptor requires two converging signals: A sufficient membrane depolarization to expel Mg^{++} from the channel at the same time that L-glutamate opens the channel by binding to the NMDA receptor. Cooperativity is explained by a threshold level of depolarization required to reduce the Mg^{++} block of the channel. This depolarization can be provided by a synchronious stimulation of the same or different excitatory afferents, explaining the associative phase. The input-specificity requirement is provided by the need of presynaptic terminals to provide a sufficient glutamate concentration to activate the NMDA receptors. Once the NMDA channel opens, calcium ions flow into the target cell.

This model explains in part the inductive phase of LTP in the postsynaptic cell. However many aspects of the induction and maintenance of this phenomenon remain unknown. For instance, recent work in the hippocampus has shown that NMDA receptor stimulation does not induce LTP exclusively, but may lead to other forms of synaptic plasticity such as short term potentiation (STP) and long term depression [LTD] (Ascher and Nowak, 1988). The mechanistic relationship between these phenomena appears to be a complex function of the magnitude, duration and history of NMDA receptor activity. Moreover, pharmacological studies using agonists and antagonists of metabotropic glutamate receptors have shown that the activation of these receptors is necessary for the induction of NMDA-dependent and -independent forms of LTP (Bashir et al., 1993). Another critical component of NMDA-induced LTP is the elevation of the postsynaptic calcium (Ca^{++}) concentration. Early studies injecting Ca^{++} chelators in the postsynaptic cell blocked the induction of LTP. These Ca^{++} increases may involve a combination of extracellular sources (entry through NMDA channels and voltage-gated Ca^{++} channels), as well as release from intracellular stores. Changes in Ca^{++} levels may be critical in triggering the signal transduction pathways that are important for the maintenace of LTP. Several calcium-sensitive enzymes have been proposed to mediate some of these events such as proteases (calpain), phosphatases (calcineurin), phospholipases, protein serine kinases (Ca^{++}/phospholipid-protein kinase - PKC, Ca^{++}/calmodulin-dependent protein kinase - CaMKII, and cAMP-dependent protein kinase -PKA) (reviewed in Bliss and Collingridge, 1993). Knockout of the gene encoding α-CaMKII severely impairs the ability of slices to exhibit LTP (Silva et al., 1992a). Interestingly, these transgenic mice show a specific spatial memory deficit, correlating molecular observations with behavior (Silva et al., 1992b). The involvement of protein tyrosine kinases in LTP has been suggested on the basis of inhibitor studies, and it may be relevant that NMDA receptor activation leads to tyrosine phosphorylation of a serine threonine kinase, MAP-2 kinase (Bading and Greenberg, 1991). Other areas of intense study are changes in gene expression after NMDA receptor activation, the nature of the retrograde messenger that influences the presynaptic cell, and activity-induced synaptic remodelling of neural connections, among others. Finally, recent findings have shown the upregulation of neurotrophins and their receptors in activity-dependent processes in the CNS, implicating these factors in mediating some of these processes (see below).

NEUROTROPHINS AND THEIR HIGH AFFINITY RECEPTORS

Neurotrophin Family Members

Extrinsic signals have been shown to play critical roles in regulating developmental events such as fate choice, migration and survival that are necessary to produce a functional hippocampus. The widespread expression of the neurotrophins and their high affinity receptors in the developing and adult hippocampus has suggested their involvement in regulating some of these processes and in supporting the maintenance of this brain region in the adult. Nerve growth factor (NGF) is the prototypical neurotrophic factor in the peripheral and central nervous systems (Levi-Montalcini, 1987). Recently, other neurotrophic factors structurally similar to NGF have been cloned, defining a neurotrophin gene family. The reported members of this family are brain-derived neurotrophic factor (BDNF) (Barde et al., 1982; Leibrock et al., 1989), hippocampal-derived neurotrophic factor (HDNF) (Ernfors et al., 1990a) also known as neurotrophin-3 (NT-3) (Hohn et al., 1990; Jones and Reichardt, 1990; Maisonpierre et al., 1990b; Rosenthal et al., 1990) and NGF-2 (Kaisho et al., 1990), and neurotrophin-4 (NT-4). NT-4 was first cloned from *Xenopus laevis*, but a putative mammalian homologue was subsequently cloned and named neurotrophin-5 (NT-5) (Berkemeier et al., 1991; Hallbook and Persson, 1991). The high degree of structural homology between NT-4 and NT-5, the similarities in receptor specificities, and biological activities have led some groups to propose that the nomenclature should be changed to NT-4 (Ip et al., 1992).

NGF, BDNF, NT-3 and NT-4 are short (approximately 120 amino acids), basic (pI about 10) proteins that share about 50% sequence identity (Hohn et al., 1990; Maisonpierre et al., 1990). All four neurotrophins seem to be initially synthesized as longer precursors that are proteolytically cleaved to release the mature neurotrophin. Phylogenetic trees of the neurotrophins reveal that NT-4 is more structurally related to BDNF than to NGF and NT-3. NGF is more related to NT-3 than to the other two neurotrophins. The structure of NT-3 seems to be equally related to NGF, BDNF and NT-4 (Ibanez and Persson, 1993; Ibanez et al., 1993). Interestingly, these structural similarities are reflected in their specificities of receptor activation and biological activities, as discussed below.

Structure-Function Studies

Structural comparisons among NGF, BDNF and NT-3 show several regions of identical amino acids which seem to correspond to residues playing structural roles in these proteins. These identities include six cysteine residues, which have been shown to form disulfide bridges in the active NGF. In addition, the amino acids flanking these cysteine residues are also highly conserved (Maisonpierre et al., 1990b). The biologically active form of the neurotrophins is a dimer of two identical subunits. Recently, the crystal structure of the NGF dimer was resolved to a 2.3 A resolution. The NGF structure reveals a rod-shaped molecule consisting of three antiparallel pairs of β-strands, which together form a flat dimerizing surface (McDonald et al., 1991). Four exposed loop regions are present on the surface of the molecule. Interestingly, these loops are the least conserved among neurotrophins, suggesting that these regions may determine their biological specificities. Experimental evidence supporting this view has been obtained using site-directed mutagenesis. Variable sequences in NGF (regions III, IV and V) were replaced with the corresponding BDNF sequences, resulting in chimeric molecules with neurotrophic activities from both BDNF and NGF (Ibanez et al., 1991). In an additional study, lysine residues -32, 34 and 95 within the loop regions of NGF were shown to provide a positively charged interface that is critical for NGF binding to its low affinity receptor (Ibanez et al., 1992a). Additional work on mutated forms of the neurotrophins is required to provide a detailed view on the points of contact between these molecules and their receptors.

trk Family Members And Ligand Specificities

It has been shown that NGF interacts with receptors of both high and low affinities (kd = 10^{-11} and 10^{-9} M, respectively) present in a variety of neuronal and nonneuronal cells (Ernfors, et al., 1990a; 1990b; Hallbook, et al., 1991; Heuer et al., 1990; Yan and Johnson, 1988). The biological effects of NGF are associated primarily with high affinity binding, which constitutes about 10% of the total binding sites on NGF-responsive cells (Chao et al., 1986). The previously cloned low affinity NGF receptor encodes a transmembrane glycoprotein of 75,000 Da (p75NGFR) that lacks a kinase domain (Johnson et al., 1986; Radeke et al., 1987). p75NGFR belongs to a family of diverse cell surface proteins which share similar cysteine-rich repeats in the extracellular domains. These repeats have been found in a number of receptors on hematopoietic cells, including TNF receptors, Fas antigen, OX40, CD30 and CD40 (Durkop et al., 1991; Mallet and Barclay, 1991; Itoh et al., 1991; Durkop et al., 1992). This receptor has been shown to bind all neurotrophins with similar affinities [kd = 10^{-9} M] (Rodriguez-Tebar et al., 1992). The characteristics of p75NGFR do not explain the selective, high affinity neurotrophin binding, or the fact that the high affinity NGF complex was shown to contain phosphorylated tyrosine residues (Meakin and Shooter, 1991). Recently, the product of the *trk* proto-oncogene, a tyrosine kinase receptor of 140,000 Da (p140trk or TrkA) has been shown to constitute an essential component of the high affinity NGF receptor. This receptor shows rapid tyrosine phosphorylation in response to NGF in the rat pheochromocytoma cell line PC12 (Kaplan et al., 1991a; Kaplan et al., 1991b; Klein et al., 1991b).

The human *trk* (tropomyosin receptor kinase) locus was first identified as a highly transforming gene resulting from a somatic rearrangement event in a colon carcinoma. This rearrangement forms a constitutively active hybrid molecule, which contains the transmembrane and catalytic domains of the *trk* gene fused to the non-muscle tropomyosin gene (Martin-Zanca et al., 1986). Isolation and molecular characterization of the chimeric oncogene led to the identification of its normal allele, the *trk*A proto-oncogene (Martin-Zanca et al., 1989). Further screening of mammalian cDNA libraries led to at least two related loci designated *trkB* (gp145trkB or TrkB) and *trkC* (gp145trkC or TrkC) (Klein et al., 1990a; Lamballe et al., 1991). The *trk*A proto-oncogene is a prototype of a small subfamily of tyrosine protein kinase cell surface receptors with closer homology to other cell surface receptors than to src-like tyrosine kinases. Although the TrkA protein is not closely related to other tyrosine kinase receptors, it has a slightly higher homology to the insulin and IGF-1 receptors and the *c-ros* proto-oncogene than to other tyrosine kinase receptors (Martin-Zanca, et al., 1989). Trk receptors have a long signal peptide, a ligand binding domain rich in consensus N-glycosylation sites, a single transmembrane domain, a juxtamembrane domain (which bridges the transmembrane region with the catalytic domain), a split tyrosine kinase catalytic domain and a short carboxy-terminal tail (reviewed in Barbacid et al., 1991).

Trk receptors contain several unique structural motifs that may provide clues to their specificity of neurotrophin binding and function as outlined briefly. Two principal motifs are present in the extracellular domains of the Trk receptors. Three tandem leucine-rich repeats flanked by cysteine-rich motifs, which have been proposed to function in protein-protein interactions. These leucine rich motifs are followed by two motifs characteristic of the extracellular domains of the immunoglobulin superfamily (immunoglobulin type C2). The juxtamembrane domain contains a highly conserved hexapeptide (ENQYF), which bears some similarity to the sequence (NPXY) found in other tyrosine kinase receptors such as EGF, c-kit, c-fms and insulin receptors among others. Disruption of this sequence has been reported to affect ligand-induced receptor internalization of the insulin receptor. Finally, the tyrosine kinase domain is highly conserved among Trk receptors, showing the consensus GlyXGlyXXGlyX that forms part of the ATP-binding site, and other subdomains found in all receptor tyrosine kinases (subdomains VI, VII and VIII). The insert that breaks the conserved region in the kinase domain is considerably shorter than those found in the kinase insert sequences of the PDGF or CSF-1 receptors. These inserts have been implicated in specifying phosphorylation targets and substrates.

Both *trkB* and *trkC* are produced by loci with intricate transcription patterns. Multiple trancripts have been identified in rat for *trkB* and *trkC* (at least nine transcripts for *trk*B, four for *trk*C). Some of these transcripts encode the full-length protein, whereas others code a second class of non-catalytic variants. These variants have identical extracellular and transmembrane regions as the full-length receptor, but have short and distinct cytoplasmic extensions. Additional uncharacterized transcripts have also been reported containing the tyrosine kinase or extracellular domain only, raising the possibility that these transcripts may code for additional cytoplasmic kinase and/or secreted extracellular domains (Valenzuela et al., 1993). Recent reports have identified *trkC* isoforms containing variable-sized amino acid insertions within the tyrosine kinase domain that split the central core of this region between subdomains VII and VIII (Hanks, 1991; Hanks et al., 1988). The *trkC* isoforms (kinase insert 14 and 25 amino acids, ki-14 and -25) show distinct biological properties and signal transduction pathways from receptors lacking the inserts (see below) (Lamballe et al., 1994; Valenzuela, et al., 1993).

When transfected into fibroblast cell lines, the full-length Trk receptors are capable of high affinity neurotrophin binding, signal transduction via a tyrosine phosphorylation cascade, and mitogenic responses (Bothwell, 1991). The in vitro assays routinely used to study neurotrophins and their receptors are proliferation and survival responses in NIH 3T3 cells, and neuronal differentiation of PC12 cells. A summary of the results from these systems has led to the following consensus: NGF and BDNF are the most effective ligands for TrkA and TrkB, respectively, although both of these receptors bind NT-3 with lower affinities (Cordon-Cardo et al., 1991; Glass et al., 1991; Kaplan, et al., 1991a; Kaplan, et al., 1991b; Klein et al., 1991a; Soppet et al., 1991). There is not a one to one correspondence between ligands and receptors, since NT-4 has been shown to bind and activate TrkB with similar affinities as BDNF (Ip et al., 1993b). TrkC has been identified as the high affinity receptor for NT-3 (Lamballe et al., 1991). The precise function of the low affinity NGF receptor (p75NGFR) in conjunction with the Trks mediating these responses is controversial (Kaplan et al., 1991a; 1991b; Bothwell, 1991). Initial studies reported that high affinity binding required a complex of p140trk and p75NGFR (Hempstead et al., 1991). In addition, PC12 cells expressing a chimeric receptor consisting of the extracellular domain of the epidermal growth factor receptor and the cytoplasmic domain of the low affinity NGF receptor undergo neuronal-like differentiation in response to EGF (Yan et al., 1991). However, several reports have convincingly disputed the requirements of coexpression of p75NGFR with TrkA to mediate NGF responses as outlined below. It has been shown that p140trk in the absence of p75NGFR binds NGF with low and high affinities in fibroblasts, and that TrkA alone mediates NGF responses in fibroblasts, Xenopus oocytes, PC12 cells and the immortal sympathoadrenal progenitor cell line, MAH (Birren et al., 1992; Cordon-Cardo, et al., 1991; Loeb et al., 1991). Moreover, receptor crosslinking experiments in fibroblast cell lines showed the presence of both p75NGFR and TrkA homodimers, but no heterodimers between p75NGFR and TrkA (Jing et al., 1992). Finally, NGF mutations that disrupt p75NGFR binding, but retain binding to p140trk are biologically active, demonstrating a functional dissociation between these two receptors (Ibanez, et al., 1992a; 1992b).

Distribution Of Neurotrophins And trk Receptors In The Developing And Adult Nervous System

The initial expression of NT-3, BDNF and NGF mRNAs occurs between E11-E12, roughly coincident with the onset of neurogenesis in the nervous system (Maisonpierre et al., 1990b). NT-4 mRNA expression is maximal at E13 (the earliest time point reported), with reduced levels at later times of development (Timmusk et al., 1993a; 1993b). Studies on the distribution of NGF mRNA (1.3 kb) have shown that this factor is synthesized in the target-regions of NGF-dependent neurons in the PNS and CNS, with highest levels of expression in the brain, heart and spleen. NT-3 mRNA (1.4 kb) is broadly distributed in peripheral tissues and in the CNS (Hohn, et al., 1990; Maisonpierre, et al., 1990a). In contrast, BDNF transcripts (1.6 and 4 kb) display the most restricted pattern of expression with highest levels found in the brain, although significant expression was detected in heart, lung and muscle (Maisonpierre et al., 1990b). NT-4 transcripts (4.4, 2.4, 1.5 and 1.3 kb) show the most widespread distribution among all the neurotrophins with detectable expression in many rat peripheral tissues, endocrine organs, early brain, and reproductive tissues (Timmusk et al., 1993a).

NT-3, BDNF and NGF mRNAs are expressed in neurons in the developing and adult rat hippocampus (Ernfors et al., 1990a, 1990b; Hohn et al., 1990; Hofer et al., 1990; Maisonpierre et al., 1990b). Northern blot studies have shown that NT-3 mRNA levels peak in the early postnatal hippocampus, in contrast to a late postnatal onset of NGF and BDNF expression. The final levels of NGF, NT-3 and BDNF mRNA expression in the adult rat hippocampus are roughly the same (Maisonpierre et al., 1990a; 1990b). *In situ* studies in the E18 hippocampal formation have revealed that BDNF mRNA-expressing cells were predominantly localized in the pyramidal cell layer, while NT-3 expression showed a scattered distribution, which localized more towards the medial hippocampus proximal to the developing granule cell layer (Ernfors et al., 1992). In the postnatal and adult hippocampus, the neurotrophins show patterns of mRNA distribution which are for the most part distinct (Ernfors et al., 1990b). In the adult rat, expression of NT-3 mRNA is restricted to the medial CA1, to CA2, and to the granule cell layer of the dentate gyrus (Ernfors et al., 1990a, 1990b; Phillips et al., 1990). Early postnatal distribution of NT-3 mRNA is essentially identical to the adult, except for the absence of NT-3 expression in the caudal hippocampus and its transient transcription in the hilar region of the dentate gyrus during the first and second postnatal weeks (Friedman et al., 1991a; 1991b). In contrast, the highest levels of BDNF message in the adult rat hippocampus are found in the pyramidal cell layer of the CA2 and CA3 regions and in the dentate hilus, with lower levels in region CA1 and the granular cell layer (Ernfors, et al., 1990a; Ernfors, et al., 1990b; Hofer et al., 1990; Phillips, et al., 1990; Wetmore et al., 1990). NGF mRNA is expressed in the adult rat primarily in scattered neurons throughout the pyramidal cell layer and stratum oriens, as well as in the hilar region (Ayer-LeLievre et al., 1988; Ernfors, et al., 1990b). The spatiotemporal differences in neurotrophin expression suggest that these proteins may act at different times of brain development, specifically supporting distinct neuronal populations in the CNS.

During mouse embryonic development trkA mRNA (4.3 kb) is first detected at E9.5 with maximal level at E13.5, followed by a decline to low levels in the adult (Martin-Zanca et al., 1990). At E13.5 trkA mRNA expression is mostly confined to neural-crest derived sensory cranial and dorsal root ganglia (Martin-Zanca et al., 1990). Lower levels of trkA expression have also been reported in the medial septum and diagonal band of Broca of the E18 basal forebrain (Ernfors et al., 1992).

Multiple trkB transcripts have been detected in the embryonic and adult rodent brain (Middlemas et al., 1991). Analysis of the trkB transcript distribution have demonstrated mRNA expression initially at E9.5 within a subset of undifferentiated neuroepithelial cells in the CNS and PNS. Some of the structures showing an early expression include the forebrain, caudal midbrain, hindbrain, spinal cord, the trigeminal ganglion and differentiating neural crest cells (Klein et al., 1990: Ernfors et al., 1992). In the embryonic pyramidal cell layer of the hippocampus, moderate labelling intensity for trkB is detected at E18. In the adult mouse, trkB mRNA expression is found in most structures of the cerebrum including the cortical layers, the thalamus, the cerebellum, the hippocampus and choroid plexus. However, lower levels of expression have also been reported in peripheral tissues (Klein et al., 1990). Probes specific for transcripts encoding the catalytic (gp145trkB) and the noncatalytic (gp95trkB) trkB gene products showed a highly specialized pattern of expression in the adult mouse. Transcripts encoding the catalytic form are present in the cerebral cortex, and the pyramidal and granular cell layers of the hippocampus, but absent in the choroid plexus or the ependymal lining of the cerebral ventricles. In contrast, transcripts for the truncated form of trkB were detected only in the choroid plexus and the ependymal lining of the ventricles (Klein et al., 1990).

In the developing mouse, the trkC gene is expressed in the central, peripheral and enteric nervous system, starting from E10.5. In addition, trkC transcripts have been seen in nonneuronal tissues such as the mesenchymal cells of the arterial wall, adipose tissue and the acini of the submaxillary gland (Lamballe, et al., 1991; Lamballe, et al., 1994; Tessarollo et al., 1993). Low levels of trkC mRNA in the hippocampus have been reported as early as E16. At E18 a moderate intensity can be seen throughout the hippocampal formation, including the pyramidal cell layer, and over the developing dentate gyrus (Ernfors et al., 1992). In the adult mouse brain, trkC is expressed in the cerebral cortex, hippocampus, certain thalamic and hypothalamic nuclei, and the granule cell layer of the cerebellum among others (Ernfors et al., 1992; Lamballe, et al., 1991; Lamballe, et al., 1994; Merlio et al., 1992a; 1992b; Tessarollo, et al., 1993). Transcripts for full-length and truncated trkC have been detected in the pyramidal cell layer and the dentate gyrus of the adult rodent hippocampus (Klein et al., 1989; Klein et al., 1990; Lamballe et al., 1991; Valenzuela et al., 1993). Transcripts for some of the trkC (ki 14 and 25) variants have been identified in various regions of the rat brain such as the cerebral hemispheres, midbrain, cerebellum and hippocampus. Glial cells from the cerebral hemispheres express mRNAs for the ki25 form. Cultured hippocampal neurons expressed these transcripts, albeit at lower levels than the unmodified form (Lamballe et al., 1993). Evidence of the insert-containing protein is very scarce, but a ki-14-specific antiserum has been shown to specifically recognize a 145-150 kD band in brain and cerebellar extracts (Lamballe et al., 1994).

In summmary, studies on the temporal and spatial pattern of expression of the neurotrophins and their high affinity receptors have found evidence consistent with target-derived as well as local modes of neurotrophin action. In a target-derived model, the neurotrophins are expressed in the target fields of trk-expressing neurons. Some examples of this pattern of expression are the NGF and NT-3 expression in the iris which is a target field for *trk-* and *trk*C-expressing sympathetic neurons of the superior cervical ganglion and sensory neurons from the trigeminal ganglion. In contrast, in other cases the neurotrophin and *trk* receptors are coexpressed in the same tissue, implying a local paracrine or autocrine mode of action. Examples of these are dorsal root, geniculate, superior, jugular, petrose and nodose ganglia, as well as in the hippocampus, frontal cortical plate and pineal recess (Ernfors et al., 1992).

Neurotrophin Receptor Signal Transduction

Insights into neurotrophin receptor signal transduction have come from studies with mitogenic growth factor receptors. Activation of tyrosine kinase receptors by their ligands is mediated by an oligomerization step which results in trans-phosphorylation of specific tyrosine residues (Ullrich and Schlessinger, 1990). Trk receptor phosphorylation on specific tyrosine residues has been shown in response to neurotrophin binding, and it is critical for receptor activation (Cordon-Cardo et al., 1991). Evidence of Trk receptor homodimer formation in response to ligand binding has recently been demonstrated (Jing et al., 1992). Specific phosphotyrosine residues of the receptor interact with various signalling molecules containing src-homology 2 (SH2) sequences (Schlessinger and Ullrich, 1992). The coupling of the activated receptor to SH2 containing proteins initiates intracellular signals through multiple pathways. SH2-containing proteins have been divided into two classes: those with enzymatic functions such as phospholipase C-γ (PLC-γ), p21ras GTPase-activating protein (GAP) and Src, and those that are thought to act as adapters such as v-crk, GRB2, SHC and the p85 subunit of phosphotidylinositol-3 (PI-3) kinase (Koch et al., 1991). To date, TrkA has been demonstrated to interact with PLC-γ and the p85 subunit of PI-3 kinase.

The involvement of the Src tyrosine kinase and the GTP-binding protein, p21c-ras, in Trk receptor signal transduction has been suggested by several observations. Expression of activated *ras* or *src* genes in PC12 cells mimics NGF-induced differentiation (Alema et al., 1985; Bar and Feramisco, 1985; Noda et al., 1985). In addition, NGF-induced neurite outgrowth in PC12 cells is blocked by microinjection of neutralizing antibodies specific for Src and Ras proteins (Hagag et al., 1986; Kremer et al., 1991). It is not yet clear how Src and Ras proteins are activated by NGF. Src activation may occur through a direct association with Trk via its SH2 domain, as is the case with platelet-derived growth factor receptor (Koch et al., 1991). For Ras activation, NGF treatment of PC12 cells induces two components that may indirectly control the activity of p21c-ras, the guanine nucleotide exchange factor and the SH2-containing guanosine triphosphatase activating protein (GAP) (Li et al., 1992). Other SH2-containing proteins, such as SHC, have been shown to be phosphorylated on tyrosine residues in response to NGF. The SH2-containing protein, GRB2, which links an EGF-like receptor tyrosine kinase to Ras activation in *C. elegans*, has been shown to interact with SHC (Rozakis-Adcock et al., 1992). This observation led some groups to propose that SHC/GRB2 may connect TrkA to Ras activation. However, it has been recently shown that TrkA does not associate directly with GRB2 (Suen et al., 1993). In contrast, a direct interaction of the activated EGF receptor with GRB2 is reported in this study.

NGF stimulation of PC12 cells stimulates the activity of several cytoplasmic protein kinases including Raf serine/threonine, mitogen-activated protein kinases (MAP also known as Erks - extracellular-signal regulated kinase), the MAP kinase activators, and RSK/S6 kinase (Boulton et al., 1991; Thomas, 1992; Wood et al.,

1992). Most of these with have been shown to act in a Ras-dependent signalling pathway. NGF stimulates hyperphosphorylation and possibly activation of Raf-1 and B-Raf in PC12 cells (Ohmichi et al., 1992a; 1992b; Wood, et al., 1992). Raf activation by NGF lies downstream of Ras, but a requirement for Raf-1 or B-Raf in NGF action has not been established (Troppmair et al., 1992). Activation of MAP kinases requires phosphorylation on both threonine and tyrosine residues (Thomas, 1992). Recently, a kinase known as MAP kinase activator was isolated which phosphorylates both threonine and tyrosine residues (Nakielny et al., 1992). These MAP kinase activators phosphorylate the serine-threonine protein kinases, MAP kinases, via a p21c-ras-dependent pathway. Finally, NGF-stimulated phosphorylation of RSK kinase by MAP kinases is also dependent on Ras (Wood, et al., 1992). Another potential substrate involved in NGF signal transduction is protein kinase C (Carter and Downes, 1992; Hama et al., 1986; Kim et al., 1991; Maher, 1988). A summary of Trk receptor signalling via Ras activation is as follows: Trk > PLC-g1/PI-3 kinase > RAS > Raf kinase > MAP kinase activator > MAP kinase > RSK kinase.

An interesting aspect of signal transduction by receptor tyrosine kinases is the determination of which structural elements in these receptors define the substrate specificity and subsequent biological outcome. For some tyrosine kinase receptors the specificity appears to be provided by the amino acids residues surrounding the substrate binding sites. For instance, PLC-γ activation by TrkA and TrkB receptors requires phosphorylation of specific tyrosine residues present in the carboxy-termini, Y785 and Y816, respectively (Vetter et al., 1991). These tyrosines are all part of the sequence YLDX, where X= a hydrophobic residue. This sequence is present in all Trk receptors. Similar interactions between TrkA and the PI-3 kinase regulatory subunit have been reported (Soltoff et al., 1992). However, there is conflicting evidence on whether PI-3 kinase phosphorylation by TrkA results from a direct or an indirect mechanism (Ohmichi et al., 1992).

It has been shown that the kinase insert variants of TrkC do not phosphorylate PLCg-1 or PI-3 kinase, in spite of containing these critical tyrosine residues that are phosphorylated in response to NT-3 (Lamballe et al., 1993). These findings suggest that the additional sequences present in these isoforms result in the elimination of the ability to phosphorylate their substrates. These variants have been shown to be biologically inactive when assayed for the induction of mitogenesis in NIH3T3 and differentiation in PC12 cells, suggesting that their lack of activity may be due, in part, to their lack of association with some signal transduction pathways.

NGF treatment of PC12 cells stimulates the transcription of numerous immediate-early genes, including c-fos, NGFIA and NGFIB (Greenberg and Ziff, 1985; Milbrandt, 1987; 1988). Within 2 hours of NGF treatment to PC12 cells, immediate-early gene induction decreases and is substituted by a second wave of neuronal gene induction (Halegoua et al., 1991). Protein kinases may play a regulatory role in the activity of some of these transcription factors. For instance, MAP kinases phosphorylate a broad range of substrate in vitro which include transcription factors such as c-fos, c-myc and c-jun (Alvarez et al., 1991; Chen et al., 1992; Pulver et al., 1991). These observations suggest that MAP kinases might mediate an integration of signals from the cell surface to the nucleus.

Neurotrophin Biological Activities

Studies on the distribution of the neurotrophins and their high affinity receptors suggest a broad function of these signalling systems in supporting neural innervation in the developing and adult brain. A particular function has been ascribed to these ligands in some of these systems in vitro, and in a few cases in vivo, as will be mentioned below. However, based on the ligand-receptor distribution, additional uncharacterized neuronal systems have been identified that should be explored further (Ernfors et al., 1992; Ibanez and Persson, 1992). NGF and BDNF support selected neuronal populations in the peripheral and central nervous systems in vivo and in vitro (Hefti, 1986; Levi-Montalcini, 1987; Hofer, et al., 1988; Rosenberg et al., 1988). NGF is a survival factor for neural crest-derived sensory and sympathetic neurons as well as for septal cholinergic neurons (Levi-Montalcini, 1987). BDNF supports each of these neuronal types, with the exception of the sympathetic ganglia and with the addition of the nodose ganglion derived-sensory neurons. In the CNS, BDNF supports the dopaminergic neurons of the substantia nigra, the GABAergic neurons of the forebrain, the cholinergic neurons of the hippocampus, the granule cells of the cerebellum, and modulates neuropeptide expression in cortical neurons (Alderson et al., 1990; Hyman et al., 1991; Ip, et al., 1992; Ip et al., 1993a; Knusel et al., 1991; Lindsay et al., 1985). The effects of NT-3 have been characterized in the PNS by the increased survival of cultured neural crest- and nodose ganglion-derived sensory neurons (Ernfors et. al., 1990; Hohn et. al., 1990; Maisonpierre et. al., 1990a; Rosenthal et. al., 1990). In the CNS, NT-3 has been shown to increase the expression of the calcium-binding protein, calbindin-D28k in a subpopulation of hippocampal neurons (Collazo et al., 1992; Ip, et al., 1993a). Recently, NT-4 has been shown to be a target-derived trophic factor for neurons of the trigeminal ganglia (Ibanez et al., 1992b).

Regulation Of Neurotrophins And Their Receptors By Neuronal Activity And Brain Insults

In many regions of the CNS, in particular in the hippocampus, mRNAs for the neurotrophins and their receptors are highly regulated by neuronal activity. Initial studies in cultured hippocampal neurons showed that kainic acid, a glutamate receptor agonist, induced BDNF and NGF mRNA levels within 3 hours of treatment (Zafra et al., 1990). This increase was mediated specifically by non-NMDA receptors, since it was blocked by non-NMDA receptor antagonists such as cyano-nitroquinoxaline-dione, CNQX. Likewise, depolarization induced by high potassium or the sodium channel agonist veratridine evoked increases in neurotrophin mRNA levels in hippocampal cultures (Zafra et al., 1990). In contrast, blockade of the glutamate receptors and/or stimulation of the GABAergic system decreases the level of NGF and BDNF in the brain (Zafra et al., 1991). The levels of BDNF and NGF mRNAs, but not NT-3 or NT-4, increase in the hippocampus, amygdala and neocortex after seizures evoked by electrolytic lesions (Gall and Isackson, 1989; Isackson et al., 1991), injections of kainic acid (Zafra et al., 1990; Timmusk et al., 1993), or electrical kindling stimulations in the hippocampus (Ernfors et al., 1991). These mRNA increases are transient, returning to pre-seizure levels after 24 hours (Ernfors et al., 1991). Seizure activity increases neurotrophin mRNA levels only in those areas that already express the factor at low levels. The upregulation of BDNF differs from NGF in showing a faster timecourse and broader distribution. Recently, seizures induced by hippocampal kindling led to upregulation of both truncated and full-length trkB mRNA, and kinase domain containing-protein with a similar spatial and temporal pattern as BDNF. This effect is specific for trkB, since no changes were detected with trk or trkC mRNAs (Merlio et al., 1992).

The rapid and large increase in neurotrophin production by neuronal activity implicates neurotrophins in activity-dependent processes in the CNS. In this regard, BDNF and NT-3, but not NGF mRNAs have been shown to increase modestly in response to stimulation inducing LTP in the CA1 pyramidal cell layer of the hippocampus, indicating that low levels of electrical activity may regulate neurotrophin expression in the hippocampus (Patterson et al., 1992). These results suggest that under normal conditions, activity-dependent processes may involve neurotrophins in some forms of synaptic plasticity such as learning and memory, but under abnormal circumstances, these processes may contribute to the progression of neurological disorders such as epilepsy.

Increases in BDNF and NGF, but not NT-3 mRNAs have also been demonstrated in the granule cells of the dentate gyrus following cerebral ischemia and insulin-induced hypoglycemic coma. *trk*B mRNA is also selectively and transiently increased in the dentate gyrus following these insults (Merlio et al., 1992). NGF has been reported to prevent neuronal damage in human cerebral cortical and rat hippocampal cell cultures from hypoglycemic insults by reducing intracellular calcium levels (Cheng and Mattson, 1991). The concomitant regulation of BDNF and *trk*B in brain regions such as the dentate gyrus that coexpress this neurotrophin and its receptor could provide a local trophic support that protects neurons against the necrosis following epileptic, ischemic and hypoglycemic insults.

TISSUE CULTURE SYSTEMS OF THE CNS

Hippocampal Primary Cultures

Tissue culture systems have been widely used to define the responses and requirements of developing neurons for factors that influence their proliferation, differentiation, and survival. Such studies have been successful with peripheral neurons (sensory, sympathetic, and parasympathetic) where the location of the cells in discrete ganglia facilitates their purification. For example, the use of cultured dorsal root ganglia has been instrumental in the identification of trophic roles for NGF and BDNF (Barde et al., 1982; Leibrock, et al., 1989; Levi-Montalcini, 1987). Similar in vitro studies in the CNS have generally been more problematic because CNS neurons can be obtained only as heterogeneous cultures of neuronal and nonneuronal cells. For this reason, the identification of specific neuronal populations in the CNS requires specific immunocytochemical markers. The importance of the hippocampus in learning processes and its degeneration in neuropathies has made it one of the most studied brain regions in vitro.

Primary or dissociated cell cultures are prepared from single cell suspensions obtained by dissociation of neuronal tissues. Immature (embryonic or early postnatal) tissues are frequently used in these preparations, since they show best survival after dissociation. Neurons in dissociated cultures retain the morphological and physiological properties of the cell populations present in the tissue of origin. Studies by Banker and Cowan (1977, 1979) established conditions for maintaining embryonic hippocampal neurons at low density in culture. These studies were possible after the establishment of serum-free conditions, in which conventional serum was substituted by a defined mixture of hormones, vitamins and cell growth additives (Bottenstein and Sato, 1979). Subsequent work has focussed on morphological and electrophysiological aspects of hippocampal neurons, allowing detailed observations of the growth and branching of axons and dendrites, and the formation of synapses (Bartlett and Banker, 1984a; Dotti et al., 1988; Furshpan and Potter, 1989). Hippocampal cultures have been used to explore the neurotoxicity of β-amyloid and excitatory neurotransmitters and to search for neuroprotective agents (Cheng and Mattson, 1991; Mattson and Kater, 1989a;

Mattson and Rychlik, 1990; Yankner et al., 1990a; Yankner et al., 1990b). Recently, the trophic effects of growth factors, such as basic fibroblast growth factor (bFGF), ciliary neurotrophic factor (CNTF) and the neurotrophins have been demonstrated in hippocampal cultures (Collazo, et al., 1992; Ip et al., 1991; Ip, et al., 1993a; 1993b; Walicke et al., 1986). Hippocampal cultures have been important in characterizing the regulation of neurotrophin expression by electrical activity as described previously (Zafra et al., 1990; 1991). A drawback to the dissociated cultures is that this system is less well suited for traditional biochemical approaches because of the limited quantity of material and heterogeneous culture composition. In spite of these limitations, biochemical studies of hippocampal cultures have provided important information about several signal transduction pathways such as growth factor and excitatory amino acid receptors (Walicke et al., 1990; Bading and Greenberg, 1991).

An important variation to dissociated cultures is the clonal microculture in which individual CNS precursors can divide at low density, allowing the clones that arise from single cells to be analyzed. This approach has been used successfully with E14 rat forebrain cells, showing that precursors can give rise to purely neuronal or glial clones, as well as mixtures of neurons and glia (Temple, 1989).

Organotypic Slice Cultures

Acute slice cultures of the adult rat and embryonic dissociated cultures are commonly used in vitro preparations for physiological, developmental and pharmacological studies. Acute slices provide the advantage of preserving the organotypic and anatomical properties of the tissue but their short term survival limits the experimental applications of this technique. Dissociated cultures provides longer cell survival, with the disadvantage that the organization and connectivity of the tissue is disrupted. Organotypic slice culture is a relatively newly established technique that combines some of the advantages of the acute slice and dissociated culture techniques. Thin slices (400 μm thick) can be cultured in a roller tube for a period of several weeks, retaining their histological organization (Zimmer and Gahwiler, 1984). This approach has proved fruitful in the study of cortical structures such as the hippocampus in which the lamellar structure can be captured in a two-dimensional culture. Specific cell types can often be identified based on their location and cell body size. Subsequent detailed studies have shown that these slice cultures undergo a considerable amount of synaptic reorganization and sprouting during the time of incubation (Caeser and Aertsen, 1991). Simplified protocols have been recently developed which also preserve the overall organotypic structure with short term survival (on the order of days instead of weeks) (Stoppin et al., 1991).

Neural Cell Lines

Cell lines are homogeneous, clonal populations of cells that can be grown indefinitely in vitro. Cell lines provide several advantages over primary cultures in terms of their clonal nature, the availability of large number of cells, and most important, their susceptibility to genetic manipulations. The advantages of cell lines must be considered against the fact that many aspects of neuronal differentiation are not accomplished by cell lines in vitro. For example, specific properties of neuronal subpopulations, or mature differentiated properties of neurons like synaptic connections are rarely seen in cell lines.

Tumor cell lines have been studied frequently by neurobiologists because, under certain conditions, these cells express some of the generic properties of neurons or glia. An example of a cell line with characteristics of undifferentiated, pluripotent cells are the embryonic carcinoma cells (EC). These cells are derived from teratoma tumors of fetal germ cells, and can differentiate in vitro and in vivo into neuronal and nonneuronal fates. Tumor cell lines have been derived from spontaneous or chemically-induced neuroblastomas, gliomas and pheochromocytomas. The pheochromocytoma cell line PC12 is one of the most extensively characterized cell lines in neurobiology. PC12 cells were isolated from a tumor of the adrenal medulla that arose following X-ray irradiation. This cell line exhibits many of the properties of adrenal chromaffin cells. When exposed to NGF, PC12 cells cease to divide and differentiate into sympathetic neuronal-like fates (Greene, 1976). PC12 cell lines have been crucial in understanding some of the molecular mechanisms underlying NGF-induced neuronal differentiation, such as signal transduction, changes in gene expression, and cellular responses. For example, the availability of PC12 mutants unresponsive to NGF have been important in identifying both the low and high affinity NGF receptors (Hempstead et al., 1989; Loeb, et al., 1991).

Recent reports have demonstrated the establishment in vitro of clonal cell lines from several regions of the developing brain. The advantage of this approach in comparison to traditional methods is that cells can be derived from specific regions, rather than from tumors that develop at random sites. One strategy involves the somatic cell fusion of tumor cells with dissociated cells. This procedure is based on the principles of the hybridoma technology frequently used to generate monoclonal antibodies. This approach has been used with success by several groups; however, the phenotype of the hybrid cell is often a complex product of nuclear and cytoplasmic factors present in the fused cells (Blusztajn et al., 1992; Greene et al., 1975; Han et al., 1991; Klee and Nerenberg, 1974). A second approach makes use of transgenic mice that express an oncogene under the control of cell-type specific promoters. Clonal cells are derived from tumors that occur in these transgenic tissues (Hammang et al., 1990; Mellon et al., 1990). Both of these approaches produce fully transformed cells. A third, nontransforming method is based on infecting primary cultures with an immortalizing oncogene using a retroviral vector, which also contains the neomycin resistance gene. Retroviruses expressing v-myc, neu and the tsA58 allele of the SV40 T antigen have been used to generate cell lines from the developing CNS, including the cerebellum, hippocampus, optic nerve, and from sympathoadrenal lineages in the PNS (Almazan and McKay, 1992; Birren and Anderson, 1990; Eves et al., 1992; Frederiksen et al., 1988b; Renfranz, et al., 1991; Ryder et al., 1990; Snyder, Deitcher et al., 1992). These studies have shown that transduction of immortalizing oncogenes into cultured cells can produce essentially unlimited mitotic activity in vitro without inducing other properties associated with tumorigenesis such as loss of contact inhibition (Land et al., 1983; Ruley, 1983). One important limitation of this approach is that retroviral infection requires at least one round of DNA synthesis; therefore, these cell lines can be made only from proliferating precursors and not from postmitotic neurons.

Work in our laboratory has focussed on T antigen-immortalized cell lines generated using retroviral vectors because the tsA58 allele encodes a protein that is stable and active at the permissive temperature (33°C) but is degraded at the non-permissive temperature (39°). An additional screen used to categorize the G418-resistant clones is the expression of the intermediate filament protein, nestin. Nestin (neuroepithelial stem cell intermediate filament) is expressed in neuroepithelial stem cells in the brain (Frederiksen and McKay, 1988a; Lendahl et al., 1990). The use of nestin has helped in identifying cell lines derived from neurepithelial precursor cells. As an example, the nestin-immunopositive cell line, HiB5, has been immortalized from E16 rat hippocampal primary cultures using the tsA58 allele of SV40 T antigen (Renfranz et al., 1991). Upon transplantation into the neonatal hippocampus and cerebellum, HiB5 cells integrate into the host tissue, acquiring morphological and antigenic properties of neurons and glial cells born at the time and location of the implant site. These results suggest that immortal cell lines can respond to environmental signals to produce differentiated cell types antigenically and functionally indistinguishable from host cells. .

Advances in the understanding of the mechanisms of neurotrophin action have been greatly facilitated by using cell lines. For example, studies that led to the biochemical characterization of the tyrosine kinase activity of gp140trk following NGF binding were possible due to the availability of the PC12 cell line (Kaplan, et al., 1991a; Klein, et al., 1991). As a second example, fibroblast cell lines have been important in identifying neurotrophins as primary or secondary ligands for Trk receptors (Cordon-Cardo et al., 1991; Glass et al., 1991; Klein et al., 1991; Soppet et al., 1991; Squinto et al., 1991). Recently, the v-myc immortalized, sympathoadrenal progenitor cell line MAH has demonstrated the role of extracellular signals in regulating TrkA receptor expression and neurotrophin responsiveness (Birren and Anderson, 1990; Birren, et al., 1992). Results from all of these systems emphasize the importance of the cell type context of Trk receptor expression in defining cellular outcome, whether proliferation in nonneuronal cells or cessation of growth and differentiation in neural systems. In addition, these cell lines are susceptible to genetic manipulations, which are crucial in dissecting neurotrophin function. Similar studies in the CNS have been hampered by the lack of cell lines from the developing and adult brain. At present, the characterization of neurotrophin effects in the brain, and in particular in the hippocampus has been limited to embryonic primary cultures (Collazo et al., 1992; Ip et al., 1993).

REFERENCES

Alderson R, Alterman A, Barde Y and Lindsay R (1990): Brain-derived neurotrophic factor increases survival and differentiated functions of rat septal cholinergic neurons in culture. Neuron 5: 297-306.

Alema S, Casalbore P, Agostini E and Tato F (1985): Differentiation of PC12 cells induced by v-src oncogene. Nature 316: 557-559.

Almazan G and McKay R (1992) An oligodendrocyte precursor cell line from rat optic nerve. Brain Res 579: 234-245.

Altman J and Bayer S (1990a): Mosaic organization of the hippocampal neuroepithelium and the multiple germinal sources of dentate granule cells. J Comp Neurol 301: 325-342.

Altman J and Bayer S (1990b): Prolonged sojourn of developing pyramidal cells in the intermediate zone of the hippocampus and their settling in the stratum pyramidale. J Comp Neurol 301: 343-364.

Altman J and Bayer S (1990c): Migration and distribution of two populations of hippocampal granule cell precursors during the perinatal and postnatal periods. J Comp Neurol 301: 365-381.

Altman J and Das G (1965): Autoradiographic and histological evidence of postnatal hippocampal neurogenesis in rats. J Comp Neurol 124: 319-336.

Alvarez E, Northwood I, Gonzales F, Latour D, Seth A, Abate C, Curran T and Davis R (1991): Pro-Leu-/Ser/Thr-Pro is a consensus primary sequence for substrate protein phosphorylation. J Biol Chem 266: 15277-15285.

Amaral D and Kurz J (1985): The time of origin of cells demonstrating glutamic acid decarboxylase-like immunoreactivity in the hippocampal formation of the rat. Neurosci Lett 59: 33-39.

Ascher P, and Nowak L (1988): Quisqualate and kainate-activated channels in mouse central neurones in culture. J Physiol (Lond) 399: 247-266.

Austin C and Cepko C (1990): Cellular migration patterns in the developing mouse cerebral cortex. Development 110: 713-732.

Ayer-LeLievre C, Olson L, Ebendal T, Seiger A and Persson H (1988): Expression of β-nerve growth factor in hippocampal neurons. Science 240: 1339-1441.

Bar S and Feramisco J (1985): Microinjection of the ras oncogene protein into PC12 cells induces morphological differentiation. Cell 42: 841-848.

Barde Y (1989): Trophic factors and neuronal survival. Neuron 2: 1525-1534.

Barde Y, Edgar D and Thoenen H (1982): Purification of a new neurotrophic factor from the mammalian brain. EMBO J 1: 549-553.

Bartlett W and Banker G (1984): Electron microscopic studies of axonal and dendritic development by hippocampal neurons in culture. I. Cells which develop without intercellular contacts. J Neurosci 4: 1944-1953.

Bartus R, Dean R, Beer B and Lippa A (1982): The cholinergic hypothesis of geriatric memory dysfunction. Science 217: 408-417.

Bashir Z, Bortolotto Z, Davies C, Berreta N, Irving A, Seal A, Henley J, Jane D, Watkins J and Collingridge G (1993): Induction of LTP in the hippocampus needs synaptic activation of glutamate metabotropic receptors. Nature 363: 347-350.

Bayer S (1980): Development of the hippocampal region in the rat: neurogenesis examined with 3H-Thymidine autoradiography. J Comp Neurol 190: 87-114.

Bekkers J and Stevens C (1990): Presynaptic mechanisms for longterm potentiation in the hippocampus. Nature 346: 724-729.

Berkemeier L, Winslow J, Kaplan D, Nikolics K, Goeddel D and Rosenthal A (1991): Neurotrophin-5: a novel neurotrophic factor that activates trk and trkB. Neuron 7: 857-866.

Birren S and Anderson D (1990): A v-myc-immortalized sympathoadrenal progenitor cell line in which neuronal differentiation is initiated by FGF but not NGF. Neuron 4: 189-201.

Birren S, Verdi J and Anderson D (1992): Membrane depolarization induces p140trk and NGF responsiveness, but not p75LNGFR, in MAH cells. Science 257: 395-397.

Blackstad T (1956): Commisural connections of the hippocampal region in the rat with special reference to the mode of termination. J Comp Neurol 105: 417-537.

Bliss T and Collingridge G (1993): A synaptic model of memory: long-term potentiation in the hippocampus. Nature 361: 31-39.

Bliss T and Lomo T (1973): Long-lasting potentiation of synaptic transmission in the dentate area of the anaesthetized rabbit following stimulation of the perforant path. J Physiol (London) 232: 331-356.

Blusztajn J, Venturini A, Jackson D, Lee H and Wainer B (1992): Acetylcholine synthesis and release is enhanced by dibutyryl c-AMP in a neuronal cell line derived from mouse septum. J Neurosci 12: 793-799.

Bothwell M (1991): Keeping track of the neurotrophin receptors. Cell 65: 915-918.

Bottenstein J and Sato G (1979): Growth of a rat neuroblastoma cell line in serum-free supplemented medium. Proc Natl Acad Sci USA 76: 514-517.

Boulton T, Nye S, Robbins D, Ip N, Radziejewska E, Morgenbesser S, DePinho R, Panayotatos N, Cobb M and Yancopoulos G (1991): ERKs: a family of protein serine-threonine kinases that are activated and tyrosine phosphorylated in response to insulin and NGF. Cell 65: 663-675.

Brodal A (1978): Neurological anatomy: In relation to clinical medicine. New York: Oxford University Press.

Caeser M and Aertsen A (1991): Morphological organization of rat hippocampal slice cultures. J Comp Neurol 307: 87-106.

Carter A and Downes C (1992): Phosphatidylinositol 3-kinase is activated by nerve growth factor and epidermal growth factor in PC12 cells. J Biol Chem 267: 14563-14567.

Chao M, Bothwell M, Ross A, Koprowski H, Lanahan A, Buck C and Sehgal A (1986): Gene transfer and molecular cloning of the human NGF receptor. Science 232: 518-521.

Chen R, Sarnecki C and Blenis J (1992) Nuclear localization and regulation of erk-and rsk-encoded protein kinases. Mol Cell Biol 12: 915-927.

Cheng B and Mattson M (1991) NGF and bFGF protect rat hippocampal and human cortical neurons against hypoglycemic damage by stabilizing calcium homeostasis. Neuron 7: 1031-1041.

Collazo D, Takahashi H and McKay R (1992): Cellular targets and trophic functions of neurotrophin-3 in the developing rat hippocampus. Neuron 9: 643-656.

Cordon-Cardo C, Tapley P, Jing S, Nanduri V, O' Rourke E, Lamballe F, Kovary K, Klein R, Jones K, Reichardt L and Barbacid M (1991): The *trk* tyrosine protein kinase mediates the mitogenic properties of nerve growth factor and neurotrophin-3. Cell 66: 173-183.

Dotti C, Sullivan C and Banker G (1988): The establishment of polarity by hippocampal neurons in culture. J Neurosci 8: 1454-1468.

Durkop H, Latza U, Hummel M, Eitelbach F, Seed B, and Stein H (1992): Molecular cloning and expression of a new member of the nerve growth factor receptor family that is characteristic for Hodgkin's disease. Cell 68: 421-427.

Ernfors P, Ibanez C, Ebendal T, Olson L and Persson H (1990a): Molecular cloning and neurotrophic activities of a protein with structural similarities to nerve growth factor: developmental and topographical expression. Proc Natl Acad Sci USA 87: 5454-5458.

Ernfors P, Merlio J and Persson H (1992): Cells expressing mRNAs for neurotrophins and their receptors during embryonic rat development. Eur J Neurosci 4: 1140-1158.

Ernfors P, Wetmore C, Olson L and Persson H (1990b): Identification of cells in the rat brain and peripheral tissues expressing mRNA for members of the nerve growth factor family. Neuron 5: 511-526.

Eves E, Tucker M, Roback J, Downen M, Rich M and Wainer B (1992): Immortal rat hippocampal cell lines exhibit neuronal and glial lineages and neurotrophin gene expression. Proc Natl Acad Sci USA 89: 4373-4377.

Frederiksen K and McKay R (1988): Proliferation and differentiation of rat neuroephitelial precursor cells in vivo. J Neurosci 8: 1144-1151.

Frederiksen K, Jat P, Valtz N, Levy D and McKay R (1988): Immortalization of precursor cells from the mammalian CNS. Neuron 1: 439-448.

Friedman W, Ernfors P and Persson H (1991a): Transient and persistent expression of NT-3/BDNF mRNA in the rat brain during postnatal development. J Neurosci 11: 1577-1584.

Friedman W, Olson L and Persson H (1991b): Cells that express brain-derived neurotrophic factor in the developing postnatal brain. Eur J Neurosci 3: 688-697.

Furshpan E and Potter D (1989): Seizure-like activity and cellular damage in rat hippocampal neurons in culture. Neuron 3: 199-207.

Galileo D, Gray G, Owens G, Majors J and Sanes J (1990): Neurons and glia arise from a common progenitor in chicken optic tectum: demonstration with two retroviruses and cell type-antibodies. Proc Natl Acad Sci USA 87: 458-462.

Gall C and Isackson P (1989): Limbic seizures increase neuronal production of mRNA for nerve growth factor. Science 245: 758-761.

Glass D, Nye S, Hantzopoulos P, Macchi M, Squinto S, Goldfarb M and Yancopoulos G (1991): TrkB mediates BDNF/NT-3-dependent survival and proliferation in fibroblasts lacking the low affinity NGF receptor. Cell 66: 405-413.

Gray G, Glover J, Majors J and Sanes J (1988): Radial arrangement of clonally related cells in the chicken optic tectum: Lineage analysis with a recombinant retrovirus. Proc Natl Acad Sci USA 85: 7356-7360.

Greenberg M, Greene L and Ziff E (1985): Nerve growth factor and epidermal growth factor induce rapid transient changes in proto-oncogene transcription in PC12 cells. J Biol Chem 260: 14101-14110.

Greene L (1976): Establishment of a nonadrenergic clonal line of rat adrenal pheochromocytoma cells which responds to nerve growth factor. Proc Natl Acad Sci USA 73: 2424-2428.

Greene L, Shain W, Chalazonitis A, Breakfield X, Minna J, Coon H and Nerenberg M (1975): Neuronal properties of hybrid neuroblastoma X sympathetic ganglion cells. Proc Natl Acad Sci USA 72: 4923-4927.

Grove E, Kirkwood T and Price J (1992): Neuronal precursor cells in the rat hippocampal formation contribute to more than one cytoarchitectonic area. Neuron 8: 217-229.

Hagag N, Halegoua S and Viola M (1986): Inhibition of growth factor-induced differentiation of PC12 cells by microinjection of antibody to Ras p21. Nature 319: 680-682.

Halegoua S, Armstrong R and Kremer N (1991): Dissecting the mode of action of neuronal growth factor. Curr Top Microbiol Immunol 165: 119-170.

Hallbook F, CF, I., and Persson, H. (1991). Evolutionary studies of the nerve growth factor family reveal a novel member abundantly expressed in Xenopus ovary. Neuron 6: 845-858.

Hama T, Huang K and Guroff G (1986): Protein kinase C is a component of a nerve growth factor sensitive phosphorylation system in PC12 cells. Proc Natl Acad Sci USA 87: 2353-2357.

Hamburger V, Brunsco-Bechtold J and Yip J (1981): Neuronal death in the spinal ganglia of the chick embryo and its reduction by nerve growth factor. J Neurosci 1: 60-71.

Hammang J, Baetge E, Behiringer R, Brinster R, Palmiter R and Messing A (1990): Immortalized retinal neurons derived from SV40 T-antigen-induced tumors in transgenic mice. Neuron 4: 775-782.

Han H, Nichols R, Rubin M, Bahler M and Greenegard P (1991): Induction of formation of presynaptic terminals in neuroblastoma cells by Synapsin IIb. Nature 349: 697-700.

Hanks S (1991): Eucaryotic protein kinases. Current Opin Struct Biol 1: 369-383.

Hanks S, Quinn A and Hunter T (1988): The protein kinase family: conserved features and deduced phylogeny of the catalytic domains. Science 241: 42-52.

Hayes T and McKay R (1993): Differentiation Of Cell Types In The Central Nervous System. Wiley-Liss Inc.

Hefti F (1986): Nerve growth factor promotes survival of septal cholinergic neurons after fimbrial transections. J Neurosci 6: 845-858.

Hempstead B, Martin-Zanca D, Kaplan D, Parada L and Chao M (1991): High affinity NGF binding requires co-expression of the trk proto-oncogene and the low affinity NGF receptor. Nature 350: 678-683.

Hempstead B, Schleifer L and Chao M (1989): Expression of a functional gene for nerve growth factor after gene transfer. Science 243: 373-375.

Heuer J, Fatemie-Nainie S, Wheeler E and Bothwell M (1990): Structure and development of the chicken NGF receptor. Dev Biol 137: 287-304.

Hirano S, Fuse S and Sohal G (1991): The effect of the floor plate on pattern and polarity in the developing central nervous system. Science, 251, 310-313.

Hofer M, Pagliusi S, Hohn A, Leibrock J and Barde Y (1990): Regional distribution of brain-derived neurotrophic factor mRNA in the adult mouse brain. EMBO J 9: 2459-2464.

Hohn A, Leibrock J, Bailey K and Barde Y (1990): Identification and characterization of a novel member of the nerve growth factor/brain-derived neurotrophic factor family. Nature 344: 339-341.

Holt C, Bertsch T, Ellis H and Harris W (1988): Cellular determination in the Xenopus retina is independent of lineage and birthdate. Neuron 1: 15-26.

Hyman C, Hofer M, Barde Y, Juhasz M, Yancopoulos G, Squinto S and Lindsay R (1991): BDNF is a neurotrophic factor for dopaminergic neurons of the substantia nigra. Nature 350: 230-232.

Ibanez C, Ebendal T, Barbany G, Murray-Rust J, Blundell T and Persson H (1992a): Disruption of the low affinity receptor-binding site in NGF allows neuronal survival and differentiation by binding to the trk gene product. Cell 69: 329-341.

Ibanez C, Ebendal T and Persson H (1991): Chimeric molecules with multiple neurotrophic activities reveal structural elements determining the specificities of NGF and BDNF. EMBO J 10: 2105-2110.

Ibanez C, Ernfors P, Timmusk T, Ip N, Yancopoulos G and Persson H (1993): Neurotrophin-4 is a target-derived neurotrophic factor for neurons of the trigeminal ganglion. Development 117: 1345-1353..

Ibanez C, Ilag L, Murray-Rust J and Persson H (1993): An extended surface of binding to Trk tyrosine kinase receptors in NGF and BDNF allows the engineering of a functional pan-neurotrophin. EMBO J 12: 2281-2293.

Ip N, Ibanez C, Nye S, McClain J, Jones P, Gies D, Belluscio L, Le Beau M, Espinosa EI, Squinto S, Persson H and Yancopoulos G (1992): Mammalian Neurotrophin-4: structure, chromosomal localization, tissue distribution and receptor specificity. Proc Natl Acad Aci USA 89: 3060-3064.

Ip N, Li Y, Van de Stadt I, Panayotatos N, Alderson R and Lindsay R (1991): Ciliary neurotrophic factor enhances neuronal survival in embryonic rat hippocampal cultures. J Neurosci 11: 3124-3134.

Ip N, Li Y, Yancopoulos G and Lindsay R (1993a): Cultured hippocampal neurons show responses to BDNF, NT-3, and NT-4, but not NGF. J Neurosci 13: 3394-3405.

Ip N, Stitt T, Tapley P, Klein R, Glass D, Fandl J, Greene L, Barbacid M and Yancopoulos G (1993b): Similarities and differences in the way neurotrophins interact with the Trk receptors in neuronal and nonneuronal cells. Neuron 10: 137-149.

Isackson P, Huntsman M, Murray K and Gall C (1991): BDNF mRNA expression is increased in the adult forebrain after limbic seizures: temporal patterns of induction distinct from NGF. Neuron 6: 937-948.

Ithoh N, Yonehara S, Ishii A, Yonehara M, Mizushima S, Sameshima M, Hase A, Seto Y and Nagata S (1991): The polypeptide encoded by the cDNA for human cell surface antigen Fas can mediate apoptosis. Cell 66: 233-243.

Jing S, Tapley P and Barbacid M (1992): Nerve growth factor mediates signal transduction through trk homodimer receptors. Neuron 9: 1067-1079.

Johnson D, Lanahan A, Buck C, Sehgal A, Morgan C, Mercer E, Bothwell M and Chao M (1986): Expression and structure of the human NGF receptor. Cell 47: 545-554.

Jones K and Reichardt L (1990): Molecular cloning of a human gene that is a member of the nerve growth factor family. Proc Natl Acad Sci USA 87: 8060-8064.

Kaisho Y, Yoshimura K and Nakahama K (1990): Cloning and expression of a cDNA encoding a novel human neurotrophic factor. FEBS Lett 266: 187-191.

Kaplan D, Hempstead B, Martin-Zanca D, Chao M and Parada L (1991a): The trk proto-oncogene product: a signal transducing receptor for nerve growth factor. Science 252: 554-558.

Kaplan D, Martin-Zanca D and Parada L (1991b): Tyrosine phosphorylation and tyrosine kinase activity of the trk proto-oncogene product induced by NGF. Nature 350: 158-160.

Kim U, Fink D, Kim H, Park D, Contreras M, Guroff G and Rhee S (1991): Nerve growth factor stimulates phosphorylation of phospholipase C-γ in PC12 cells. J Biol Chem 266: 1359-1362.

Klee W and Nerenberg M (1974): A neuroblastoma X glioma hybrid cell line with morphine receptors. Proc Natl Acad Sci USA 69: 3474-3477.

Klein R, Conway D, Parada L and Barbacid M (1990): The trkB tyrosine protein kinase gene codes for a second neurogenic receptor that lacks the tyrosine kinase domain. Cell 61: 647-656.

Klein R, Nanduri V, Jing S, Lamballe FPT, Bryant S, Cordon-Cardo C, Jones K, Reichardt L and Barbacid M (1991a): The trkB tyrosine protein kinase is a receptor for brain-derived neurotrophin factor and neurotrophin-3. Cell 66: 395-403.

Klein R, Jing S, Nanduri V, O'Rourke E and Barbacid M (1991b): The trk protooncogene encodes a receptor for nerve growth factor. Cell 65: 189-197.

Knusel B, Winslow J, Rosenthal A, Burton L, Seid D, Nikolics K and Hefti F (1991): Promotion of central cholinergic and dopaminergic differentiation by brain-derived neurotrophic factor but not neurotrophin-3. Proc Natl Acad Sci USA 88: 961-965.

Koch C, Anderson D, Moran M, Ellis C and Pawson T (1991): SH2 and SH3 domains: Elements that control interactions of cytoplasmic signaling proteins. Science 252: 668-674.

Kremer N, D'Arcangelo G, Thomas S, DeMarco M, Brugge J and Halegoua S (1991): Signal transduction by nerve growth factor and fibroblast growth factor in PC12 cells requires a sequence of Src and Ras actions. J Cell Biol 115: 809-819.

Lamballe F, Klein R and Barbacid M (1991) trkC, a new member of the trk family of tyrosine protein kinases, is a receptor for neurotrophin-3. Cell 66: 967-979.

Lamballe F, Smeyne R and Barbacid M (1994): Developmental expression of trkC, the Neurotrophin-3 receptor, in the mammalian nervous system. J Neurosci 14: 14-28.

Land H, Parada L and Weinberg R (1983): Tumorigenic conversion of primary embryo fibroblast requires at least two cooperating oncogenes. Nature 304: 596-602.

Lander A (1987): Molecules that make axons grow. Mol Neurobiol 1: 213-245.

Leibrock J, Lottspeich F, Hohn A, Hofer M, Hengerer B, Masiakowski P, Thoenen H and Barde Y (1989): Molecular cloning and expression of brain-derived neurotrophic factor. Nature 341: 149-152.

Lendahl U, Zimmerman L and McKay R (1990): CNS stem cells express a new class of intermediate filament proteins. Cell 60: 585-595.

Levi-Montalcini R (1987): The nerve growth factor: thirty-five years later. Science 237: 1154-1162.

Li B, Kaplan D, Kung H and Kamata T (1992): Nerve growth factor stimulation of Ras-guanine nucleotide exchange factor and GAP activities. Science 256: 1456-1459.

Lindsay R, Thoenen H and Barde Y (1985): Placode and neural crest-derived sensory neurons are responsive at early developmental stages to brain-derived neurotrophic factor. Dev Biol 112: 319-328.

Loeb D, Maragos J, Martin-Zanca D, Chao M, Parada L and Greene L (1991): The trk proto-oncogene rescues NGF responsiveness in mutant NGF-nonresponsive PC12 cell lines. Cell 66: 961-966.

Lorente de Nó R (1934): Studies on the structure of the cerebral cortex. II. Continuation of the study of the ammonic system. J Psychol Neurol 46: 113-177.

Maher P (1988): Nerve growth factor induces protein-tyrosine phosphorylation. Proc Natl Acad Sci USA 85: 6788-6791.

Maisonpierre P, Belluscio L, Friedman B, Alderson R, Wiegand S, Furth M, Lindsay R and Yancopoulos G (1990a): NT-3, BDNF, and NGF in the developing rat nervous system: parallel as well as reciprocal patterns of expression. Neuron 5: 501-509.

Maisonpierre P, Belluscio L, Squinto S, Ip N, Furth M, Lindsay R and Yancopoulos G (1990b): Neurotrophin-3: a neurotrophic factor related to NGF and BDNF. Science 247: 1446-1451.

Malinow R and Tsien R (1990): Presynsaptic enhancement shown by whole cell recordings of long-term potentiation in hippocampal slices. Nature 346: 177-180.

Mallet S and Barclay A (1991): A new superfamily of cell surface proteins related to nerve growth factor receptor. Immunol. Today 12: 220-223.

Manabe T, Renner P and Nicoll R (1992): Postsynaptic contribution to long-term potentiation revealed by the analysis of miniature synaptic currents. Nature 355: 50-55.

Martin-Zanca D, Hughes S and Barbacid M (1986): A human oncogene formed by the fusion of truncated tropomyosin and protein kinase sequences. Nature 319: 743-748.

Martin-Zanca D, Oskam R, Mitra G, Copeland T and Barbacid M (1989): Molecular and biochemical characterization of the human trk protooncogene. Molec Cell Biol 1989: 24-33.

Mattson M, and Kater S (1989): Development and selective neurodeneration in cells cultured from different hippocampal regions. Brain Res 490: 110-125.

Mattson M and Rychlik B (1990): Glia protect hippocampal neurons against excitatory amino acid-induced degeneration: involvement of fibroblast growth factor. Int J Dev Neurosci 8: 399-416.

McConnell S (1988): Fates of visual cortical neurons in the ferret after isochronic and heterochronic transplantation. J Neurosci 8: 945-974.

McConnell S and Kaznowski C (1991): Cell cycle dependence of laminar determination in developing neocortex. Science 254: 282-285.

McDonald N, Lapatto R, Murray R, Gunning J, Wlodawer A and Blundell T (1991): New protein fold revealed by a 2.3 Å resolution crystal structure of nerve growth factor. Nature 354: 411-414.

Meakin S and Shooter E (1991): Molecular investigations on the high-affinity nerve growth factor receptor. Neuron 6: 153-163.

Mellon P, Windle J, Goldsmith P, Padola C, Roberts J and Weiner R (1990): Immortalization of hypothalamic GnRH neurons by genetically targeted tumorigenesis. Neuron 5: 1-10.

Merlio J, Ernfors P, Jaber M and Persson H (1993): Molecular cloning of rat trkC and identification of cells expressing mRNAs for members of the trk family in the rat Central Nervous System. Neurosci 51: 513-532.

Merlio J, Ernfors P, Kokaia Z, Middlemas D, Bengzon J, Kokaia M, Smith M, Siesjo B, Hunter T, Lindvall O and Persson H (1993): Increased production and activation of the tyrosine kinase receptor trkB after brain insults. Neuron 10: 151-164.

Middlemas D, Lindberg R and Hunter T (1991): trkB, a neural receptor protein-tyrosine kinase: evidence for a full-length and two truncated receptors. Mol Cell Biol 11: 143-153.

Milbrandt J (1987): Nerve growth factor induced gene encodes a possible transcription regulatory factor. Science 238: 797-799.

Milbrandt J (1988): Nerve growth factor induces a gene homologous to the glucocorticoid receptor gene. Neuron 1: 183-188.

Morris R, Davis S and Butcher S (1990): Hippocampal synaptic plasticity and NMDA receptors: a role in information storage. Phil.Trans R. Soc 329: 187-204.

Nakielny S, Cohen P, Wu J and Sturgill T (1992): MAP kinase activator from insulin-stimulated skeletal muscle is a protein threonine/tyrosine kinase. EMBO J 11: 2123-2129.

Noda M, Ko M, Ogura A, Liu D, Amano T, Takano T and Ikawa Y (1985): Sarcoma viruses carrying the ras oncogene induce differentiation-associated properties in a neuronal cell line. Nature 318: 73-75.

Ohmichi M, Decker S and Saltiel A (1992a): Activation of phosphatidylinositol-3 kinase by nerve growth factor involves indirect coupling of the *trk* proto-oncogene with *src* homology 2 domains. Neuron 9: 769-777.

Ohmichi M, Pang L, Decker S and Saltiel A (1992b): Nerve growth factor stimulates the activities of the Raf-1 and the mitogen-activated protein kinases via the trk protooncogene. J Biol Chem 267: 14604-14610.

Phillips H, Hains J, Laramee G, Rosenthal A and Winslow J (1990): Widespread expression of BDNF but not NT-3 by target areas of basal forebrain cholinergic neurons. Science 250: 290-294.

Price J and Thurlow L (1988): Cell lineage in the rat cerebral cortex: a study using retroviral-mediated gene transfer. Development 104: 473-482.

Pulver B, Kyriakis J, Avruch J, Nikolakaki E and Woodget J (1991): Phosphorylation of c-Jun mediated by Map kinases. Nature 353: 670-674.

Purves D (1988): "A Molecular Basis For Trophic Interactions In Vertebrates" Cambridge, Massachusetts: Harvard University Press.

Radeke M, Misko T, Hsu C, Herzenberg L and Shooter E (1987): Gene transfer and molecular cloning of the rat nerve growth factor receptor. Nature 9: 593-597.

Rakic P (1988): Specification of cerebral cortical areas. Science 241: 170-176.

Ramón y Cajal S (1990): "New Ideas On The Structure Of The Nervous System In Man And Vertebrates" Ist Edition (Swanson N, Swanson LW Trans.) Cambridge, MA: MIT Press.

Renfranz P, Cunningham M and McKay R (1991): Region-specific differentiation of the hippocampal stem cell line HiB5 upon transplantation into the developing mammalian brain. Cell 66: 713-729.

Rodriguez-Tebar A, Dechant G, Gotz R and Barde Y (1992): Binding of neurotrophin-3 to its neuronal receptors and interactions with nerve growth factor and brain-derived neurotrophic factor. EMBO J 11: 917-922.

Rohrer H, Hofer M, Hellweg R, Korsching S, Stehle A, Saadat S and Thoenen H (1988): Antibodies against mouse nerve growth factor interfere in vivo with the development of avian sensory and sympathetic neurons. Development 103: 545-552.

Rosenthal A, Goeddel D, Nguyen T, Lewis M, Shih A, Laramee G, Nikolics K and Winslow J (1990): Primary structure and biological activity of a novel human neurotrophic factor. Neuron 4: 767-773.

Rozakis-Adcock A, McGlade J, Mbamalu G, Pelicci G, Daly R, Li W, Batzer A, Thomas S, Brugge J, Pelicci P, Schlessinger J and Pawson T (1992): Association of SHC and GRB2/sem-5 SH2 containing proteins is implicated in activation of the Ras pathway by tyrosine kinases. Nature 360: 689-692.

Rozenberg F, Robain O, Jardin L and Ben-Ari Y (1989): Distribution of GABAergic neurons in late fetal and early postnatal rat hippocampus. Dev Brain Res 50: 177-187.

Ruley H (1983): Adenovirus early region 1A enables viral and cellular transforming genes to transform primary cells in culture. Nature 304: 602-606.

Ryder E, Snyder E and Cepko C (1990): Establishment and characterization of multipotent neural cell lines using retrovirus vector-mediated oncogene transfer. J Neurobiol 21: 356-375.

Schlessinger J and Ullrich A (1992): Growth factor signaling by receptor tyrosine kinases. Cell 9: 383-391.

Silva A, Paylor R, Wehner J and Tonegawa S (1992b): Impaired spatial learning in α-Calcium-Calmodulin Kinase II Mutant Mice. Science 257: 206-211.

Silva A, Stevens C, Tonegawa S and Wang Y (1992a): Deficient hippocampal long-term potentiation in α-Calcium-Calmodulin Kinase II mutant mice. Science 257: 201-205.

Snyder E, Deitcher D, Walsh C, Arnold-Aldea S, Hartwieg E and Cepko C (1992): Multipotent neural cell lines can engraft and participate in development of mouse cerebellum. Cell 68: 33-51.

Soltoff S, Rabin S, Cantley L and Kaplan D (1992): Nerve growth factor promotes the activation of phosphatidylinositol 3-kinase and its association with the trk tyrosine kinase. J Biol Chem 267: 17472-17477.

Soppet D, Escandon E, Maragos J, Middlemas D, Reid S, Blair J, Burton L. Stanton B, Kaplan D, Hunter T, Nikolics K and Parada L. (1991): The neurotrophic factor brain-derived neurotrophic factor and neurotrophin-3 are ligands for the trkB tyrosine kinase receptor. Cell 65: 895-903.

Spemann, H. (1938). "Embryonic development and induction". New Haven, Conn.: Yale University Press.

Stoppin L, Buchs P and Muller D (1991): A simplified method for organotypic cultures of nervous tissue. J Neurosci Meth 37: 173-182.

Swanson L and Cowan W (1977): An autoradiographic study of the organization of the efferent connections of the hippocampal formation in the rat. J Comp Neurol 17: 49-84.

Temple S (1989): Division and differentiation of isolated CNS blast cells in microculture. Nature 340: 471-473.

Tessarollo L, Tsoulfas P, Martin-Zanca D, Gilbert D, Jenkins N, Copeland N and Parada L (1993): trkC, a receptor for neurotrophin-3, is widely expressed in the developing nervous system and in non-neuronal tissues. Development 118: 463-475.

Tessier-Lavigne M, Placzek M and Lumsden A (1988): Chemotropic guidance of developing axons in the mammalian central nervous system. Nature 336: 775-778.

Thomas G (1992): MAP kinase by any other name smells just as sweet. Cell 68: 3-6.

Timmusk T, Belluardo N, Metsis M and Persson H (1993a): Widespread and developmentally regulated expression of Neurotrophin-4 in rat brain and peripheral tissues. Eur. J. Neurosci., 5, 605-613.

Timmusk T, Palm K, Metsis M, Reintam T, Paalme V, Saarma M and Persson H (1993b): Multiple promoters direct tissue-specific expression of the rat BDNF gene. Neuron 10: 475-489.

Troppmair J, Bruder J, App H, Cai H, Liptak L, Szeberenyi J, Cooper G and Rapp U (1992): Ras controls coupling of growth factor receptors Protein Kinase C in the membrane to Raf-1 and B-Raf protein serine kinases in the cytosol. Oncogene 7: 1867-1873.

Tsoulfas P, Soppet D, Escandon E, Tessarollo L, Rosenthal A, Nikolics K and Parada L (1993): The trkC locus encodes multiple neurogenic receptors that exhibit differential response to NT-3 in PC12 cells. Neuron 10: 975-990.

Turner D and Cepko C (1987): A common progenitor for neurons and glia persists in rat retina late in development. Nature 328: 131-136.

Turner D, Snyder E and Cepko C (1990): Lineage-independent determination of cell type in the embryonic mouse retina. Neuron 4:.833-845.

Ullrich A and Schlessinger J (1990): Signal transduction by receptors with tyrosine kinase activity. Cell 61: 203-212.

Valenzuela D, Maisonpierre P, Glass D, Rojas E, Nunez L, Kong Y, Gies D, Stitt T, Ip N and Yancopoulos G (1993): Alternative forms of rat TrkC with different functional capabilities. Neuron 10: 963-974.

Vetter M, Martin Z, Parada L, Bishop J and Kaplan D (1991): Nerve growth factor rapidly stimulates tyrosine phosphorylation of phospholipase C-gamma 1 by a kinase activity associated with the product of the trk protooncogene. Proc Natl Acad Sci USA 88: 5650-5654.

Walicke P (1988): Basic and acidic fibroblast growth factors have trophic effects on neurons from multiple CNS regions. J Neurosci 8: 2618-2627.

Walicke P, Cowan W, Ueno N, Baird A and Guillermin R (1986): Fibroblast growth factor promotes survival of dissociated hippocampal neurons and enhances neurite extension. Proc Natl Acad Sci USA 83: 817-821.

Wetmore C, Ernfors P, Persson H and Olson L (1990): Localization of brain-derived neurotrophic factor mRNA to neurons in the brain by in situ hybridization. Exp Neurol 109: 141-152.

Wetts R and Fraser S (1988): Multipotent precursors can give rise to all major cell types of the frog retina. Science 239: 1142-1145.

Witter M (1989): "Connectivity Of The Rat Hippocampus" (3 ed.). New York: Alan R. Liss, Inc.

Wood K, Sarnecki C, Roberts T and Blenis J (1992): Ras mediates nerve growth factor receptor modulation of three signal transducing protein kinases: MAP kinase, Raf-1, and RSK. Cell 68: 1041-1050.

Yan Q and Johnson E (1988): An immunohistochemical study of the nerve growth factor receptor in developing rats. J Neurosci 8: 3481-3498.

Yankner B, Caceres A and Duffy L (1990a): Nerve growth factor potentiates the neurotoxicity of beta amyloid. Proc Natl Acad Sci USA 87: 9020-9023.

Yankner B, Duffy L and Kirschner D (1990b): Neurotrophic and neurotoxic effects of the amyloid beta protein are reversed by tachykinin neuropeptides. Science 250: 279-282.

Zafra F, Hengerer B, Leibrock J, Thoenen H and D, L. (1990). Activity dependent regulation of BDNF and NGF mRNAs in the rat hippocampus is mediated by non-NMDA glutamate receptors. EMBO J 9: 3545-3550.

Zimmer J and Gahwiler B (1984): Cellular and connective organization of slice cultures of the rat hippocampus and fascia dentata. J Comp Neurol 228: 432-446.

FROM PRECURSOR CELL BIOLOGY TO TISSUE REPAIR IN THE O-2A LINEAGE

Andrew K. Groves[1] and Mark Noble[2]

[1]Division of Biology, 216-76, California Institute of
Technology, Pasadena, CA 91125, U.S.A. [2]Ludwig
Institute for Cancer Research, 91 Riding House Street,
London W1P 8BT, U.K.

INTRODUCTION

One of the preoccupations of our laboratory in recent years has been to achieve a detailed understanding of the mechanisms that control the proliferation and differentiation of glial cells and their precursors in the developing and adult central nervous systems. A long term goal of such studies is to determine the role of glial cells in aspects of disease and injury in the CNS, and to transplant well-characterised populations of precursor cells back into the CNS to assess their capacity for repair. In this review, we describe recent studies of glial precursor cells derived from the neonatal and adult CNS that have suggested possible strategies for the repair of demyelinated CNS lesions. In addition, we discuss how a novel approach for generating conditionally immortal cell lines has allowed us to develop an in vitro model that may be of use in the study of glial scarring.

THE O-2A*perinatal* PROGENITOR

The first step in our studies on oligodendrocyte development was the discovery that cultures derived from white matter tracts of the CNS contained two distinct astrocyte populations, termed type-1 and type-2 astrocytes (Raff et al., 1983a). These two cell types could be readily distinguished from each other on the basis of morphology, antigenic phenotype and response to growth factors. Most importantly, we found that optic nerves of perinatal rats contained a population of glial precursors which did not express glial fibrillary acidic protein (GFAP) at the time of isolation, but which could be induced to become GFAP[+] by growth in tissue culture.

Shortly after the identification of type-2 astrocytes, we discovered that the precursor cells which we had shown could become type-2 astrocytes in vitro could also be induced to differentiate into oligodendrocytes (Raff et al., 1983b).

Neural Cell Specification: Molecular Mechanisms and Neurotherapeutic Implications
Edited by Juurlink *et al.*, Plenum Press, New York, 1995

171

Differentiation of these cells into oligodendrocytes occurred when these precursor cells, termed oligodendrocyte-type-2 astrocyte (O-2A) progenitor cells, were grown in chemically-defined medium and in the absence of inducing factors. In contrast, differentiation into astrocytes required the presence of appropriate inducing factors, at least one of which is found in foetal sera from a number of different species (Raff et al., 1983b, 1983c).

Initial studies on oligodendrocyte differentiation of O-2A progenitors isolated from perinatal rat optic nerve revealed an unexpected paradox. The perinatal optic nerves in our studies were isolated at a time of maximal division of the oligodendrocyte lineage in vivo (Skoff et al., 1976a,b), yet O-2A progenitor cells (termed O-2A^{perinatal} progenitors to distinguish them from their adult counterpart; see below) did not divide in tissue culture. Resolution of this paradox began with the discovery that cortical astrocytes promoted O-2A^{perinatal} progenitor division *in vitro* (Noble and Murray, 1984). The astrocytes used in these studies expressed a phenotype similar to that of type-1 astrocytes of the optic nerve, which are the first identifiable glial cells to appear in the nerve (Miller et al., 1985). The similarity of these two populations led us to suggest that these astrocytes were responsible for supplying one or more diffusible mitogens necessary to keep O-2A progenitor cells in division. Moreover, populations of O-2A^{perinatal} progenitors grown in the presence of purified cortical astrocytes were capable of undergoing extensive division while also continuing to generate more oligodendrocytes (Noble and Murray, 1984), a pattern of behaviour similar to that occurring in vivo. Thus, the failure of O-2A^{perinatal} progenitors to divide in our initial in vitro studies was due to the lack of necessary mitogens, which appeared to be supplied by another glial cell type.

Further studies demonstrated that purified cortical astrocytes could also promote the correctly timed differentiation in vitro of O-2A^{perinatal} progenitors isolated from optic nerves of embryonic rats (Raff et al., 1985). The mechanisms by which this timing is controlled remain a mystery, although all evidence to date indicates that it is the O-2A^{perinatal} progenitors themselves which are measuring elapsed time (Temple and Raff, 1986; Raff et al., 1988; Bögler and Noble, 1994). A potential link between the measurement of elapsed time by dividing cells and the control of differentiation has also been observed, *inter alia*, for fibroblasts and haematopoetic stem cells (for review, see Groves et al., 1991). In the case of O-2A^{perinatal} progenitors, it appears that this biological clock causes clonally related dividing progenitors to synchronously differentiate into oligodendrocytes within a limited number of cell divisions (Temple and Raff, 1986; Raff et al., 1988; Wren et al., 1992). However, it is not yet known whether the mechanism which underlies this synchronous differentiation of clones of dividing cells is also responsible for the first appearance of oligodendrocytes in the rat optic nerve at the day of birth in vivo, or the equivalent time in vitro.

The effects of purified cortical astrocytes, and of type-1 astrocytes from the optic nerve, on O-2A^{perinatal} progenitor division in vitro appear to be mediated by platelet-derived growth factor (PDGF; Noble et al., 1988; Raff et al., 1988; Richardson et al., 1988; Pringle et al. 1989). O-2A^{perinatal} progenitors exposed to either PDGF or astrocyte-conditioned medium exhibited a bipolar morphology, migrated extensively (with an average migration rate of 21.4 ± 1.6 µm/hr) and divided with an average cell cycle length of 18 h. PDGF was also as potent as type-1 astrocytes at promoting the correctly timed differentiation in vitro of embryonic O-2A progenitors into oligodendrocytes (Raff et al., 1988). Moreover, antibodies to PDGF blocked the mitogenic effect of type-1 astrocytes on embryonic O-2A progenitor cells, causing these cells to cease division and to differentiate prematurely, even when growing on monolayers of type-1 astrocytes. Thus, this single mitogen was able to elicit a complex behavioural phenotype from O-2A^{perinatal} progenitors that included normal functioning of the cellular mechanisms involved in the measurement of elapsed time. More recent studies have indicated that neurons, which also promote division of O-2A^{perinatal} progenitors in vitro (Gard and Pfeiffer, 1990; Hunter and Bottenstein, 1991), may also be a source of PDGF. At present, however, the specific contributions of either neuron- or astrocyte-derived PDGF to the development of the O-2A lineage in vivo is not yet known.

O-2A progenitor cells can also express a variety of developmental programs other than the one elicited by exposure to PDGF (Bögler et al., 1990). For example, O-2A^{perinatal} progenitors induced to divide by basic fibroblast growth factor (bFGF) were multipolar and showed little migratory behaviour. In addition, cells induced to divide by bFGF had a cell-cycle length of 45 ± 12 hr, in contrast with the 18 ± 4 hr cell cycle length elicited by exposure to PDGF. These results indicate that PDGF and bFGF function in the O-2A lineage as modulators of differentiation as well as functioning as promoters of cell division. PDGF and bFGF also differ in their effects on oligodendrocytes themselves, in that only bFGF is able to promote division of these cells (Eccleston and Silberberg, 1985; Saneto and de Vellis, 1985; Bögler et al., 1990).

The effects of bFGF on the differentiation of O-2A progenitors are dependent on the context in which bFGF is seen by these cells. In our initial report we found that in optic nerve cultures exposed to bFGF the progenitor cells differentiated into oligodendrocytes prematurely. In contrast McKinnon et al. (1990) found in cultures prepared from cortical preparations that bFGF inhibited oligodendrocyte generation. More recently (Mayer et al., 1993), we found that purified cortical astrocytes secrete a factor (or factors) that overides the inhibitory effects of bFGF. Thus purified O-2A progenitors are inhibited from generating oligodendrocytes by exposure to bFGF while in heterogenous cultures such inhibition does not occur reproducibly. We have also found that the inhibitory effects of bFGF can be overridden by ciliary neurotrophic factor, a potential promoter of oligodendrocyte generation and maturation, thus raising the possibility that this is the active factor produced by cortical astrocytes (Mayer et al., 1994).

The most intriguing result of our studies with PDGF and bFGF was the discovery that O-2A^{perinatal} progenitors exposed simultaneously to these two mitogens continued to divide without differentiating into oligodendrocytes (Bögler et al., 1990). For example, cultures of optic nerves of 19d old rat embryos began to generate oligodendrocytes after 2 days when established in the presence of PDGF alone (Raff et al., 1988; Bögler et al., 1990), yet remained free of oligodendrocytes even after 10 d of growth in the presence of PDGF + bFGF (Bögler et al., 1990). Further studies have demonstrated that O-2A^{perinatal} progenitors can be continually grown for a year or more in vitro so long as cells are continuously exposed to both of these mitogens (S.C. Barnett and M. Noble, unpublished observations).

The discovery that co-operation between growth factors can cause prolonged self-renewal of precursors revealed a previously unknown means of regulating self-renewal in a precursor cell population. Such cooperation may, however, represent a more general phenomenon, as indicated by the importance of growth factor co-operation in promoting the extended division in vitro of other neural precursor cells (Weiss et al., this volume), and haematopoetic stem cells (Cross and Dexter, 1991).

Remyelination by Transplanted O-2A^{perinatal} Progenitors

The discovery that co-operativity between PDGF and bFGF allows extensive self renewal of O-2A^{perinatal} progenitor cells enabled us to prepare large numbers of highly purified precursor cells for the first time. Such populations allowed us to perform transplantation experiments to determine the capacity of this precursor cell type for regeneration and repair in animal models of demyelinating disease (Groves et al., 1993).

To investigate the possibility that populations of purified oligodendrocyte precursors could be used to achieve remyelination of demyelinating CNS lesions, we transplanted a variety of preparations of purified O-2A^{perinatal} progenitors into demyelinating lesions in the spinal cords of adult rats. These lesions were generated by injection of ethidium bromide into spinal cord white matter that had been exposed to 40 Grays of X-irradiation three days previously. This lesion model has been shown to consist of demyelinated axons residing in a glia-free environment which cannot be remyelinated by cells derived from the host animal (Blakemore and Crang, 1988; 1989; Crang et al., 1992).

In initial experiments, O-2A progenitor cells derived from optic nerves of 7 day old rats were grown in culture in the presence of PDGF and bFGF. Cells were passaged after one week in culture, and again after a further two weeks. This procedure routinely yielded cultures containing > 95% O-2A progenitor cells. A suspension of this cell population was injected into spinal lesions of ten syngeneic adult rats, with a series of non-transplanted lesioned animals and animals injected with medium alone serving as controls. Eight out of the ten lesioned animals receiving injections of O-2A progenitor cells displayed extensive remyelination by oligodendrocytes after three weeks, in contrast to the complete lack of remyelination observed in control lesions. In certain areas of the lesions, up to 90% of the demyelinated axons were reinvested with myelin sheaths possessing ultrastructural features and periodicity characteristic of CNS myelin.

We next demonstrated that the remyelinating oligodendrocytes in our experimental lesions were unequivocally derived from the transplanted cells by expressing the bacterial ß-galactosidase gene in O-2A progenitor cells whose purity had been enhanced to >99.5% by antibody-mediated cell capture (Barres et al., 1992; Mayer et al., 1993). Cells infected with the BAG retrovirus (Price et al., 1987) produced remyelination after the same survival period, although to a somewhat lesser degree than that exhibited by uninfected cells. Nevertheless, ß-gal expressing oligodendrocytes clearly contributed to the repair of the lesion.

The above experiments demonstrate that highly purified populations of O-2A progenitor cells may be used to achieve extensive remyelination of demyelinated CNS lesions. These populations could be expanded in vitro for several weeks simply by the application of two co-operating growth factors, without the necessity for the introduction of immortalising oncogenes, and could also be genetically modified to express a foreign protein in the lesion site. The present studies thus extend previous demonstrations of remyelination by transplantation of glial cells (e.g. Gumpel et al., 1983; Blakemore and Crang, 1988; 1989; Crang et al., 1992) by the utilization of purified precursor cells, by the use of co-operating growth factors to expand these populations and by the demonstration that these cells can be genetically modified and still retain function. In our study, such modification extended only to the genetic marking of transplanted cells with ß-galactosidase. These experiments show, however, that it may be possible to engineer O-2A progenitors to deliver therapeutic molecules such as members of the neurotrophin or cytokine families. Finally, our experiments also offer one of the first direct demonstrations that O-2A progenitor cells can give rise to myelinating oligodendrocytes in vivo.

THE O-2A*adult* PROGENITOR - A STEM CELL OF THE ADULT CNS

The experiments described above represent an example of attempts to bring about repair in the CNS by the transplantation of populations of embryonic or neonatal precursor cells. In general such precursor cells have only limited capacity for self-renewal, being instead destined to give rise to large numbers of differentiated progeny within a small period of time. Alternatively, it is conceivable that populations of undifferentiated stem cells may exist in the adult CNS, and that such populations could be recruited to effect repair without recourse to transplantation strategies. We describe below the characterisation of one such cell type, the adult form of the O-2A progenitor cell.

At the outset of studies in our laboratory on adult optic nerves, we found that O-2A progenitors isolated from adult animals (O-2A*adult* progenitors hereafter) differed from their perinatal counterparts in several ways. O-2A*adult* progenitors have a unipolar morphology *in vitro* (Wolswijk and Noble, 1989), whereas O-2A*perinatal* progenitors are usually bipolar (Small et al., 1987; Wolswijk and Noble, 1989). In addition, O-2A*adult* progenitors have a longer average cell-cycle time in vitro than O-2A*perinatal* progenitors (65 ± 18 h v. 18 ± 4 h; Noble et al., 1988; Wolswijk and Noble, 1989), migrate more slowly (4.3 ± 0.7 μm/h v. 21.4 ± 1.6 μm/hr; Small et al., 1987; Wolswijk and Noble, 1989) and take longer to differentiate (5 days versus 2 days for 50% differentiation; Wolswijk and Noble, 1989). Furthermore, O-2A*adult* progenitors stimulated to divide by type-1 astrocytes are labelled with the O4 antibody, whilst dividing O-2A*perinatal* progenitors are O4-negative (Wolswijk and Noble, 1989; I. Sommer and M. Noble, unpublished observations).

The discovery of an adult-specific glial precursor population raised a series of questions which were addressed in our subsequent studies. These questions were: (i) When do O-2A*adult* progenitors first appear? (ii) What is the developmental origin of these cells? (iii) How is a population of O-2A progenitors maintained in the optic nerve of adult animals? (iv) What are the molecular mechanisms responsible for the differences in behavior between O-2A*perinatal* progenitors and O-2A*adult* progenitors?

Small numbers of O-2A*adult* progenitor-like cells may be identified in rat optic nerve cultures from as early as 7 days after birth, although such cells do not appear to be present in optic nerve cultures isolated from newborn animals (Wolswijk et al., 1990). O-2A*adult* progenitors become the dominant progenitor population in the optic nerve by 1 month after birth. However, replacement of the perinatal progenitor population by the adult one occurs gradually, and both O-2A*perinatal* and O-2A*adult* progenitors can be isolated from optic nerves during the period between 7 days and 1 month after birth.

The source of O-2A*adult* progenitors during development appears to be a subpopulation of O-2A*perinatal* progenitors (Wren et al., 1992). This hypothesis is supported by time-lapse microcinematographic observations of cultures of dividing progenitors isolated from optic nerves of 3-week old rats. In these experiments, we observed families of cells in which the founder member expressed characteristics of O-2A*perinatal* progenitors and initially gave rise to further cells with the characteristics of O-2A*perinatal* progenitors (i.e., bipolar morphology, rapid rates of migration and division), but subsequent progeny in the family expressed the characteristics of O-2A*adult* progenitors (i.e., unipolar morphology, slow rates of division and migration). In addition, continued passaging of perinatal progenitors on monolayers of purified cortical astrocytes was associated with the generation of adult progenitor-like cells and loss of perinatal-progenitor like cells, a transition superficially similar to that which appears to occur in vivo (Wolswijk et al., 1990).

In contrast to the events thought to be involved in the initial appearance of O-2A^{adult} progenitors, we believe that maintenance of these cells in the adult optic nerve may be the result of asymmetric division and differentiation of O-2A^{adult} progenitors themselves (Wren et al., 1992). Analysis of the composition of individual colonies derived from O-2A^{adult} progenitors, grown at clonal densities, indicates that these cells can produce colonies containing both oligodendrocytes and dividing progenitor cells. Such a colony composition is strikingly different from that observed in colonies derived from O-2A^{perinatal} progenitors, in which the generation of oligodendrocytes is associated with cessation of progenitor division and initiation of differentiation in all members of a clonal family (Temple and Raff, 1986; Wren et al., 1992).

The capacity of O-2A^{adult} progenitors to undergo asymmetric division and differentiation is consistent with the hypothesis that these cells are stem cells (Wren et al., 1992). O-2A^{adult} progenitors also express other properties of stem cells, such as long cell-cycle times (Wolswijk and Noble, 1989; Wolswijk et al., 1991; Wren et al., 1992). The composition of clonal colonies of O-2A^{adult} progenitors growing *in vitro* on purified cortical astrocytes for 25 days also suggests that some of these cells may be able to also exist in a virtually quiescent state, with cell cycle lengths in excess of 150h, without differentiating (Wren et al., 1992). Moreover, O-2A^{adult} progenitors seem to be capable of undergoing prolonged self-renewal, at least *in vitro* (Wren et al., 1992). Finally, these cells have the important quality of being maintained in the optic nerve throughout life (Wolswijk and Noble, 1989; Wolswijk et al., 1990). Thus, at this stage of our research, we would suggest that it is most appropriate to regard O-2A^{adult} progenitors as stem cells (i.e., these cells are capable of functioning as a self-renewing population throughout life) and to regard O-2A^{perinatal} progenitors as true progenitor cells (i.e., cells which are programmed to differentiate within a limited time-span).

The derivation of O-2A^{adult} progenitors from O-2A^{perinatal} progenitors has interesting implications in respect to the possible developmental origin of stem cell populations. The results of our studies suggest that precursor cells with properties appropriate for early development can give rise not only to terminally differentiated end-stage cells, but also to a second generation of precursors with properties appropriate for later developmental periods (see discussions in Wolswijk and Noble, 1989; Noble et al., 1991). In the O-2A lineage, it is this second group of precursors which represents the self-renewing stem cells. Whether the ancestral relationship between stem cells and pre-stem cells in other lineages is similar to that seen in the O-2A lineage is not yet known.

We have only recently begun to investigate the molecular mechanisms which allow O-2A^{perinatal} and O-2A^{adult} progenitors to express their different behaviours, and know very little about the physiological alterations which underlie the differences between these cells. We have seen that O-2A^{perinatal} and O-2A^{adult} progenitors can express their characteristic properties when grown on the same monolayer of purified cortical astrocytes (Wolswijk et al., 1990). We have also found that both types of progenitors can be isolated from optic nerves of perinatal rats of between 7 days and 1 month of age, suggesting that these populations also co-exist in vivo. Moreover, we have found that the specific properties of O-2A^{adult} progenitors are elicited by growth in the presence of

PDGF (Wolswijk et al., 1991), just as exposure of O-2A^{perinatal} progenitors cells to PDGF is associated with expression of the specific properties which characterize these cells (Noble et al., 1988). Thus, these two cell types respond to a single signalling molecule by expressing different behaviours. Taken together, these results indicate that the differences between the O-2A^{adult} and O-2A^{perinatal} progenitors are intrinsic to the cells themselves and are not likely to be due to, for example, changes in the microenvironments found in the perinatal versus adult CNS, or due to responsiveness to different growth factors found within the same environment.

O-2A Lineage Cells and Repair of Demyelinating Damage in Multiple Sclerosis

Studies of the O-2A^{adult} progenitor cell in our laboratory have raised interesting questions concerning the properties of these cells and certain features of multiple sclerosis (MS). We originally suggested that the slow division rate of O-2A^{adult} progenitors induced to divide by type-1 astrocytes or PDGF might limit the ability of these cells to participate in lesion repair (Wolswijk and Noble, 1989; Noble et al., 1991). In addition, we found that O-2A^{adult} progenitors and immature oligodendrocytes bind and activate complement in the apparent absence of specific antibody, thus leading to their own destruction (Wren and Noble, 1989; see also Scolding et al., 1989). This property was not shared by O-2A^{perinatal} progenitors or other glial cells. Both of these properties are consistent with observations that repair of demyelinating lesions in children may occur more effectively than in adults (Kriss et al., 1988; for discussion of these issues see also Wren and Noble, 1989; Compston et al., 1991).

One of our most exciting recent findings concerning O-2A^{adult} progenitors is the discovery that simultaneous exposure of these cells to PDGF + bFGF converts many of them to a rapidly dividing and highly motile phenotype (Wolswijk and Noble, 1992). The potential relevance of such findings to the repair of lesions in vivo is indicated by the increased production of FGFs and PDGF in CNS lesions (Logan, 1990 Nietro-Sampedro et al., 1988; Lotan and Schwarz, 1992) and indications that repair of virally-induced demyelination in vivo appears to be preceded by increases in the numbers of O-2A^{adult} progenitor-like cells (Godfraind et al., 1989). It is particularly intriguing that our studies also suggest that the ability of O-2A^{adult} progenitors to maintain a rapidly dividing phenotype is not maintained beyond a small number of divisions, suggesting intrinsic limitations may exist in the extent to which these cells are capable of contributing to myelin repair (Wolswijk and Noble, unpublished observations). Such a possibility is reminiscent of claims that MS lesions are initially repaired, but eventually become permanently demyelinated.

THE H-2K^{b}tsA58 TRANSGENIC MOUSE - A SOURCE OF CONDITIONALLY IMMORTAL CELL LINES

Our studies on the biology of the perinatal and adult forms of the O-2A progenitor cell have been assisted immeasurably by our observations that co-operativity between growth factors can be utilised to maintain precursor cells in a proliferative state in the absence of differentiation. However, there are still only a small number of cases where growth factor co-operativity may be used to maintain precursor cells in an undifferentiated state. Many laboratories have instead sought to immortalise precursor cells by the introduction of immortalising oncogenes (see Noble et al., 1992 for review). Although this approach has undoubtedly been of immense importance, it has several disadvantages. Firstly, the efficiency of integration of the immortalising gene into the host cell genome following retroviral infection is rather low, and drops still further by several orders of magnitude when primary cells are infected instead of established cultures. This is a particularly intractable problem when populations of very rare cells are to be targeted for immortalisation. Secondly, there is a general requirement for the host cell to undergo at least one round of DNA synthesis in order to successfully integrate the immortalising gene into its genome. Thirdly, the requirement for reverse transcription of the retrovirus, followed by integration of the provirus into the host genome means that up to 72 hours can elapse between retroviral infection of a cell and expression of the immortalising gene in that cell. Lastly, the seemingly random integration of the provirus into the host cell genome means that different cell lines isolated from the same cell type using the same immortalising oncogene cannot necessarily be compared with one another.

In order to circumvent some of these difficulties, we developed a line of transgenic mice (H-2K^{b}tsA58) that harbour the tsA58 temperature sensitive mutant of the SV40 large T antigen (Tag) gene under the control of the H-2K^b Class I antigen promoter (Weiss et al., 1983; Israel et al., 1986). The H-2K^{b}tsA58 construct exerts two levels of control over the expression of the SV40 large T antigen in transgenic mice. The oncogene is not expressed functionally in vivo due both to the generally low level of transcription from the H-2K^b promoter, and to the high body temperature of the mouse. However, the immortalising activity of the T antigen can be activated in cells derived from the animal by growing them at 33°C in the presence of interferon (Jat et al., 1991). The transgenic animals are able to develop and breed normally in virtually all respects, although older animals do tend to develop thymic hyperplasia (but not neoplasia) with time (Jat et al., 1991). Cells derived from H-2K^{b}tsA58 transgenic mice are capable of being grown indefinitely at 33°C in the presence of interferon. However, no growth is seen when interferon is removed from the cultures, or when the cultures are shifted to 39.5°C, or both interferon withdrawal and temperature shift-up is imposed (Jat et al., 1991).

The combination of existing studies on the effects of TAg in tissue culture and in transgenic animals indicates clearly that this gene expresses its immortalising function in a very wide variety of cell types, including neuroblasts of the rodent CNS (Renfranz et al., 1991). In addition, expression from the H-2K^b promoter can be induced, or enhanced, in almost all cell types by exposure of cells to interferons (Wallach et al., 1982; Israel et al., 1986). Thus, in cells or tissues (such as brain) which normally express little or no Class I antigens, exposure to interferons activates transcription from this promoter. Moreover, cells which constitutively express high levels of Class I antigens can still be induced to express even higher levels of expression by exposure to interferons.

Astrocyte Cell Lines Derived from H-2K^{b}tsA58 Transgenic Mice Express Properties of Glial Scar Tissue

Our interest in interactions between neurons and glia in development, injury and disease led us to derive a series of cortical astrocyte lines from the H-2K^{b}tsA58 transgenic mouse line (Groves et al., 1993b). We examined four astrocyte lines in detail, all of which displayed similar properties.The four astrocyte lines expressed GFAP, albeit at lower levels than that seen in primary mouse cortical astrocyte cultures, and supported the proliferation and differentiation of O-2A progenitor cells through the production of PDGF. In addition, the four lines supported neurite outgrowth of both CNS and PNS neurons in the absence of aggregation or fasciculation.

In the course of our studies of neurite outgrowth on the astrocyte lines, we were surprised to observe that the extent of neurite outgrowth was significantly less than that seen on monolayers of primary cortical astrocytes. For example, mean postnatal cerebellar granule neurite lengths on the four astrocyte lines ranged from 120 µm to 150 µm, compared with a value of 370 µm seen on primary astrocytes. We hypothesised that this markedly less permissive substrate might be due to the expression of inhibitory molecules on the surface of the astrocyte lines. Characterisation of the expression of cell membrane and extracellular matrix molecules indicated that the astrocyte lines expressed a phenotype associated with glial scar tissue. All four lines expressed members of the J1/tenascin family, laminin and chondroitin sulphate proteoglycans, all of which are present in some glial populations during development and are particularly expressed in CNS lesions. (Liesi et al., 1983; Liesi, 1985; McKeon et al., 1991; Laywell and Steindler, 1991; Laywell et al., 1992). It was of interest to observe that the four astrocyte cell lines were also less effective at supporting the growth of P7 dorsal root ganglion neurons than primary astrocytes, although inhibition of neurite outgrowth from dorsal root ganglion neurons derived from of 18 day old embryos was not seen. In this respect also, the astrocyte cells lines appeared to behave like scar tissue, which is markedly less inhibitory for the growth of immature neurons as compared with mature neurons (Fawcett et al., 1989)

A final surprising observation was made when we examined the migration of O-2A progenitor cells on the astrocyte cell lines. As mentioned above, the astrocyte cell lines produced PDGF and promoted the division of O-2A progenitors in vitro. However, the astrocyte cell lines differed markedly from primary astrocytes in their ability to support O-2A progenitor migration. O-2A progenitor cells formed small, tight colonies on monolayers of all four astrocyte lines after seven days of growth, whereas on primary cortical astrocytes, O-2A progenitor cells were distributed much more evenly over the entire monolayer.

To determine whether the astrocyte lines were inhibiting the migration of O-2A progenitors (as compared with failing to promote such migration), we performed two confrontation experiments. O-2A progenitors were either plated onto primary astrocytes and allowed to migrate onto the astrocyte cell lines, or were plated onto the astrocyte cell lines and allowed to migrate onto primary astrocytes. In experiments where O-2A progenitor cells growing on primary astrocytes were challenged with a monolayer of transgenic astrocyte lines, very few O-2A progenitor cells were observed to cross the interface between the primary and transgenic astrocytes. Instead, the progenitor cells appeared to migrate to the astrocyte interface and stop, frequently aligning their processes along the interface. The apparent failure of O-2A progenitor cells to cross the interface from primary to transgenic astrocytes was not the result of a failure to cross an astrocyte interface per se, since in the majority of cases where O-2A progenitor cells were plated onto transgenic astrocyte monolayers and challenged with monolayers of primary astrocytes, progenitor cells were able to cross the astrocyte interface and migrate considerable distances over the primary astrocyte monolayer.

The inhibition of O-2A progenitor migration by our astrocyte cell lines is of particular interest in considering the possible properties of astrocytes involved in scar formation. Circumstantial evidence has suggested that certain specialised populations of optic nerve astrocytes from the lamina cribrosa may act as a barrier to the migration of O-2A progenitor cells in vivo, and these cells have been suggested to express properties like astrocytes seen in glial scar tissue (ffrench-Constant et al., 1988).

It is perhaps surprising that the four astrocyte lines examined all express markers characteristic of lesioned adult astrocytes although they were originally isolated from neonatal animals. This antigenic phenotype did not appear to be the result of expression of T antigen, as primary cortical astrocytes isolated from H-2K^{b}tsA58 transgenic mice and grown in permissive conditions express the properties we, and others, have previously documented for neonatal astrocyte populations examined in vitro. We do not yet know whether the cells we have grown represent a subset of astrocytes which simply grew particularly well. An alternative explanation for our observations, however, is that these astrocytes continued to mature in vitro despite the expression of low levels of TAg. Such a possibility is consistent with a variety of other data in our laboratory (unpublished observations).

The availability of clonal cell lines which exhibit properties much like that of glial scar tissue offers us a simple in vitro system for analyzing the biochemical cues which hinder neurite outgrowth from mature neurons and for identifying factors which might inhibit migration of O-2A progenitor cells. In relation to our interest in repair of demyelinating damage, it is this latter question which is of particular interest, as chronically demyelinated lesions are characterised by the presence of a large number of scar-like astrocytes (e.g. Ludwin, 1981). Our present results, together with the purported function of the scar-like astrocytes of the lamina cribrosa (ffrench-Constant et al., 1988), raise the possibility that one reason for the failure of these lesions to repair is due to an exclusion of O-2A progenitors from the lesion site.

CONCLUDING REMARKS

In this review, we have shown how a detailed study of CNS glial cell lineages in vitro has allowed us to make the first tentative steps towards proposing how glial cells and their precursors may be used to bring about repair in the CNS. In particular, we wish to stress that an understanding of the fundamental biological difference between perinatal and adult forms of the O-2A progenitor cell has suggested that these two cell types could conceivably be utilised in very different ways to contribute to repair of demyelinating lesions. In addition, the H-2K^{b}tsA58 transgenic mice offer us a means of immortalising other precursor cell populations, or to provide us with simplified cell culture systems with which to analyse CNS development and injury.

ACKNOWLEDGEMENTS

We wish to thank our colleagues at the Ludwig Institute, both past and present, for their support and advice over many years. Paris Ataliotis, Sue Barnett, Kishore Bhakoo, Oliver Bögler, Parmjit Jat, Margot Mayer, Guus Wolswijk and Damian Wren all participated in the research described in this review. In addition, it is a pleasure to thank Bill Blakemore, Robin Franklin and John Crang of the Cambridge School of Veterinary Medicine for an immensely enjoyable and fruitful series of collaborations, and to Dimitris Kioussis of the NIMR, Mill Hill for his help in generating the H-2K^{b}tsA58 transgenic mice.

REFERENCES

Barres BA Hart IK Coles HSR Burne JF Voyvodic JT Richardson WD and Raff MC (1992): Cell death and control of cell survival in the oligodendrocyte lineage.Cell 70: 31-46.

Blakemore WF and Crang AJ (1988): Extensive oligodendrocyte remyelination following injection of cultured central nervous system cells into demyelinating lesions in adult central nervous system.Dev Neurosci 10: 1-11.

Blakemore WF and Crang AJ (1989): The relationship between type-1 astrocytes, Schwann cells and oligodendrocytes following transplantation of glial cell cultures into demyelinating lesions in the adult rat spinal cord. J Neurocytol 18: 519-528.

Bögler O and Noble M (1994): Measurement of time in oligodendrocyte-type-2 astrocyte progenitors is a cellular process distinct from differentiation or division. Dev Biol 162: 525-538.

Bögler O Wren DR Barnett SC Land H and Noble MD (1990): Co-operation between two growth factors promotes extended self-renewal, and inhibits differentiation, of O-2A progenitor cells.Proc Natl Acad Sci USA 87: 6368-6372.

Compston DAS Scolding NJ Wren DR and Noble M (1991): The pathogenesis of demyelinating disease: insights from cell biology.Trends Neurosci 14: 175-182.

Crang AJ Franklin RJM Blakemore WF Noble M Barnett SC Groves A Trotter J and Schachner M (1992): The differentiation of glial cell progenitor populations following transplantation into non-repairing CNS glial lesions in adult animals. J Neuroimmunol 40: 243-254.

Cross M and Dexter TM (1991): Growth factors in development, transformation and tumorigenesis. Cell 64: 271-280.

Eccleston A and Silberberg DR (1985): Fibroblast growth factor is a mitogen for oligodendrocytes in vitro. Dev Brain Res 21: 315-318.

Fawcett JW Housden E Smith-Thomas and L Meyer RL (1989): The growth of axons in three-dimensional astrocyte cultures. Dev Biol 135: 449-458.

ffrench-Constant C Miller RH Burne JF and Raff MC (1988): Evidence that that migratory oligodendrocyte-type-2 astrocyte (O-2A) progenitor cells are kept out of the rat retina by a barrier at the eye-end of the optic nerve. J Neurocytol 17: 13-25.

Gard AL and Pfeiffer SE (1990): Two proliferative stages of the oligodendrocyte lineage (A2B5+O4- and O4+GalC-) under different mitogenic control. Neuron 5: 615-625.

Godfraind C Friedrich VL Holmes KV and Dubois-Dalcq M (1989): In vivo analysis of glial cell phenotypes during a viral demyelinating disease. J Cell Biol 109: 2405-2416.

Groves AK Bögler O Jat PJ and Noble M (1991): The cellular measurement of time. Curr Opin Cell Biol 3: 224-229.

Groves AK Barnett SC Franklin RJM Crang AJ Mayer M Blakemore WF and Noble M (1993a): Repair of demyelinated lesions by transplantation of purified O-2A progenitor cells. Nature 362: 453-455.

Groves AK Entwistle A Jat PS and Noble M (1993b): The characterisation of astrocyte cell lines that display properties of glial scar tissue. Dev Biol 159: 87-104.

Gumpel M Baumann N Raoul M and Jacque C (1983): Survival and differentiation of oligodendrocytes from neural tissue transplanted into newborn mouse brain. Neurosci Lett 37: 307-311.

Hunter SF and Bottenstein JE (1991): O-2A glial progenitors from mature brain respond to CNS neuronal cell line-derived growth factors. J Neurosci Res 28: 574-582.

Israel A Kimura A Fournier A Fellous M and Kourilsky P (1986): Interferon response sequence potentiates activity of an enhancer in the promoter of a mouse H-2 gene. Nature 322: 743-746.

Jat PS Noble MD Ataliotis P Tanaka Y Yannoutsos N Larssen L and Kioussis D (1991): Direct derivation of conditionally immortal cell lines from an H-2K^b-tsA58 transgenic mouse. Proc Natl Acad Sci USA 88: 5096-5100.

Kriss A Francis DA Cuendet F Halliday AM Taylor DS Wilson J Keast-Butler J Batchelor JR and McDonald WI (1988): Recovery after optic neuritis in childhood. J Neurol Neurosurg Psychiat 51: 1253-1258.

Laywell E and Steindler D (1991): Boundaries and wounds, glia and glycoconjugates: Cellular and molecular analyses of developmental partitions and adult brain lesions. Ann NY Acad Sci 633: 122-141.

Laywell E Dörries U Bartsch U Faissner A. Schachner M and Steindler D (1992): Enhanced expression of the developmentally regulated extracellular matrix molecule tenascin following adult brain injury. Proc Natl. Acad Sci USA 89: 2634-2638.

Liesi P (1985): Laminin-immunoreactive glia distinguish regenerative adult CNS systems from non-regenerative ones. EMBO J 4: 2505-2511.

Liesi P Dahl D Vaheri A (1983): Laminin is produced by early rat astrocytes in primary culture. J Cell Biol 96: 920-924.

Logan A (1990): The role of fibroblast growth factors in the central nervous system. Trends Endocrin Metab 1: 149-154.

Lotan M and Schwarz M (1992): Postinjury changes in platelet-derived growth factor-like activity in fish and rat optic nerves. J Neurochem 58: 1637-1642.

Ludwin SK (1981): Pathology of demyelination and remyelination. In Waxman SG and Ritchie JM (eds): "Demyelinating Disease: Basic And Clinical Electrophysiology," Raven Press: New York, pp 123-168.

Mayer M Bögler O and Noble M (1993): The inhibition of oligodendrocytic differentiation of O-2A progenitors caused by basic fibroblast growth factor is overridden by astrocytes. Glia 8: 12-19.

Mayer M Bhakoo K and Noble M (1994): Ciliary neurotrophic factor and leukaemia inhibitory factor promote the generation, maturation and survival of oligodendrocytes in vitro. Development 120: 143-153.

McKeon RJ Schreiber RC Rudge JS and Silver J (1991): Reduction of neurite outgrowth in a model of glial scarring following CNS injury is correlated with the expression of inhibitory molecules on reactive astrocytes. J Neurosci 11: 3398-3411.

McKinnon RD Matsui T Dubois-Dalcq M and Aaronson SA (1990): FGF modulates the PDGF-driven pathway of oligodendrocyte development. Neuron 5: 603-614.

Miller RH David S Patel ER and Raff MC (1985): A quantitative immunohistochemical study of macroglial cell development in the rat optic nerve: in vivo evidence for two distinct astrocyte lineages. Dev Biol 111: 35-43.

Nieto-Sampedro M Lim R Hicklin DJ and Cotman CW (1988): Early release of glia maturation factor and acidic fibroblast growth factor after rat brain injury. Neurosci Lett 86: 361-365.

Noble M (1991): Points of controversy in the O-2A lineage: Clocks and type-2 astrocytes. Glia 4: 157-164.

Noble M Groves AK Ataliotis P and Jat PS (1992): From chance to choice in the generation of neural cell lines. Brain Pathol 2: 39-46.

Noble MD and Murray K (1984): Purified astrocytes promote the in vitro division of a bipotential glial progenitor cell. EMBO J 3: 2243-2247.

Noble MD Murray K Stroobant P Waterfield MD and Riddle P (1988): Platelet-derived growth factor promotes division and inhibits premature differentiation of the oligodendrocyte/type-2 astrocyte progenitor cell. Nature 333: 560-562.

Price J Turner D and Cepko CL (1987): Lineage analysis in the vertebrate nervous system by retrovirus-mediated gene transfer. Proc Natl Acad Sci USA 84: 156-160.

Pringle N Collarini EJ Mosley MJ Heldin C-H Westermark B and Richardson WD (1989): PDGF A chain homodimers drive proliferation of bipotential (O-2A) glial progenitor cells in the developing rat optic nerve. EMBO J 8: 1049-1056.

Raff MC Abney ER and Fok-Seang J (1985): Reconstitution of a developmental clock in vitro: a critical role for astrocytes in the timing of oligodendrocyte differentiation. Cell 42: 61-69.

Raff MC Abney ER Cohen J Lindsay R and Noble M (1983a): Two types of astrocytes in cultures of developing rat white matter: differences in morphology, surface properties and growth characteristics. J Neurosci 3: 1289-1300.

Raff MC Miller RH and Noble M (1983b): A glial progenitor cell that develops in vitro into an astrocyte or an oligodendrocyte depending on the culture medium. Nature 303: 390-396.

Raff MC Miller RH and Noble M (1983c): Glial cell lineages in the rat optic nerve. Cold Spring Harbor Symp Quant Biol 48: 569-572.

Raff MC Lillien LE Richardson WD Burne JF Noble MD (1988): Platelet-derived growth factor from astrocytes drives the clock that times oligodendrocyte development in culture. Nature 333: 562-565.

Renfranz PJ Cunningham MJ and McKay RDG (1991): Region-specific differentiation of the hippocampal stem cell line HiB5 upon implantation into the developing brain. Cell 66: 713-729.

Richardson WD Pringle N Mosley M Westermark B and Dubois-Dalcq M (1988): A role for platelet-derived growth factor in normal gliogenesis in the central nervous system. Cell 53: 309-319.

Saneto RP and de Vellis J (1985): Characterisation of cultured rat oligodendrocytes proliferating in a serum-free chemically defined medium. Proc Natl Acad Sci USA 82: 3509-3513.

Scolding NJ Morgan BP Houston A Campbell AK Linington C and Compston DAS (1989): Normal rat serum cytotoxicity against syngeneic oligodendrocytes. J Neurol Sci 89: 289-300.

Skoff RP Price DL Stocks A (1976a): Electron Microscopic autoradiographic studies of gliogenesis in rat optic nerve. 1. Cell proliferation. J Comp Anat 169: 291-311.

Skoff RP Price DL and Stocks A (1976b): Electron Microscopic autoradiographic studies of gliogenesis in rat optic nerve. 2. Time of origin. J Comp Anat 169: 313-323.

Small RK Riddle P and Noble MD (1987): Evidence for migration of oligodendrocyte-type-2 astrocyte progenitor cells into the developing rat optic nerve. Nature 328: 155-157.

Temple S and Raff MC (1986): Clonal analysis of oligodendrocyte development in culture: Evidence for a developmental clock that counts cell divisions. Cell 44: 773-779.

Wallach D Fellous M and Revel M (1982): Preferential effect of g-interferon on the synthesis of HLA antigens and their mRNAs in human cells. Nature 299: 833-836.

Weiss EH Mellor A Golden L Fahrner K Simpson E Hurst J and Flavell RA (1983): The structure of the mutant H-2 gene suggests that the generation of polymorphism in H-2 genes may occur by gene conversion-like events. Nature 301: 671-674.

Wolswijk G and Noble M (1989): Identification of an adult-specific glial progenitor cell. Development 105: 387-400.

Wolswijk G and Noble M (1992): Cooperation between PDGF and FGF converts slowly dividing O-2A^{adult} progenitor cells to rapidly dividing cells with characteristics of O-2A^{perinatal} progenitor cells. J Cell Biol 118: 889-900.

Wolswijk G Riddle PN and Noble M (1990): Co-existence of perinatal and adult forms of a glial progenitor cell during development of the rat optic nerve. Development 109: 691-698.

Wolswijk G Riddle PN and Noble M (1991): Platelet-derived growth factor is mitogenic for O-2A^{adult} progenitor cells. Glia 4: 495-503.

Wren DR and Noble M (1989): Oligodendrocytes and oligodendrocyte/type-2 astrocyte progenitor cells of adult rats are specifically susceptible to the lytic effects of complement in absence of antibody. Proc Natl Acad Sci USA 86: 9025-9029.

Wren DR Wolswijk G and Noble M (1992): In vitro analysis of the origin and maintenance of O-2A^{adult} progenitor cells. J Cell Biol 116: 167-176.

NEURONAL CELL SPECIFICATION FROM CNS STEM CELLS

S. Weiss, S. Ahmed, A. Vescovi[1] and B.A. Reynolds[2]

Neuroscience Research Group, Departments of Anatomy and
Pharmacology & Therapeutics, University of Calgary Faculty of
Medicine, Calgary, AB, T2N 4N1, CANADA; [1]Instituto Neurologica
"C Besta", Via Celoria, Milan, ITALY; [2]NeuroSpheres Ltd, Calgary,
AB, CANADA

INTRODUCTION

Little is known about the factors and pathways leading to the production of neurons of the mammalian central nervous system (CNS). Although many in vitro and in vivo studies have described the presence of both unipotent and multipotent CNS progenitor cells and their regulation by extrinsic signaling molecules (reviewed in McKay, 1989), the pathways from primitive precursor to differentiated neuron have not been identified. This is contrasted by cells of the lympho-hematopoietic system, whereby a number of pathways leading from the primitive hematopoietic precursors to fully differentiated blood and immune cells have been described (Metcalf, 1989). Our lack of understanding regarding CNS cell production is due, partially, to a paucity of descriptions of primitive CNS precursors. Recently, we have isolated a cell from the embryonic (Reynolds et al., 1992) and adult (Reynolds and Weiss, 1992) mammalian CNS that fulfills the criteria representative of a stem cell. In this chapter, we review a series of studies which suggest that the pathway from CNS stem cell to differentiated neurons may involve the sequential actions of growth factors that act to influence undifferentiated neural precursors towards the neuronal lineage.

EGF-RESPONSIVE CNS STEM CELLS

Stem cells have been best described in studies of isolated cells from tissues that are self-renewing, such as the hematopoietic system, the skin and the intestinal crypts (Hall and Watt, 1989). Stem cells derived from these tissues share several features, including the ability to: proliferate, exhibit self maintenance, generate a large number of differentiated progeny characteristic of the tissue, retain their multi-lineage potential over time, and participate in the repair of the tissue after injury or

Neural Cell Specification: Molecular Mechanisms and Neurotherapeutic Implications
Edited by Juurlink *et al.*, Plenum Press, New York, 1995

185

disease (Potten and Loeffler, 1990). We have identified an epidermal growth factor (EGF)-responsive proliferating cell that is present in the embryonic (Reynolds et al., 1992) to adult (Reynolds and Weiss, 1992) mouse brain, which fulfills many of the criteria described above. This cell, when exposed to EGF, proliferates and forms a sphere of undifferentiated cells that can be induced, by serum and/or additional factors, to differentiate into neurons and glia (Reynolds et al., 1992; Reynolds and Weiss, 1992). We have recently shown (Reynolds and Weiss, submitted for publication) that, in addition to proliferation, this cell is capable of self renewal, the production of neuron, astrocyte and oligodendrocyte precursors and the retention of multipotency both in vitro and in vivo. In addition, we have found that in the adult mammalian forebrain, this cell is found in the subependymal zone and participates in subependymal zone re-population (Morshead et al., 1994). Therefore we propose that the EGF-responsive CNS cell is a stem cell capable of producing undifferentiated cells that can be differentiated into neurons and glia.

bFGF-RESPONSIVE NEURONAL PROGENITOR CELLS

Previous studies have identified basic fibroblast growth factor (bFGF) as a mitogen for neuronal precursors (Gensburger et al., 1987; Kilpatrick and Bartlett, 1993, Ray and Gage, 1994). However, there were no reports of the identity of the individual cells that responded to bFGF. After finding that the undifferentiated cells within the EGF-responsive stem cell-generated spheres expressed FGFR1 (the bFGF receptor), we examined the actions of bFGF on single cells dissociated from these spheres. This study (Vescovi et al., 1993) showed that bFGF acted on two distinct cell populations by stimulating their *limited* proliferation - a characteristic of progenitor cells. One of these bFGF-responsive progenitor cells is a unipotent neuroblast, a cell that divides to produce 3-17 neuronal precursors. The second bFGF-responsive cell is a bipotent neuronal/astroglial progenitor, a cell that produces a limited number of neuron and astrocyte precursors. These cells, found in almost equal proportions, make up approximately 9% of all the cells produced by the EGF-responsive stem cell. For both of the bFGF-responsive progenitors, however, differentiation of the neurons, as determined by morphologic and antigenic criteria, required the addition of serum Thus, the production of neuronal precursors from CNS stem cells may involve the stimulation of stem cell proliferation by EGF, followed by expansion of neuronal progenitors by bFGF. Additional factors, however, would be required for neuronal differentiation.

BDNF-RESPONSIVE NEURONAL PRECURSORS

Neurotrophins have been described as growth factors that specifically enhance the survival and process outgrowth of neurons of both the peripheral nervous system and CNS. We probed EGF-generated spheres for the expression of *trk* neurotrophin receptors. These analyses revealed that the undifferentiated cells expressed *trk* B, the receptor for brain derived neurotrophic factor (BDNF). When single spheres were exposed to BDNF, a greater number of neuronal precursors developed the morphologic and antigenic characteristics of differentiated neurons (Ahmed et al., submitted for publication). In particular, BDNF dramatically enhanced the outgrowth and branching of neuritic processes. Paradigms that involved addition and removal of the factor over extended periods yielded data suggesting that BDNF acts on all neuronal precursors, generated by the EGF-responsive stem cell. Moreover, these data suggest that BDNF enhances their differentiation/maturation but not their long term survival. Additional factors are likely required for the long term maintenance and survival of the neurons derived from CNS stem cells.

SUMMARY

Analogous to stem cells of other tissues, the EGF-responsive stem cells are capable of producing all the cells of the CNS. In order to do so, the stem cell produces undifferentiated cells, themselves capable of proliferation and ultimate terminal differentiation. With respect to neuronal production, the EGF-responsive stem cell produces at least two types of neuronal progenitors. These progenitors proliferate in response to bFGF to produce undifferentiated neurons that can be differentiated in response to BDNF and other undetermined factors. Hence, it is reasonable to conclude that the sequential actions of growth factors may be associated with neuronal cell specification from CNS stem cells.

ACKNOWLEDGEMENTS

The research referred to in this paper was supported by grants from the Medical Research Council of Canada (MRC) and Ciba-Geigy Canada Ltd. to S.W. During the course of this work, B.A.R. was recipient of an Alberta Heritage Foundation for Medical Research (AHFMR) Studentship. S.W. is an AHFMR Scholar and an MRC Scientist.

REFERENCES

Ahmed S, Reynolds BA and Weiss S (1994): BDNF induces the differentiation of neuronal precursors derived from EGF-responsive CNS stem cells. (submitted for publication).

Gensburger C, Labourdette G and Sensenbrenner M (1987): Brain basic fibroblast growth factor stimulates the proliferation of rat neuronal precursor cells in vitro. FEBS Lett 217: 1-5.

Hall PA and Watt FM (1989): Stem cells: the generation and maintenance of cellular diversity. Development 106: 619-633.

Kilpatrick TJ and Bartlett PF (1993): Cloning and growth of multipotential neural precursors: requirements for proliferation and differentiation. Neuron 10: 255-265.

McKay RDG (1989): The origins of cellular diversity in the mammalian central nervous system. Cell 58: 815-821.

Metcalf D (1989): The molecular control of cell division, differentiation commitment and maturation in haemopoietic cells. Nature 339: 27-30.

Morshead CM, Reynolds BA, Craig CM, McBurney MW, Staines WA, Morassutti D. Weiss S and van der Kooy D (1994): Neural stem cells in the adult mammalian forebrain: a relatively quiescent subpopulation of subependymal cells. Neuron 13: 1071-1082.

Potten CS and Loeffler M (1990): Stem cells: attributes, cycles, spirals, pitfalls and uncertainties. Lessons for and from the Crypt. Development 110: 1001-1020.

Ray J and Gage FH (1994) Spinal cord neuroblasts proliferate in response to basic fibroblast growth factor. J Neurosci 14: 33548-3564.

Reynolds BA and Weiss S (1994): Isolation and characterization of a mammalian embryonic CNS stem cell. (submitted for publication).

Reynolds BA and Weiss S (1992): Generation of neurons and astrocytes from isolated cells of the adult mammalian central nervous system. Science 255: 1707-1710.

Reynolds BA, Tetzlaff W and Weiss S (1992): A multipotent EGF-responsive striatal embryonic progenitor cell produces neurons and astrocytes. J Neurosci 12: 4565-4574.

Vescovi A, Reynolds BA, Fraser DD and Weiss S (1993): Basic fibroblast growth factor regulates the proliferative fate of both unipotent (neuronal) and bipotent (neuronal/astroglial) epidermal growth factor-generated progenitor cells. Neuron 11: 951-966.

COMMUNITY LIVING IN FOREBRAIN: NEURAL PRECURSORS MODIFY THEIR REGIONAL SPECIFICITY AFTER HETEROTOPIC TRANSPLANTATION IN THE TELENCEPHALON

Gordon J. Fishell

The Skirball Institute, 550 1st Avenue, New York, NY, 10016, U.S.A.

INTRODUCTION

It is at present unclear at what point during neural development neurons become committed to a particular regional fate. I have examined this question in forebrain by transplanting embryonic precursors from the basal (striatal) telencephalon to the adjacent dorsal (cortical) telencephalon. Striatal ventricular zone cells transplanted to a striatal environment adopt morphologies and axonal projections characteristic of striatal cells. In contrast, striatal ventricular zone cells transplanted to a cortical environment acquired morphologies and axonal projections specific to cortex. These findings suggest that position specific cues play an instructive role in determining critical aspects of regional phenotype. Presently, we are examining whether regional markers of basal forebrain such as Dlx-2, are shut-off and regional markers of dorsal forebrain, such as Emx-1 are expressed in cells moved from the striatal to the cortical VZ. Regardless of the outcome, the present experiments emphasize that restricted fate is not equivalent to restricted potential. This suggests that while the process of regional specification within these cells has begun at the time of transplantation, they are not yet regionally determined. Hence even after dorsal versus basal regions of VZ can be distinguished by both their characteristic shape and a variety of molecular markers, critical aspects of regional phenotype remain responsive to environmental cues.

REGIONALIZATION AND DETERMINATION IN THE TELENCEPHALON

Central to the question of how unique regional identities are achieved within the neuroaxis are the concepts of specification, fate and determination (Slack, 1983). Understanding the difference between these terms and their application to cellular

Neural Cell Specification: Molecular Mechanisms and Neurotherapeutic Implications
Edited by Juurlink *et al.*, Plenum Press, New York, 1995

189

patterning is critical to any examination of the mechanisms underlying commitment of cells to a mature phenotype. As it relates to forebrain, considerable evidence supports that the regional structure which neural precursors contribute can be predicted from the earliest time when regional proliferative zones of the forebrain are specified (Grove et al., 1993; Luskin et al., 1988; Walsh and Cepko, 1992, 1993; Price et al., 1992; Simone et al., 1992; Bulfone et al., 1993). These findings suggest that specification and fating of neurons to precise regions of the neuraxis is a relatively early developmental event. The question of whether these cells are regionally determined, however, has not been as thoroughly addressed. To do this requires an experimental approach where the fate of a precursor is challenged by placing it in a novel environment. To examine regional determination in forebrain, I have investigated the limits of cellular potential of neural cells through transplantation.

Although the neurons that populate both dorsal and basal forebrain become postmitotic over a similar time course their regional organization and cellular diversity in mature nervous system are strikingly different. The major basal structure of the forebrain, the striatum, has a nuclear organization and is largely composed of neurons with a single neuronal morphology, the medium spiny cell (Kemp and Powell, 1971). In contrast, the cortex, which forms the dorsal aspect of the forebrain, has a laminar organization and contains a diverse variety of neuronal morphologies, including pyramidal, stellate and fusiform cells (Shatz, 1992). These regions of forebrain also have highly specific patterns of axonal outgrowth. Cortical projection neurons consist of three basic classes: i) subcortically projecting neurons, which supply innervation to the brainstem and spinal cord, ii) commissural projecting neurons, which project to the contralateral cortex and iii) associational projecting neurons, which supply the intrinsic connections between various cortical areas and laminae. In contrast, the majority of projection neurons within the striatum are directed towards two subcortical structures, the globus pallidus and the substantia nigra. The close physical proximity of these structures combined with their divergent developmental patterning and axonal connections makes this area of the forebrain ideal for addressing questions of how regional patterning is established within the CNS.

To understand how this process occurs within the telencephalon, it is necessary to examine the earliest time in development when the regional divergence in patterning is evident. In contrast, to the segmental organization seen in hindbrain (Lumsden and Keynes, 1989: also see review by Lumsden et al, in this issue) regions within the forebrain are not separated by a 'traditional' neuromeric boundary and in fact share a common developmental history for much of their early development. The first indicators of their divergent regional fates appear a few days prior to the first wave of neurogenesis, with the emergence of discrete dorsal (cortical) and basal (striatal) ventricular zones (VZs) (Johnston, 1923; Smart and Sturrock, 1979), each having a characteristic pattern of gene expression and proliferation. The dorsal cortical VZ has a sheet-like epithelial structure and is molecularly characterized by the expression of a variety of genes including Emx-1 and 2, and Wnt-7b (reviewed by Puelles and Rubenstein, 1993). In contrast, the basal striatal VZ has a bulb-like structure (often referred to as the lateral ganglionic eminence) and is molecularly characterized by the expression of genes including Dlx-1 and 2, MASH-1 and Nkx-2.

Previous experiments have examined, through fate mapping, the extent of the domains that individual progenitors can populate with their progeny. These studies suggest that individual clones can populate surprisingly large areas of cortex (Walsh and Cepko, 1992) but only rarely cross the boundary region dividing dorsal and ventral telencephalon (Walsh and Cepko, 1993). However these observations must be tempered by the caveat that, due to the technical difficulties involved in marking specific progenitors early in development, these studies have been restricted to periods after the establishment of discrete dorsal and ventral regions of forebrain. This leaves open the possibility that earlier in development dorsal and ventral telencephalon share common precursors. Recently, direct visualization of the movement of neural progenitors within forebrain has been done (Fishell et al., 1993). This study has offered at least a partial explanation as to the cellular dynamics underlying the observed fate maps. This work has shown that while telencephalon progenitors are able to move within both the dorsal and basal VZs, they are unable to cross the border region that separates them. This raises the question of whether the cells situated on either side of this boundary area are restricted in terms of their regional identity or whether they are simply restricted in terms of their spatial movements. To assess whether the establishment of this border signifies that the cells on either side of it are regionally determined, I have transplanted striatal precursors to examine whether they are able to integrate and differentiate within the adjacent cortical environment (Fishell, 1994). The lipophylic dye PKH 26 was used to vitally label cells prior to transplantation. This allowed the detailed morphology and to some extent the projection pattern of these cells to be visualized.

TRANSPLANTATION STUDIES

In Vitro Transplantation Of Striatal Precursors Into Cortical VZ

To investigate the behavior of basal telencephalic precursors heterotopically moved to the dorsal telencephalic proliferative zone, in vitro heterotopic transplants were performed. Precursors from the striatal ventricular zone were then placed onto the cortical VZ surface of age matched forebrain explants. Although the cortical ventricular zone was intact, heterotopically transplanted cells were able to integrate into the cortical ventricular zone within four to eight hours after transplantation. Examination of the labeled cells at a variety of time points indicated that the cells integrated into the host tissue through a series of discrete steps. After a period of as little as four hours, cells placed on the ventricular surface of cortex extended a neuritic process into the host tissue. Within eight hours many of the cell bodies of donor cells had followed this process into the ventricular zone itself. By twelve hours some of the transplanted cells were indistinguishable from cells within the host tissue. To examine whether homotopic transplants had a higher affinity for reintegration than the heterotopically placed striatal precursors, the experiment was repeated using E15 cortical cells as the donor population. Surprisingly, although cortical cells are derived from the host environment, their ability to integrate into the host ventricular zone was neither quicker nor more efficient.

In Vivo Transplantation Of Striatal Precursors Into Cortical VZ

Given the ability of precursors to integrate either heterotypically or homotypically into in vitro explants, similar in vivo transplants were performed on embryonic animals by transplanting cells into the cerebral ventricles. Most animals receiving transplants had numerous cells integrated into both the striatal and cortical VZs. This allowed the fate of cells which integrated within the cortex to be directly compared with those that incorporated into the striatum.

Homotopic Integration Of Striatal Precursors. Transplanted striatal VZ cells that reincorporated homotopically within basal forebrain differentiated into neurons with a morphology typical of medium spiny striatal cells (Fig. 1). To test whether these neurons also made axonal projections appropriate to striatum, the retrograde neuronal tracer Fast Blue was injected into the substantia nigra, an area normally innervated by the striatum. In accordance with their position and morphology, many of transplanted cells were double-labeled with Fast Blue and PKH-26, suggesting that the transplanted cells had established efferent projections to the substantia nigra.

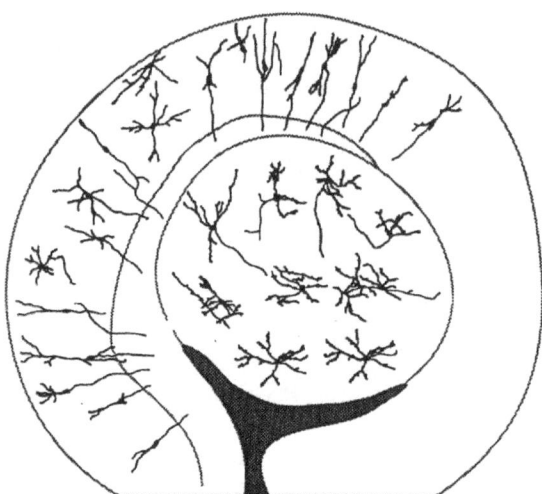

Figure 1. The morphology of striatal precursors after transplantation to the embryonic (E16.5) forebrain of recipient hosts. Shown are representative examples of cells after a fifteen day survival period following incorporation into either the cortex or the striatum. Transplanted cells were labeled with the lipophylic dye PKH 26. Using camera lucida the morphology of transplanted cells could be reconstructed. The resulting figure is a collage of transplanted cells from a total of eighteen different host animals.

Heterotopic Integration Of Striatal Precursors. Not only are the organization and morphology of the mature neuronal population within the cortex distinctly different than those seen in the striatum, so are the developmental events that lead to their establishment. Unlike striatal development, where postmitotic neurons coalesce into a nuclear structure, cortical development is characterized by a predictable series of steps in which the cells sequentially migrate, assemble and differentiate to form an ordered set of cortical laminae. Striatal VZ cells that incorporated into the cortex were examined during each of these phases to determine how closely their developmental progression follows normal cortical maturation.

Two days after transplantation, at E19, labeled cells were seen migrating through the intermediate zone and into the cortical plate. The profile of migrating cells was reminiscent of the appearance of neurons as they migrate either in vivo or in vitro. Five days after transplantation, at P1-P2, the transplanted striatal VZ cells that incorporated into the cortex were positioned in cortical laminae and had a morphology similar to developing pyramidal or stellate cells. At P6-7, after a ten day survival period, transplanted cells within the cortex produce a variety of typical mature "cortical" morphologies including pyramidal, stellate and fusiform (Fig. 1). By this age the appearance of the cells was appropriate to their position within the cortex. In the deeper layers of cortex (layers V and VI), most cells (>85%) resembled cortical pyramidal cells, with an apical dendrite and a descending axon. In the middle layers of cortex (layer IV), stellate cells with a radial array of dendrites and an axon projecting either to the deeper or more superficial cortical laminae predominated (65%, with the remainder having a pyramidal morphology). In the most superficial cortical laminae, transplanted cells took on a fusiform morphology (>95%, with the remainder having a stellate morphology).

Interestingly, when the projection pattern of the transplanted neurons was assessed through retrograde tracing, many of the transplanted cells that had adopted a pyramidal morphology had formed a commissural projection to the contralateral cortex. This is exactly in accord with the connections made by cortical cells during late cortical development (Koester and O'Leary, 1993). As striatal cells do not form axonal projections to either of these targets, this strongly suggests that the projection patterns of the transplanted cells are a result of these cells acquiring fates appropriate for their cortical "host" environment.

CONCLUSIONS

The present results suggest that precursors within the proliferative zones of forebrain remain responsive to regional cues after overt regionalization has occurred. Taking this as evidence that the regional fate of forebrain VZ cells is controlled by local cues, how then are these cues established within the CNS? It seems likely that local interactions between communities of cells within specific regions of the CNS act synergistically to provide positional cues (Barbe and Levitt, 1991; Bally-Cuif et al., 1992; Guthrie et al., 1992; Cohen-Tannoudji et al., 1994). This raises the question of how adjacent regions can maintain distinct sets of positional cues. It has been observed that the movement of VZ cells is restricted at the border between dorsal and basal VZs. Indeed, we have observed a glial palisade that divides these ventricular zones, which could act as a mechanical barrier to lateral movement between VZs (Fishell, 1994; Zimmerman et al., 1994). Together with the present findings, this suggests a model where boundaries between VZs act to isolate cell communities within forebrain. This in turn permits unique sets of positional cues to develop within cell communities populating specific areas of the CNS.

REFERENCES

Bally-Cuif L, Alvarado-Mallart R-M, Darnell DK and Wasse M (1992): Relationship between Wnt-1 and En-2 expression domains during early development of normal and extopic met- and mesencephalon. Development 115: 999-1009.

Barbe MF and Levitt PJ (1991): The early commitment of fetal neurons to the limbic cortex. J Neurosci 11: 519-533.

Bulfone A, Puelles L, Porteus MH, Frohma, MA, Martin GR and Rubenstein JLR (1993): Spatially restricted expression of Dlx-1, Dlx-2, (Tes-1), Gbx-2 and Wnt-3 in the embryonic day 12.5 mouse forebrain defines potential transverse and longitudinal segmental boundaries. J Neurosci 13: 3155-3172.

Cohen-Tannoudji M, Babinet C and Wassef M (1994): Early determination of a mouse somatosensory cortex marker. Nature 368: 460-463.

Fishell GJ, Mason CA and Hatte, M. (1993): Dispersion of neural precursors within the germinal zones of the forebrain. Nature 362: 636-638.

Grove EA, Williams BP, Li DQ, Hajihosseini M, Friedrich A and Price J (1993): Multiple restricted lineages in the embryonic rat cerebral cortex. Development 117: 553-561.

Johnston JB (1923): Further contributions to the study of the evolution of the forebrain. J Comp Neurol 35: 337-348.

Kemp JM and Powell TPS (1971): The structure of the caudate nucleus of the cat: Light and electron microscope. Phil Trans B 262: 383-401.

Koester S and O'Leary DDM (1993): Connectional distinction between callosal and subcortical projecting cortical neurons is determined prior to axon extension. Develop Biol 160: 1-14.

Lumsden A and Keynes R (1989): Segmental patterns of neuronal development in the chick hindbrain. Nature 337: 424-428.

Luskin MB, Pearlman AL and Sanes JR (1988): Cell lineage in the cerebral cortex of the mouse studied in vivo and in vitro with a recombinant retrovirus. Neuron 7: 685-647.

Price M., Lazzaro D, Pohl T, Mattei M-G, Ruther U, Olivo J-C, Duboule D and Di Lauro R (1992): Regional expression of the homeobox gene NKX-2.2 in the developing mammalian forebrain. Neuron 8: 241-255.

Puelles L and Rubenstein JL (1993): Expression patterns of homeobox and other putative regulatory genes in the embryonic mouse forebrain suggest a neuromeric organization. TINS 16: 472-479.

Shatz CL (1992): Dividing up the neocortex. Science 258: 237-238.

Slack JMW (1983): "From Egg To Embryo: Determinative Events In Early Devlopment." Cambridege University Press, London and New York.

Simeone A, Acampora D, Gulisano M, Stornaiuol, A,and Boncinelli E (1992): Nested expression domains of four homeobox genes in devloping rostral brain. Nature 358: 687-690.

Smart IHM and Sturrock RR (1979): The development of the medial and lateral ganglionic eminences. Ontogeny of the Neostriatum 23: 127-146.

Walsh C and Cepko CL (1992): Widespread dispersion of neuronal clones across functional regions of the cerebral cortex. Science 255: 434-440.

Walsh C and Cepko CL (1993): Clonal dispersion in proliferative layers of developing cerebral cortex. Nature 362: 633-635.

Zimmerman L, Parr B, Lendahl U, Cunningham M, McKay R, Gavin B, Mann J, Vassileva G and McMahon A (1994): Independent regulatory elements in the nestin gene direct transgene expression to neural stem cells or muscle precursors. Neuron 12: 11-24.

TARGETED NEOCORTICAL CELL DEATH GUIDES THE FATE OF TRANSPLANTED NEURAL PRECURSORS

Jeffrey D. Macklis

Department of Neurology and Program in Neuroscience,
Harvard Medical School, Neuroscience Division,
Mental Retardation Research Center, Children's
Hospital, Boston, MA 02115, U.S.A.

INTRODUCTION

The potential future transplantation of immature or genetically modified neurons to repopulate or provide molecular support to injured regions of the central nervous system may offer a future therapy for degenerative, dysgenetic, or acquired nervous system injury (Björklund et al., 1982; Gage et al., 1987; Gage and Fisher, 1991). One potential approach to such plasticity and reorganization is via control of donor cells using normal developmental mechanisms with additional cellular modifications and pharmacologic influences. Although repopulation of specific circuitry within the central nervous system offers one potential future therapeutic approach to degenerative or acquired disease, understanding the mechanisms of cellular control of neuronal differentiation and integration within various regions of the central nervous system would also be critical to therapeutic approaches in which engineered cells are transplanted to perform metabolic function or provide molecular signals or support to surrounding, intrinsic cells. Appropriate manipulations may allow grafted cells to acquire the correct phenotypes to provide molecular supplements or specific cellular repopulation. The latter would require exquisitely precise anatomic and physiologic cellular integration within the nervous system, especially within the neocortex, the most specialized and highly complex of all brain structures. Joint approaches combining cell biology with the rapidly advancing molecular understanding of nervous system pathobiology are likely to be of great importance both in possible future therapy and in modeling disease elements toward the understanding and prevention of developmental and degenerative disorders.

A model system in mice which provides selective, defined, and localized neuronal degeneration within neocortex (Macklis and Madison, 1991; 1993) allows careful dissection of key individual effects on the migration, integration, and function of immature neocortical neurons and genetically engineered neural precursor cell lines transplanted into the neocortex of selectively and anatomically

Neural Cell Specification: Molecular Mechanisms and Neurotherapeutic Implications
Edited by Juurlink et al., Plenum Press, New York, 1995

195

localized neuron-deficient hosts (Sheen and Macklis, 1992; Macklis, 1993). Selective photolytic neocortical neuron cell death can be induced using noninvasive laser energy that penetrates through the neocortex without injury to intermixed neurons, glia, axons, vascular and connective tissue, causing degeneration to desired subpopulations of neurons in vivo. Neurons are targeted retrogradely with latex nanospheres containing cytolytic chromophores activated only by 670 nm energy. This apoptotic cell death is associated with a cascade of events that parallel those found during apoptotic programmed cell death: singlet oxygen production, lipid peroxidation of lysosomal membranes, release of proteases, disruption of cytosolic proteins and cytoskeletal components, loss of calcium homeostasis by activation of L-type calcium channels and via increased membrane porosity, DNA fragmentation, and subsequent loss of cellular architecture with apoptotic bodies and phagocytic removal (Sheen et al., 1992; Sheen and Macklis, 1994). The slow, non-necrotic process of targeted neuronal cell death in vivo may activate many of the same physiological cues that are activated by programmed cell death during normal development and during organizational refinement in the adult vertebrate nervous system, explaining the directed migration and differentiation of neuronal precursors transplanted into these regions of targeted neuronal degeneration.

Neuronal transplantation in neocortex as a repopulation therapy for degenerative injury would necessitate specific migration and synaptic integration of grafted cells into positions within the neuronal network left vacant by degenerated neurons. Models for such transplantation must reflect the intricate interconnections of neurons which degenerate within cerebral cortex. Nonselective lesioning methods (e.g. ischemic, excitotoxic, surgical) do not allow for such migration and integration in either neocortex or cerebellum. The model system of targeted photolytic neuronal cell death, producing selective neocortical degeneration using noninvasive laser illumination, allows directed studies of neocortical transplantation by providing degeneration to callosally projecting pyramidal neurons that is geographically defined, slowly progressive, and cell-type specific.

Controlled neuronal degeneration allows analysis of developmental factors, normally reduced or absent in adult mice, responsible for neuronal specification, migration, and differentiation during normal neocortical development. The central hypothesis under investigation is that extremely specific degeneration to individual neuronal subpopulations will lead to transient alterations in the surrounding cell populations and extracellular matrix that promote elements of developmentally correct plasticity, even within adult or aged animals. These regulatory alterations may mimic those influenced by apoptotic programmed cell death during initial nervous system development.

We now have compelling evidence in a variety of distinct systems that three components of normal neocortical differentiation can be modified by targeted photolytic neuronal cell death: i) initial specification to neuronal phenotype, ii) migration and iii) cell-type specific neuronal differentiation. Experiments in juvenile, adolescent, and adult hosts reveal preferential migration, differentiation, and integration by transplanted embryonic neurons in regions of selective neuron degeneration (Macklis, 1993; Sheen and Mackliss, 1992). Neurons can be seen to actively migrate toward the neuron-deficient regions using time-lapse digital confocal microscopy (Hernit-Grant and Macklis, 1994). Once in position, neurons differentiate and a small number send long projections to the original distant targets in the contralateral hemisphere, even in fully adult mice (Hernit-Grant and Macklis, 1993). Other experiments (Yoon et al., 1993) using genetically immortalized, multipotent cerebellar precursor cells transplanted into adult host mice with targeted neocortical neuron death demonstrate unique specification of a subpopulation toward a neuronal phenotype (morphology, ultrastructure, and immunocytochemical expression); morphologically and ultrastructurally they appear to possess neocortical pyramidal phenotype. Such differentiation is not seen in neocortex in intact or kainic acid lesioned controls beyond the embryonic period (Snyder et al., 1993).

Studies of integration by embryonic neurons and multipotent precursors into "vacant" neuronal locations within an otherwise normal environment at various developmental stages offer an in vivo three-dimensional equivalent to studies of cellular development and interaction in tissue culture. The host brain can be viewed as a custom, neurological "mutation" of precisely defined cell type, geographic location and time after onset of neuronal degeneration. Issues of laminar commitment, cell fate, and extracellular/intracellular cues during normal developmental cell positioning can be probed in vivo by alteration of both donor cell type and host environment. Similarly, understanding the factors controlling such integration is critical to the goal of returning function by transplantation within the complex neocortex injured by a variety of dysgenetic or degenerative influences. The specific experiments described here primarily study neuronal migration and differentiation following transplantation of embryonic neocortical neurons or clonal, multipotent precursor cells into "custom-generated" pyramidal neuron-deficient neocortex of juvenile and adult mice. The events we observe occur within a highly defined and controllable host microenvironment that has been altered to attempt to re-establish certain features of the environment present during initial neocortical development. Similar paradigms may prove useful in transplantation studies involving other regions of the nervous system and other degenerative diseases. These same studies will contribute to our understanding of normal and pathological events of neuronal migration, differentiation, and integration during neocortical development.

HOST MODEL SYSTEM: PHOTOLYTIC DEGENERATION IN NEOCORTEX

Targeted Photolytic Neuronal Cell Death

We have developed a novel approach to targeted neuronal cell death involving laser-induced photolytic degeneration, in order to address questions of cell-cell interactions and specifically microenvironmental effects on neuronal migration, laminar specificity, differentiation, and integration, both during neocortical development and following transplantation. This approach was first developed in vitro (Macklis and Madison, 1985; 1991; 1993; Madison et al., 1988) and then applied to murine neocortex in vivo (Madison et al., 1990). The currently employed photolytic neuron cell death (Fig. 1) results from the photoactivation of retrogradely transported, nontoxic nanospheres (Katz et al., 1984) carrying a chromophore (chlorin e_6). Chlorin e_6 produces the cytotoxic molecule singlet oxygen when excited at near-infrared wavelengths, a toxic species with a diffusion range of approximately 0.1-0.2 microns that causes lipid peroxidation, protein oxidation, and interference with DNA transcription (Sheen and Mackliss, 1994).

Light at these wavelengths (produced by a specific 674 nm laser and beam controlling optics) penetrates deeply through nervous system tissue without absorption or cellular injury to non-targeted cells. At long wavelengths in the range of 650 nm to 850 nm, an "optical window" exists where near-infrared light energy can penetrate nervous system tissue several millimeters without significant absorption by unpigmented tissue (Anderson and Parrish, 1983). Photolysis by singlet oxygen ($^{1}O_{2}$) is a mechanism by which unfocused laser energy produces extremely selective, noninvasive degeneration to entire subpopulations of neurons containing appropriate exogenously targeted photolytic chromophores with absorption in the near-infrared range. Light activated chromophores may be selectively targeted to appropriate populations of neurons, producing no injury prior to specific wavelength illumination. It is unnecessary to aim at targeted neurons because optical and absorptive properties of these cells and neighboring tissue provide selectivity of cellular injury. Such photolytic degeneration allows a two stage selectivity via non-toxic agents that are delivered relatively

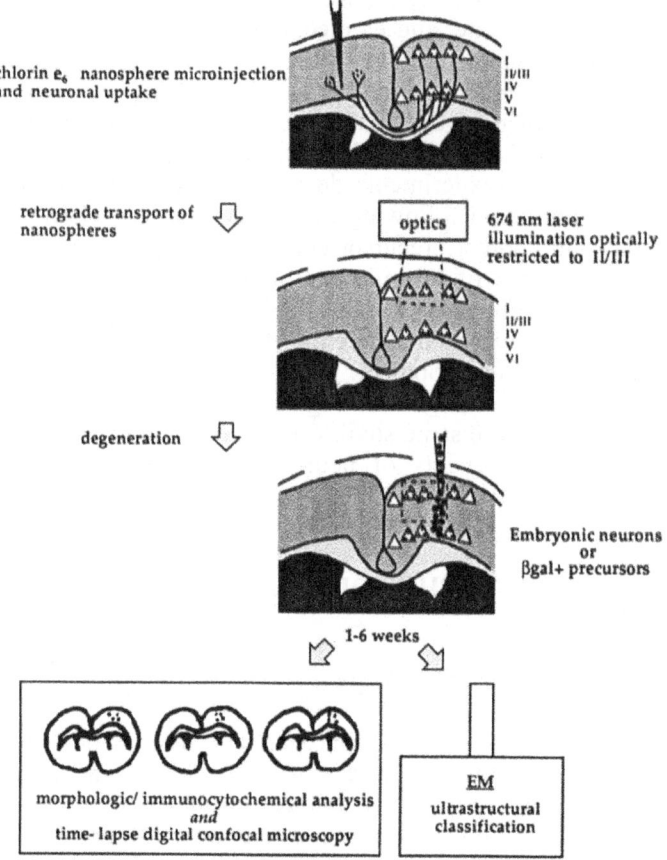

Figure 1. Sequential steps to produce regions of targeted photolytic pyramidal neuron degeneration for precursor cell transplantation. Energy dosimetry is selected to initiate slowly progressive, non-inflammatory, apoptotic neuronal cell death to targeted neurons within lamina II/III of the exposure field. Labeled neurons outside the defined region of laser illumination remain uninjured. Precursor cells are transplanted; their cell type specification, migration, final differentiation, and cellular integration may be studied.

specifically, with added specificity achieved by geographic localization of laser illumination. Targeted photolytic neuronal cell death is different from methods using laser microbeams, fluorescent dye injections, or focused fluorescent beams to rapidly injure individual neurons in vitro, or individual developmentally important cells or proteins in developing invertebrates and translucent vertebrate embryos (Miller and Selverston, 1979; Bently and Caudy, 1983; Raper et al., 1984; Eisen et al., 1989; Jay and Keshishian, 1990). Only microns of tissue penetration are possible with those methods, and only small numbers of cells or localized protein are injured with individual exposures. Using targeted photolytic degeneration, only cells selectively prelabeled with the chromophore undergo the slow, non-necrotic cell death following laser illumination; intermixed neurons, glia, axons and connective tissue are unaffected.

Unique Mechanisms of Targeted Photolytic Cell Death

Previous work (Madison and Macklis, 1993) demonstrated that loss of targeted pyramidal neurons in laminae II/III of mouse cortex is first apparent by routine histology approximately 3 to 4 weeks after laser exposure. More sensitive silver degeneration stains reveal that early stages of degeneration begin from a few hours to a few days following laser illumination, and numbers of pyknotic, degenerating cells are maximal at 8-10 days (Fig. 2). Histologically, this very slow and progressive, non-inflammatory cell death demonstrates these and other elements in common with apoptosis. The death of such cells is mediated by both calcium dependent and calcium independent mechanisms associated with the progressive loss of membrane integrity over hours to days (Macklis and Madison, 1985; Sheen et al., 1992).

We have further investigated the mechanisms of this targeted cell death and found a cascade of events that parallel those found following genetic activation of apoptotic programmed cell death, beginning with singlet oxygen production causing lipid peroxidation of lysosomal membranes, release of proteases, disruption of cytosolic proteins and cytoskeletal components, loss of calcium homeostasis by activation of L-type calcium channels and via increased membrane porosity, DNA fragmentation seen by gel electrophoresis with a ladder pattern and by 3'-OH end terminal labeling in vivo and in vitro, and subsequent loss of cellular architecture with apoptotic bodies and phagocytic removal (Sheen and Macklis, 1994). The slow, non-necrotic process of targeted neuronal cell death in vivo may activate many of the same physiological cues that are activated by programmed cell death during normal development and during organizational refinement in the adult vertebrate nervous system.

NEOCORTICAL DEGENERATION AND NEURONAL TRANSPLANTATION

Juvenile Host Mice

These approaches were modified for use in the neonatal period, to produce selective neuron deficiency in young mice beyond the normal migrational period, but during a period in which the normal developmental environment might be most generally permissive for repopulation, integration, and neuronal transplant viability. The hypothesis under investigation was that the laminae of host cortex deficient in neurons in this novel model may exert selective neurotropic action on fetal neocortical neurons, just as the Purkinje cell layer deficient in neurons in the PCD mutant has been shown to be selectively repopulated after transplantation of immature cerebellum. Neuron counts revealed a reduction in total neuron density of typically 25%-30% within targeted areas, with approximately 65% loss of larger projection neurons and no change in the number of small, presumptive interneurons. Regions of neuronal degeneration were approximately 600-800 μm wide and spanned the thickness of lamina II/III. No significant inflammatory reaction was visible and glial fibrillary acidic protein immunoreactivity was not increased.

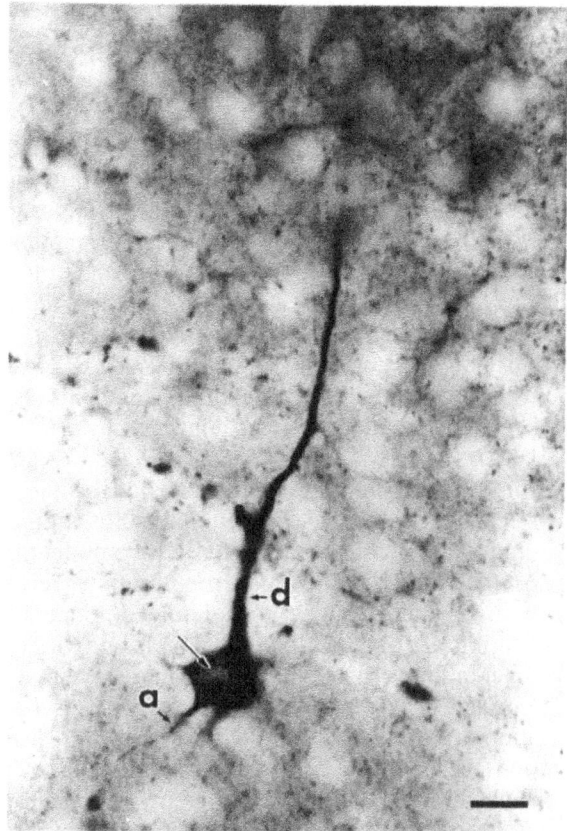

Figure 2. A callosally projecting pyramidal neuron in lamina II/III undergoing targeted photolytic cell death, visualized by silver degeneration staining. The apical dendrite ("d"), the axon ("a") projecting toward the corpus callosum, and the prominent, pale nucleus (arrow) are indicated. Scale bar = 25 μm.

Embryonic day 17 (E17) neocortical cell suspensions, containing recently postmitotic neurons destined to form lamina II/III, were transplanted lateral to these geographically defined regions of ongoing neuron degeneration at postnatal day 14 to 15 (P14-P15). Controls included several ages of donor embryonic neocortical cells transplanted into intact hosts, transplants into kainic acid lesion zones, embryonic cerebellar transplants to the photolytic degeneration zones, lysed neuron transplants, and transplants distant from the photolytic degeneration zones. Cellular injections spanned laminae II to V, to provide donor neurons both lateral and laminar choice for possible migration and integration. Donor cells were labeled in vitro with unique nanospheres that allowed distinct identification of donor cells at both light- (LM) and electron-microscopic (EM) levels. Samples of each graft suspension were followed over the first few weeks in tissue culture to assure excellent cellular viability and intracellular fluorescent nanosphere labeling.

These experiments demonstrated at the LM and EM levels that transplanted neurons undergo unique and directed long-distance migration, pyramidal neuron differentiation, and synaptic integration within the regions of cell type-specific neuronal cell death, not seen under any of the control conditions. Neurons placed near host zones of neuron deficiency within laminae II/III migrated up to 780 μm and integrated specifically within these zones. Quantitative analysis revealed that 44% of transplanted neurons migrated into the regions of photolytic degeneration beyond three standard deviations of the controls. Migration and integration did not occur in normal, unaffected deeper layers IV to VI of these experimental mice, nor in the normal lamina II/III bordering the transplantation site on the side opposite the neuron-deficient region. Control grafts revealed only minimal local spread without laminar preference. The control results were consistent with the extensive literature on grafts of neocortex, hippocampus, or cerebellum into neonatal or adult rodents. Donor neocortical neurons within the regions of targeted neuron death largely assumed a pyramidal neuron morphology and extended early processes; neurons outside these regions and in controls were small and ovoid, morphologically nonpyramidal. Electron microscopy was used to further confirm donor identity of these neurons, using standard morphologic criteria. The striking results suggested that targeted photolytic neuronal cell death produces an environment which, at the least, is permissive for cellular repopulation and may even lead to enhancement or re-expression of developmental cues used to guide normal neocortical development.

Adult Host Mice

Another set of experiments (Sheen and Macklis, 1992) considered whether potentially upregulated developmental signals may similarly guide directed repopulation in adult mice with selective neuronal degeneration, at times when normal migration and development is considered completed. Experiments using adolescent and adult hosts with selective neuron degeneration demonstrated equivalent findings within the fully adult host environment that are not seen with excitotoxic or other nonselective neuronal injury. Analysis of these experiments in vivo in the days immediately following transplantation revealed small clusters of migrating donor neurons moving away from the injection site, across normal layers V and IV, and toward the neuron-deficient regions beginning approximately four days after transplantation. These neurons entered the regions of ongoing neuronal degeneration by one week, grouped in clusters that later assumed pyramidal neuron morphology and extended processes. By two weeks following transplantation, neurons migrated 400-500 μm and began extending processes toward the corpus callosum. Approximately nineteen percent of these neurons send long projections to the contralateral homologous somatosensory cortex over the next several weeks even in such fully adult brains (Hernit-Grant and Macklis, 1993). Projections were not made to ipsilateral S2 or motor cortex, nor to thalamus, potential but incorrect targets, suggesting a high level of specificity to the newly formed projections. Parallel preliminary studies using a living slice culture system and time-lapse digital video confocal microscopy demonstrate both radial and tangential migration across intact deeper lamina V and IV by transplanted neurons on their way to lamina II/III repopulation, using axons and de-differentiated radial glia, re-expresssing the marker RC-2 (Edwards et al., 1990), as guiding structural elements (Hernit-Grant and Macklis, 1994).

Microenvironmental alterations following targeted neuronal cell death appear to be able to alter the laminar positioning fate of precursor cells, as demonstrated by analysis of transplanted precursor cells following tritiated thymidine birthdating (Sheen and Macklis, 1992). Donor E14 neurons "destined" to form deep laminae V and VI were prelabeled with combinations of fluorescent nanospheres, DiI, and ^{3}[H]thymidine to determine neuronal "birthdate". Heavily tritium-labeled E14 birthdate neurons, labeled for 6-8 hours in situ, long enough to be "determined" during the cell cycle to populate layer V and VI (McConnell and Kaznowski, 1991), migrated deep within the regions of targeted neuronal degeneration in superficial lamina II/III, along with lightly labeled or unlabeled neurons. Many neurons that migrated into these regions differentiated into large diameter pyramidal neurons, unlike the smaller, ovoid interneuron morphology of neurons that remained local to the injection site. These results suggest that E14 neuronal precursors can be directed within a range of potential differentiation choices by extrinsic microenvironmental signals after the time they would be normally "committed" in situ, and that such microenvironmental alterations are possible even in adult mice.

Multipotent Neural Precursor Cell Lines

In two ongoing collaborations, we are investigating the potential control exerted by these altered microenvironments over two recently developed, oncogene-transformed, multipotent cell lines: the cerebellar-derived C17-2 line described by Snyder et al. (1992) and the HiB5 line described by Renfranz et al (1991). We have investigated the potential differentiation and migration of the C17-2 cells transplanted from early neural tube closure (E9) through full adulthood in mice, finding broad embryonic potential that is restricted postnatally following the developmental schedule of normal ontogeny; of note, no neuronal differentiation occurred in neocortex postnatally (Snyder et al., 1993). When transplanted into 1, 2, 4, and 6 week old intact hosts, or adult mice with defined kainic acid lesions to assess effects of nonspecific excitotoxicity, C17-2 cells display only glial differentiation, both oligodendroglia and astrocytes by morphologic, immunocytochemical, and ultrastructural criteria.

In sharp contrast, in regions of photolytically-targeted neuron death in adult mice, striking and unique morphologic and ultrastructural specification to neuronal phenotype occurs in a subpopulation of transplanted progenitors, never observed in intact or kainic acid-lesioned regions (Yoon et al., 1993). Ultrastructural neuronal morphology was determined by standard criteria as well as direct identification of axosomatic synapses, neurofilaments, and microtubules. We interpret this as further support for the notion that this targeted cell death is altering microenvironmental signals toward those found during initial development, allowing or directing neocortical neuronal differentiation that normally occurs only during embryonic corticogenesis.

In parallel studies using the HiB5 line, we first investigated differentiation in embryonic through adult neocortex, both intact and kainic acid lesioned, as controls for currently ongoing experiments placing these cells into regions of photolytically targeted neuron death. We find a sharp restriction to glial phenotype (morphologic, ultrastructural, and immunocytochemical) in intact neocortex after the first postnatal day, paralleling the findings with C17-2 cells. We have also demonstrated microenvironmental effects on HiB5 cells in striatum of adult mice leading to neuronal phenotype only within restricted, periventricular regions, as opposed to glial phenotype in neocortex, corpus callosum, and the remainder of the striatum (Sheen et al., 1993).

DISCUSSION

These results suggest that there are likely to be multiple levels of hierarchical control over the migration, laminar positioning, and differentiation of neocortical neurons during development and, therefore, during attempts to experimentally repopulate regions of neocortex. During the normal developmental timetable in cortex, a specific time sequence of expression of neuronal, afferent, glial, trophic, and extracellular matrix signals may occur, and these naturally disappear after a relatively brief period (Angevine and Sidman, 1961; Geschwind and Hockfield, 1989; Austin and Cepko, 1990; Culican et al., 1990; Cohen et al., 1991). Some of these permissive or enhancing factors may be upregulated or reactivated given a specific enough degeneration of a selected subpopulation of neurons. The host model used in these studies provides control and definition over both temporal rate and extent of the neuronal death, and, thus, potential host environmental cues.

The finding of apparent preferential migration of a large fraction of transplanted E17 or E14 neurons to regions of host mouse neocortex with selective neuron degeneration provides a partial answer to the hypothesis under study regarding restoration of normal cytoarchitecture. The suggestion from the experiments to date is that, even after the normal period for appropriate neuronal migration to correct laminar positions is completed, it is possible to reactivate a specific combination of environmental cues to allow cellular integration that otherwise would not occur in normal or nonspecifically lesioned animals. It is possible that such reactivated cues are capable of at least partially superseding intrinsic neuronal signals acquired during the cell cycle within the germinal matrix, allowing repopulation by neurons otherwise "destined" to form other, deeper lamina, for example. Contributions by extracellular matrix components, glial populations (re-expression of radial glial morphology, e.g.) and neuronal populations (local-circuit neurons and more distant afferents) may all be important. Such cues may also allow earlier pluripotent progenitor cells or genetically modified, pluripotent cell lines to acquire correct phenotype when provided with appropriately specific developmental signals.

Recent studies suggest that extrinsic developmental cues are centrally important in neuronal laminar "fate" determination, but that such cues are normally available only during a brief period of active neuronal migration during neocortical development. At least two lines of inquiry suggest that integration in complexly connected systems is possible if host and donor conditions are appropriate; definition of these complex conditions and necessary modifications are not established. Studies using the cerebellum of Purkinje cell degeneration (PCD) mutant mice by Sotelo, Alvarado-Mallart, and colleagues have shown the feasibility of neural grafting procedures to restore elements of damaged neuronal systems that are connected in a "point-to-point" fashion (Sotelo and Alvarado-Mallart, 1986; 1987a, b; Gardette et al., 1990; Sotelo et al., 1990). Embryonic transplants of cerebellum partially repopulate the defective host cerebellum with Purkinje cells and establish relatively appropriate, specific afferent synaptic contacts. Grafted Purkinje cells appear to retain portions of their own "internal clock" of the normal developmental and migrational sequence and timetable. Our experiments suggest that it is also possible that migration and integration of embryonic or genetically engineered neurons transplanted into appropriately modified diseased neocortex may mimic events occurring during normal ontogeny.

McConnell and colleagues have defined the importance of the normal developing host environment to laminar specification, suggesting limitations on such migrational potential by immature neocortical neurons when grafted heterochronically within normal, unlesioned neocortex of the still-developing postnatal ferret (McConnell, 1985, 1988; McConnell and Kaznowski, 1991). These studies demonstrate limited potential to alter laminar "fate" after a brief developmental period during the cell cycle within the germinal matrix. These two sets of results are not inconsistent within an interpretation that the host microenvironment can be permissive to immature migrational phase neurons, but that exquisitely selective models of neuronal degeneration are necessary to allow appropriate cellular laminar positioning, cellular integration, and intercellular connections. One interpretation is that the uniqueness of cerebellar Purkinje cell specification offers more redirected growth to restore appropriate positioning, whereas increased inherent complexity in neocortex makes appropriate migration difficult. An alternate interpretation is that the PCD neuronal degeneration produces necessary microenvironmental cues that are absent in the intact developing ferret neocortical system. Our evidence to date argues in favor of the latter; similar migration and integration occur within neocortex within regions of targeted photolysis and slowly progressive neuronal degeneration.

The goal of ongoing experiments is to extend the investigation of the mechanisms by which transplanted neuronal precursors within a host environment beyond the normal migrational period can be "convinced" to migrate and integrate into positions lacking neurons that are normally generated and integrated much earlier within a normal, developing mouse. The specific repopulation after normal ontogeny is complete indicates microenvironmental cues can be reactivated in more mature rodents to initiate or permit reconstitution of normal cytoarchitecture. Identification and characterization of cellular and molecular controls over these events will be important in the future.

Information regarding this possible hierarchy of "cues" that guide laminar positioning and neuronal differentiation during development will allow directed studies of possible transplantation of neocortical precursor neurons or cell lines as a future therapy for degenerative, acquired, or developmental injury to neocortex (Rosenberg et al., 1988; Temple, 1989; Horellou et al., 1990; Gage and Fisher, 1991; Lee et al., 1991; Renfranz et al., 1991; Reynolds and Weiss, 1992; Snyder et al., 1992; Whittemore and White, 1993). Such defined lines could have custom modifications to provide genetic and pharmacologic control in vitro and in vivo after transplantation. Similar paradigms could prove useful in transplantation studies involving other regions of the nervous system and other degenerative diseases. These same studies will contribute to our understanding of normal and pathological events of neuronal migration, differentiation, and intergration during neocortical development.

ACKNOWLEDGEMENTS

This work was supported by NIH grants from NICHD, NINDS, and a Mental Retardation Research Center grant, and by grants from the Alzheimer's Association and the William Randolph Hearst Fund. I thank Tina Wilusz and Liem Tran for excellent technical assistance. J.D.M. is a Rita Allen Foundation Scholar.

REFERENCES

Anderson RR and Parrish JA (1983): Selective photothermolysis: precise microsurgery by selective absorption of pulsed radiation. Science 220: 524.
Angevine JB, Jr and Sidman RL (1961): Autoradiographic study of the cell migration

during the histogenesis of the cerebral cortex in the mouse. Nature 192: 766.

Austin CP and Cepko CL (1990): Cellular migration patterns in the developing mouse cerebral cortex. Development 110: 713.

Bently D and Caudy M (1983): Pioneer axons lose directed growth after selective killing of guidepost cells. Nature 304: 62.

Björklund A, Stenevi U, Dunnett SB and Gage FH (1982): Cross-species neural grafting in a rat model of Parkinson's disease. Nature 298: 652.

Cohen CS, Dreyfus CF and Black IB (1991): NGF and excitatory neurotransmitters regulate survival and morphogenesis of cultured cerebellar Purkinje cells. J Neurosci 11: 462.

Culican SM, Baumrind NL, Yamamoto M and Pearlman AL (1990): Cortical radial glia: identification in tissue culture and evidence for their transformation to astrocytes. J Neurosci 10: 684.

Edwards MA, Yamamoto M and Caviness VS Jr (1990): Organization of radial glia and related cells in the developing murine CNS. An analysis based upon a new monoclonal antibody marker. Neuroscience 36: 121.

Eisen JS, Pike SH and Debu B (1989): The growth cones of identified motorneurons in embryonic zebrafish select appropriate pathways in the absence of specific cellular interactions. Neuron 2: 1097.

Gage FH, Wolff JA, Rosenberg MB, Xu L, Yee J.-K, Shults C and Friedmann T (1987): Grafting genetically modified cells to the brain: possibilities for the future. Neurosci 23: 795.

Gage FH and Fisher LJ (1991): Intracerebral grafting: a tool for the neurobiologist. Neuron 6: 1.

Gardette R, Crepel F, Alvarado-Mallart RM and Sotelo C (1990): Fate of grafted embryonic Purkinje cells in the cerebellum of the adult "Purkinje cell degeneration" mutant mouse. II. Development of synaptic responses: an in vitro study. J Comp Neurol 295: 188.

Geschwind DH and Hockfield S (1989): Identification of proteins that are developmentally regulated during early cerebral corticogenesis in the rat. J Neurosci 9: 4303.

Hernit-Grant CS and Macklis JD (1993): Morphology and projections of embryonic neocortical cells transplanted into neuron-deficient somatosensory cortex of young adult mice. Soc Neurosci Abst 19: 357.9.

Hernit-Grant CS and Macklis JD (1994): Re-expression of the radial glial cell marker, RC-2, in young adult mouse cortex following targeted neuronal death. Soc Neurosci Abst 20: 682.6.

Horellou P, Brundin P, Kalen P, Mallet J and Björklund A (1990): In vivo release of dopa and dopamine from genetically engineered cells grafted to the denervated rat striatum. Neuron 5: 393.

Jay DG and Keshishian H (1990): Laser inactivation of fascilin I disrupts axon adhesion of grasshopper pioneer neurons. Nature 3448: 595.

Katz LC, Burkhalter A and Dreyer WJ (1984): Fluorescent latex microspheres as a retrograde neuronal marker for in vivo and in vitro studies of visual cortex. Nature 310: 498.

Lee HJ, Hoffmann PC, Kontur PJ, Won LA, Hammond DN and Wainer BH (1991): Immortalization of mesencephalic dopaminergic neurons by somatic cell fusion. Brain Res 522: 67.

Macklis JD (1993): Transplanted neocortical neurons integrate within selectively neuron-deficient regions produced by chromophore targeted laser photolysis. J Neurosci 13(9): 3848.

Macklis JD and Madison R (1985): Unfocused laser illumination kills dye-targeted mouse neurons by selective photothermolysis. Brain Res 359: 158.

Macklis JD and Madison RD (1991): Neuroblastoma grafts are noninvasively removed within mouse neocortex by selective laser activation of intracellular photolytic chromophore. J Neurosci 11: 2055.

Madison RD, Macklis JD and Frosch MP (1988): Noninvasive laser microsurgery selectively damages labeled mouse neurons: dependence on incident laser dose and absorption. Brain Res 445: 101.

Madison R, Macklis J and Thies C (1990): Latex nanosphere delivery system (LNDS): novel nanometer-sized carriers of fluorescent dyes and active agents selectively target neuronal subpopulations via uptake and retrograde transport. Brain Res 522: 90.

Madison, RD and Macklis JD (1993): Noninvasively induced degeneration of neocortical pyramidal neurons in vivo: Selective targeting by laser activation of retrogradely transported photolytic chromophore. Exp Neurol 121: 153.

McConnell SK (1985): Migration of cerebral cortical neurons after transplantation into the brains of ferrets. Science 229: 1268.

McConnell SK (1988): Fates of visual cortical neurons in the ferret after isochronic and heterochronic transplantation. J Neurosci 8: 945.

McConnell SK and Kaznowski CE (1991): Cell cycle dependence of laminar determination in developing neocortex. Science 254: 282.

Miller JP and Selverston AI (1979): Rapid killing of single neurons by irradiation of intracellularly injected dye. Science 206: 702.

Raff MC (1992): Social controls on cell survival and cell death Nature 356: 397.

Raper JA, Bastiani MJ and Goodman CS (1984): Pathfinding by neuronal growthcones in grasshopper embryos. IV. The effects of ablating the A and P axons upon the behavior of the G growthcone. J Neurosci 4: 2329.

Renfranz PJ, Cunningham MG and McKay RDG (1991): Region-specific differentiation of the hippocampal stem cell line HiB5 upon implantation into the developing mammalian brain. Cell 66: 713.

Reynolds BA and Weiss S (1992): Generation of neurons and astrocytes from isolated cells of the adult mammalian central nervous system. Science 255: 1707.

Rosenberg MB, Friedmann T, Robertson RC, Tuszynski M, Wolff JA, Breakefield XO and Gage FH (1988): Grafting genetically modified cells to the damaged brain: restorative effects of NGF expression. Science 242: 1575.

Sheen VL and Macklis JD (1992): Transplanted embryonic neocortical neurons undergo increased differentiation in selectively neuron-deficient cortex of adolescent mice. Soc Neurosci Abst 18:393.13: Targeted neocortical in adult mice guides migration and differentiation of transplanted embryonic neurons. (manuscript submitted).

Sheen VL and Macklis JD (1994): Apoptotic mechanisms in targeted neuronal cell death by chromophore-activated photolysis. Exp Neurol (in press).

Sheen VL, Dreyer EB and Macklis JD (1992): Calcium-mediated neuronal degeneration follows singlet oxygen production by chromophore-targeted photolysis. NeuroReport 3: 705.

Sheen VL, Cunningham MG, Hernit-Grant CS, McKay RDG and Macklis JD (1993): Hippocampal derived, immortalized stem cells (HiB5) undergo morphologic differentiation in adult murine striatum. Soc Neurosci Abst 19: 256.2.

Snyder EY, Deitcher DL, Walsh C, Arnold-Aldea S, Hartweig EA and Cepko CL (1992): Multipotent neural cell lines can engeaft and participate in development of mouse cerebellum. Cell 68: 1.

Snyder EY, Yandava BD, Pan Z-H, Yoon CH and Macklis JD (1993): Multipotent neural progenitor cell lines can engraft and participate in development of multiple structures at multiple stages along the mouse neuraxis. Soc Neurosci Abst 256.3.

Sotelo C, Alvarado-Mallart RM, Gardette R and Crepel F (1990): Fate of embryonic Purkinje cells in the cerebellum of the adult "Purkinje cell degeneration" mutant mouse. I. Development of reciprocal graft-host interactions. J Comp Neurol 295:

165.

Sotelo C and Alvarado-Mallart RM (1986): Growth and differentiation of cerebellar suspensions transplanted into the adult cerebellum of mice with heterodegenerative ataxia. Proc Natl Acad Sci USA 83: 1135.

Sotelo C and Alvarado-Mallart RM (1987): Reconstruction of the defective cerebellar circuitry in adult purkinje cell degeneration mutant mice by Purkinje cell replacement through transplantation of solid embryonic implants. Neurosci 20: 1.

Sotelo C and Alvarado-Mallart RM (1987): Embryonic and adult neurons interact to allow Purkinje cell replacement in mutant cerebellum. Nature 327: 421.

Temple S (1989): Division and differentiation of isolated CNS blast cells in microculture. Nature 340: 471.

Whittemore SR and White LA (1993): Target regulation of neuronal differentiation in a temperature-sensitive cell line derived from medullary raphe. Brain Res 615: 27.

Yoon CH, Snyder EY and Macklis JD (1993): Immortalized pluripotent neural precursors undergo morphologic differentiation when transplanted into selectively neuron-deficient neocortex of adult mouse. Soc Neurosci Abst 19: 256.4: Neural progenitors can replace degenerated neurons in adult mouse neocortex. (manuscript submitted).

NEURODEGENERATION INDUCED BY DE-REGULATION OF NEUROFILAMENT GENE EXPRESSION IN TRANSGENIC MICE

J.-P. Julien, F. Côté, G. Charron, and J.-F. Collard

Centre for Research in Neuroscience, McGill University,
The Montreal General Hospital, Research Institute,
Montreal, H3G 1A4, CANADA

INTRODUCTION

Neurofilaments (NFs) are formed by the copolymerization of three proteins, the light (NF-L), medium (NF-M) and heavy (NF-H) subunits (Hoffman and Lasek,1975; Liem et al., 1978). The three NF subunits are encoded by different genes and several reports have shown that the genes are under separate developmental control (Shaw and Weber, 1982; Carden et al., 1987; Cochard and Paulin, 1984; Julien et al., 1986). Direct evidence for a role of neurofilaments in the control of axonal calibers was recently provided from the recent analyses of a mutant quail (Ohara et al. 1993) and of a transgenic mouse (Eyer and Peterson 1994) with NF-deficient axons.

Abnormal accumulations of neurofilaments in distinct regions of the neuron occur in a variety of disorders including amyotrophic lateral sclerosis (ALS) (Carpenter, 1968; Hirano et al., 1984; Schmidt et al., 1987; Sasaki et al., 1989; Chou, 1992), an inherited giant axonal neuropathy (Carpenter et al., 1974) and toxic neuropathies (Graham et al., 1984; Griffin et al., 1978; Troncoso et al., 1992). Until recently, the NF accumulations in these pathologies were perceived as a secondary effect of an impairment in axonal transport (Gadjusek 1985; Griffin and Watson, 1988; Chou 1992). However, the recent finding that overexpression of either the NF-L or NF-H gene provokes in transgenic mice a neuronopathy with features highly reminescent to those found in human motor neuron diseases (Côté et al., 1993; Xu et al., 1993) provided evidence for a more direct involvement of NF proteins in neuropathologies and it emphasized the importance to elucidate the mechanisms regulating neurofilament gene expression.

In the first part of this review, we discuss the mechanisms regulating expression of the human NF-L and NF-H genes. In the second section, we will present evidence based on the analysis of transgenic mice bearing the human NF-H gene that deregulated levels of NF proteins can provoke a neurodegeneration

Neural Cell Specification: Molecular Mechanisms and Neurotherapeutic Implications
Edited by Juurlink *et al.*, Plenum Press, New York, 1995

resulting from a general disruption of axonal transport.

REGULATION OF NEUROFILAMENT GENE EXPRESSION

Sequence Requirements for Neuronal Expression of the NF-L Gene

The genes encoding the three NF proteins have been cloned and sequenced (Julien et al., 1987, 1988; Lees et al., 1988; Levy et al., 1988) but the molecular signals regulating their transcriptional activation in neurons have not yet been elucidated. In transgenic mice, the complete NF-L gene was found to be correctly expressed in the nervous system (Julien et al., 1987; Beaudet et al., 1992; Monteiro et al., 1990; Yazdanbaksh et al., 1993). To date, the combined results indicate that elements located in both upstream and intragenic NF-L regions contribute to the regulation of this gene (Beaudet et al., 1992; Yazdanbaksh et al., 1993; Ivanov et al., 1992; Nakahira et al., 1990; Schneidman et al., 1992). To identify the intragenic regions that confer neuronal expression, we tested in transgenic mice the effect of

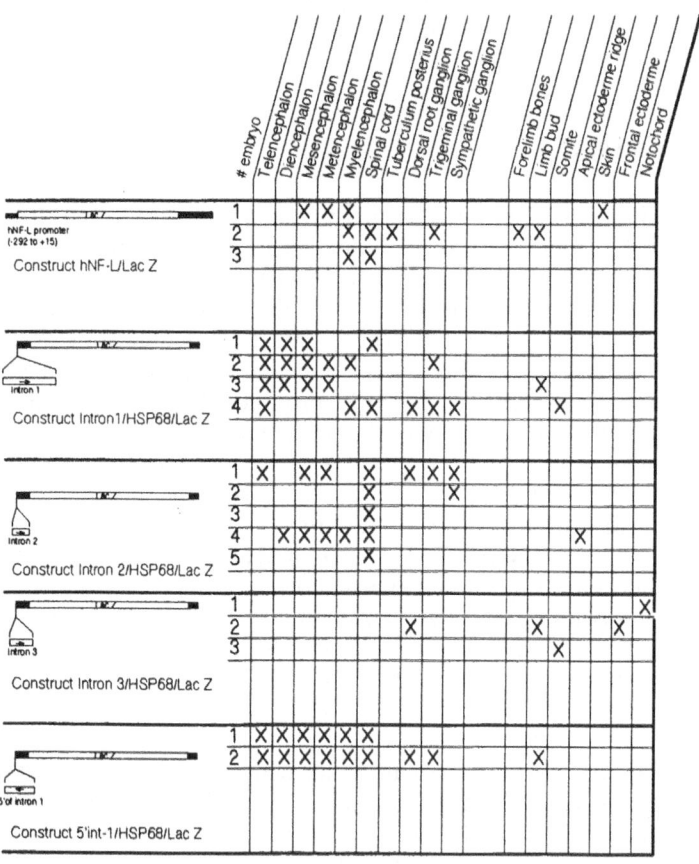

Figure 1. Sites of expression of NF-L/lacZ constructs in transgenic embryos.

various internal NF-L DNA regions on the activity of the basal heat shock promoter fused to the lacZ reporter. The spatial expression of the NF-L/hsp/lacZ constructs in whole 13-day-old embryos was determined histochemically using X-gal as a substrate. Fig. 1 summarizes the sites of expression of the different NF-L/hsp/lacZ fusion constructs. Our results indicate that introns 1 and 2 of the NF-L gene each contains regulatory elements to confer neuronal expression whereas no activity was found in intron 3. The exact sequences responsable for neuronal expression in intron 1 and 2 have not yet been identified.

The proximal NF-L promoter region (-0.3 kb) does contain regulatory elements that are capable of directing neuronal expression of lacZ in transgenic mice. In addition to TATA and Sp1 elements which may be responsible for a basal activity of the NF-L promoter, our promoter analysis using P19 embryonal carcinoma (EC) cells revealed the presence of functional AP-1 and Krox-24 regulatory elements at -48 and -139 bp, respectively (Pospelov et al., 1994). The presence of these elements provides further evidence that regulation of neurofilament gene expression can be mediated, at least in part, by pathways directly coupled to extracellular signals such as growth factors or other external stimuli (Lindenbaum et al., 1988; Tsuneishi et al., 1993).

Copy-Dependent Expression of the Human NF-H Gene

We have derived lines of transgenic mice carrying multiple copies of a 39 kb genomic fragment (Cos4NFH) containing the four exons of the human NF-H gene flanked by 9.6 kb of upstream and 13.4 kb of downstream sequences (Côté et al., 1993). Interestingly, the DNA fragment bearing the human NF-H gene was found to be expressed in a copy-dependent manner in transgenic mice both at the mRNA and protein levels (Côté et al., 1994). This represents the first example of a transgene being expressed in a copy-dependent manner in neurons and the identification of the NF-H regulatory elements responsible for copy-dependent expression may constitute a valuable tool in the development of future vectors for gene therapy in the nervous system.

NEURONOPATHY IN TRANSGENIC MICE OVER-EXPRESSING NF-H PROTEINS

Aberrant NF accumulations in spinal motor neurons

Transgenic mice expressing human NF-H proteins at levels of approximately two-fold the levels of endogenous mouse NF-H protein appear normal during the first few weeks of postnatal development but progressively manifest signs of neurological abnormality. When lifted by their tails, they reflexively contract their fore- and hind-limbs whereas normal mice extend their legs. The disease progresses with signs of weakness. A simple behavioral test, the grasping ability, can be used to monitor, during aging, the progressive motor dysfunction (Collard and Julien, 1994). We have found that at six months old the NF-H transgenics lose their ability to uphold their weight when grasping a pencil. At one year of age, severely affected mice have difficulty in ambulating.

Microscopic examination of spinal cord sections from adult NF-H transgenic mice revealed the presence of abnormal neurofilamentous swellings in the perikaryon and proximal axon of spinal motor neurons (Fig. 2). Many neurons show eccentrically localized nuclei with prominent swellings of their perikarya and proximal axons. Electron microscopy revealed that other organelles including mitochondria and vesicles are displaced into a submembranal disposition or squeezed through the densely packed 10-nm neurofilaments. Such dramatic neurofilamentous swellings was also detected in some sensory neurons of the dorsal root ganglia (DRG) which are also high producers of neurofilaments (Côté et al., 1993). Conversely, there is an atrophy of axons in the ventral root and in sciatic nerve and the progressive axonopathy is accompanied by secondary atrophy of skeletal muscle fibers (Côté et al., 1993). A plausible explanation from these data is that the excess levels of NF-H protein, which forms NF cross-linkages, can provoke an impairment of neurofilament transport resulting in the gradual piling up of newly synthesized neurofilaments in the perikarya and proximal axons.

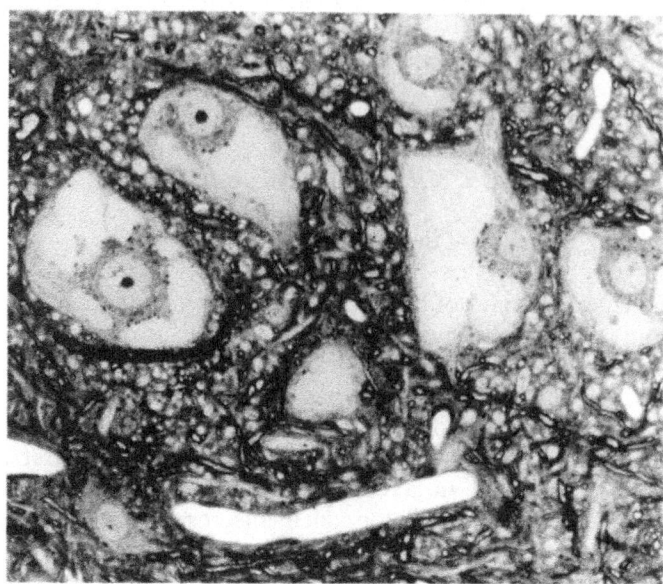

Figure 2. Neurofilament accumulations in spinal motor neurons of NF-H transgenic mice. Magnification = 150x.

Disruption of axonal transport in NF-H transgenics

To determine if an impairment of axonal transport accompanies the motor neuron disease in transgenic mice, ^{35}S-methionine was injected into the spinal cord of mice followed, after various time intervals, by SDS gel electrophoresis and fluorography of 5 mm segments of sciatic nerves. Fig. 3 illustrates the changes in the ratios of ^{35}S-labeled NF proteins which migrated in the 10 mm segment of sciatic nerve 4 weeks after injection of ^{35}S-methionine in spinal cord. Our results reveal a dramatic reduction in the rate of transport of NF proteins in motor axons of NF-H transgenics aged 3 and 18 months as compared to tubulin or actin proteins. In addition, the NF proteins are transported in abnormal ratios in the transgenics, the NF-M protein being virtually undetectable in the sciatic nerve of 18 month transgenics. Our interpretation is that exceeding levels of incorporated NF-H protein might affect intracellular transport of neurofilament proteins either by imposing additional drag due to extra cross-linkages or by impeding normal interactions of neurofilaments with a slow transport motor. Moreover, the defect in axonal transport is not limited to NF proteins in NF-H transgenics. The velocity of transport of other cytoskeletal proteins, such as tubulin and actin proteins, was affected albeit to a lower extent than NF proteins and we noted a defect in transport of some proteins associated with the fast component in NF-H transgenics (Côté, F. and Julien, J.-P., unpublished data).

NF-H Transgenics as a Model of ALS

Several features of the NF-H transgenic mice described here make them a relevant model of ALS. Human ALS is an adult-onset neurological disorder resulting from the degeneration of motor neurons in brain and spinal cord. Mutations in the copper/zinc dismutase (SOD1) gene located on chromosome 21 have been recently identified in a subset of familial ALS cases (Rosen et al. 1993) but for more than 95% of ALS patients the etiology remains unknown. A characteristic pathological finding in either familial or sporadic ALS is the presence of neurofilamentous swellings in the perikaryon and proximal axon of spinal motor neurons (Carpenter, 1968; Hirano et al., 1984; Schmidt et al., 1987; Sasaki et al., 1989; Hirano and Kato, 1992; Chou, 1992). The NF accumulations in spinal motor neurons of NF-H transgenics resemble very much those found in ALS. The pathology, which is accompanied by a progressive disruption of axonal transport, is pertinent to the "dying back" model hypothesized as a mechanism underlying ALS (Gajdusek, 1985; Griffin and Watson, 1988; Chou, 1992). The validity of the transgenic mouse models with deregulated levels of NF proteins (Cote et al., 1993; Xu et al., 1993) is further supported by the recent report of alterations in NF mRNA

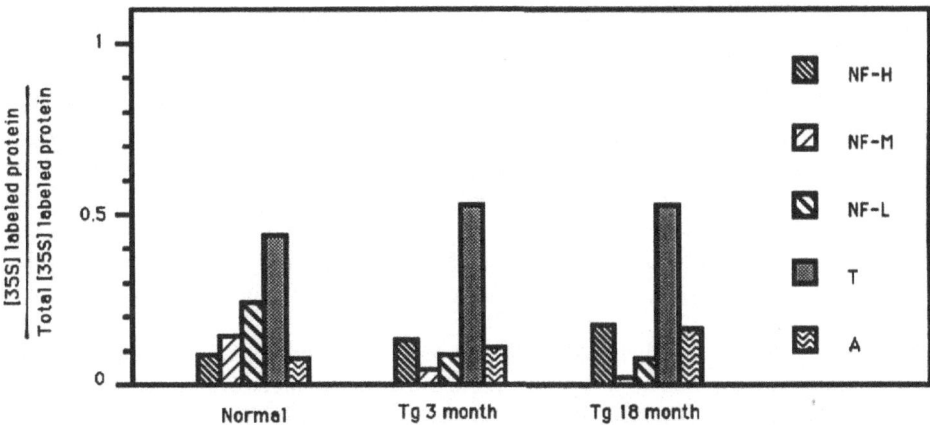

Figure 3. Quantitative analysis of labeled cytoskeletal proteins transported in sciatic axons 28 days after labeling of spinal motor neurons with ^{35}-methionine. The labeled proteins in 5-mm segments of the sciatic nerve were analysed by SDS-PAGE and fluorography. The results represent the ratios of the labeled proteins in the 10-mm segment away from the spinal cord. A=actin; NF-H= neurofilament heavy; NF-L = neyrofilament light; NF-M= neurofilament medium; T=tubulin.

ratios in hereditary canine spinal muscular atrophy, a well established animal model of human motor neuron disease (Muma and Cork, 1993). Finally, compelling evidence for NF involvement in ALS pathogenesis was provided by our recent finding of mutations in the NF-H gene of 5 out of 350 ALS patients examined (Figlewicz et al., 1994). These NF-H mutations consist of codon deletions in the domain of phosphorylation repeats. It is plausible that such mutations could alter the cross-linking properties of the NF-H protein resulting in a tendency to develop abnormal neurofilament accumulations. Work is now in progress to test in transgenic mice the in vivo effects of the NF-H variants on neurofilament organization and neuronal function.

ACKNOWLEDGEMENTS

The technical assistance of Daniel Houle and Pascale Hince is gratefully acknowledged. This work was supported by the Medical Research Council of Canada and the ALS Association (USA). The production of transgenic mice was supported by the Canadian NeuroScience Network of Centre of Excellence. J.-F.C. is a recipient of a fellowship from the Medical Research Council of Canada, F.C. has a studentship from the Rick Hansen Foundation and J.-P.J. has a senior scholarship from F.R.S.Q.

REFERENCES

Beaudet L, Charron G, Houle D, Tretjakoff I, Peterson A and Julien J.-P. (1992): Intragenic regulatory elements contribute to transcriptional control of the neurofilament light gene. Gene 116: 205-214, .

Carden MJ, Trojanowski JQ, Schlaepfer WW and Lee, VM-Y (1987): Two-stage expression of neurofilament polypeptides during rat neurogenesis with early establishment of adult phosphorylation patterns. J Neurosci 7: 3489-3504.

Carpenter S (1968): Proximal enlargement in motor neuron diseases. Neurol 18: 841-851.

Chou SM (1992): Pathology - Light microscopy of amyotrophic lateral sclerosis. In Smith RA (ed): "Handbook of Amyotrophic Lateral Sclerosis," Marcel Dekker Inc, New York, 133-181.

Carpenter S Karpati G Andermann F and Gold R (1974): Giant axonal neuropathy: a clinically and morphologically distinct neurological disease. Arch Neurol 31: 312.

Cochard P and Paulin D (1984): Initial expression of neurofilament and vimentin in the central and peripheral nervous system of the mouse embryo in vivo, J Neurosci 4: 2080-2094.

Collard JF and Julien JP (1994): A simple test to monitor motor dysfunction in a transgenic mouse model of amyotrophic lateral sclerosis. J Psychiatry and Neuroscience (in press).

Côté F, Collard J-F and Julien J-P (1993): Progressive neuronopathy in transgenic mice expressing the human neurofilament heavy gene: A mouse model of amyotrophic lateral sclerosis. Cell 73: 35-47.

Côté F, Collard J-F, Houle D and Julien J-P (1994): Copy-dependent and correct developmental expression of the human neurofilament heavy gene in transgenic mice. Molec Brain Res 26: 99-105.

Eyer J and Peterson A (1994): Neurofilament-deficient axons and perikaryal aggregates in viable transgenic mice expressing a neurofilament-beta-galactosidase fusion protein. Neuron 12: 389-416.

Figlewicz DA, Krizus A, Martinoli MG, Meininger V, Klein V, Rouleau GA and Julien J-P (1993): Variants of the heavy neurofilament gene subunit are associated with the development of amyotrophic lateral sclerosis. Hum Molec Genet 3: 1757-1761.

Gajdusek DC (1985): Hypothesis: interference with axonal transport of neurofilament as a common pathogenetic mechanism of certain diseases of the CNS. N Engl J Med 312: 714-719.

Graham DG, Szakal-Quin G, Priest JW and Anthony DC (1984): In vitro evidence that covalent crosslinking of neurofilaments occurs in gamma-diketone neuropathy. Proc Natl Acad Sci USA 81: 4979-4982.

Griffin JW, Hoffman PN, Clark AW, Carroll PT and Price DL (1978): Slow axonal transport of neurofilament proteins: Impairment by ß,ß'-iminodipropionitrile administration. Science 202: 633-635.

Griffin JW and Watson DF (1988): Axonal transport in neurological disease. Ann Neurol

23: 3-13.

Hirano A, Donnenfeld H, Sasaki S and Nakano I (1984): Fine structural observations of neurofilamentous changes in amyotrophic lateral sclerosis. J Neuropathol Exp Neurol 43: 461-470.

Hirano A and Kato S (1992): Fine structural study of sporadic and familial amyotrophic lateral sclerosis. In Smith RA (ed): "Handbook of Amyotrophic Lateral Sclerosis," Marcel Dekker Inc, New York, 183-192.

Hoffman PN and Lasek RJ (1975): The slow component of axonal transport. Identification of the major structural polypeptides of the axon and their generality among mammalian neurons. J Cell Biol 66: 351-366.

Ivanov TR and Brown IR (1992): Interaction of multiple nuclear proteins with the promoter region of the mouse 68-kDa neurofilament gene. J Neurosci 32: 149-158.

Julien J-P, Meijer D, Hurst J and Grosveld F (1986): Cloning and developmental expression of the murine neurofilament gene family. Mol Brain Res 1: 243-250.

Julien J-P, Tretjakoff I, Beaudet L and Peterson A (1987): Expression and assembly of a human neurofilament protein in transgenic mice provide a novel neuronal marking system. Genes Dev 1: 1085-1095.

Julien JP, Grosveld F, Yazdanbajksh K, Flavell D, Meijen D and Mushynski W (1987): Structure of the human neurofilament light gene (NF-L): a unique exon-intron organization in the intermediate filament gene family. Biochem Biophys Acta 909: 10-20.

Julien JP, Côté F, Beaudet L, Sidky M Flavell D, Grosveld F and Mushynski W (1988): Sequence and structure of the mouse gene coding for the largest neurofilament subunit. Gene 68: 307-314.

Lees JF, Shneidman PS, Skuntz SF, Carden MJ and Lazzarini RA (1988): The structure and organization of the human heavy neurofilament subunit (NF-H) and the gene encoding it. EMBO J 7: 1947-1955.

Levy E, Liem RKU, D'Eustachio P and Cowan NJ (1982): Structure and evolutionary origin of the gene encoding mouse NF-M, the middle molecular mass neurofilament protein. Eur J Biochem 166: 71-72.

Liem RKH, Yen S-H, Salomon GD and Shelanski ML (1978): Intermediate filaments in nervous tissues. J Cell Biol 79: 637-645.

Lindenbaum MH, Carbonetto S, Grosveld F, Flavell D and Mushynski WE (1988): Transcriptional and post-transcriptional effects of nerve growth factor on expression of the three neurofilament subunits in PC-12 cells. J Biol Chem 263: 5662-5667.

Monteiro MJ, Hoffman PN, Gearhart JD and Cleveland DW (1990): Expression of NF-L in both neuronal and nonneuronal cells of transgenic mice: increased neurofilament density in axons without affecting caliber. J Cell Biol 111: 1543-1557.

Muma NA and Cork LC (1993): Alternations in neurofilament mRNA in hereditary canine spinal muscular atrophy. Lab Invest 69: 436-442.

Nakahira K, Ikenaha K, Wada K, Tamura T, Furuichi T and Mikoshiba K (1990): Structure of the 68 kDa neurofilament gene and regulation of its expression. J Biol Chem 265: 19786-19791.

Ohara O, Gahara Y, Miyake T, Teraoka H and Kitamura T (1993): Neurofilament deficiency in quail caused by nonsense mutation in neurofilament-L gene. J Cell Biol 121: 387-395.

Rosen DR, Siddique T, Patterson D, Figlewicz DA, et al. (1993): Mutations in Cu/Zn superoxide dismutase gene are associated with familial amyotrophic lateral sclerosis. Nature 362: 59-62.

Sasaki S, Murayana S, Yamane K, Sakuma H and Takeishi M (1989): Swelling of proximal axons in a case of motor neuron disease. Ann Neurol 25: 520-522.

Schmidt ML, Carden MJ, Lee VM-Y and Trojanowski JQ (1987): Phosphate dependent and independent neurofilament epitopes in the axonal swellings of patients with motor neuron disease and controls. Lab Invest 56: 282-294.

Schneidman PS, Bruce J, Schwartz ML and Schlaepfer WW (1992): Negative regulatory regions are present upstream in the three mouse neurofilament genes. Molec Brain Res 13: 127-138.

Shaw G and Weber K (1982): Differential expression of neurofilament triplet proteins in brain development. Nature 298: 277-279.

Troncoso JC, Gilbert MR and Muma NA (1992): Neurotoxicology: light metals, In Smith RA (ed): "Handbook of Amyotrophic Lateral Sclerosis," Marcel Dekker Inc, New York, 543-558.

Tsuneishi S, Sano K and Nakamura H (1993): Serum depletion increases the neurofilament protein mRNA levels in a neuroblastoma cell line, GOTO. Molec Brain Res 17: 119-128.

Xu Z, Cork LC, Griffin JW and Cleveland DW (1993): Increased expression of neurofilament subunit NF-L produces morphological alterations that resemble the pathology of human motor neuron disease. Cell 73: 23-33.

Yazdanbakhsh K, Fraser P, Kioussis D, Vidal M, Grosveld F and Lindenbaum M (1993): Functional analysis of the human neurofilament light chain gene promoter. Nucl Acids Res 21: 455-461.

STRATEGIES FOR NEURAL CELL REPLACEMENT IN NEURODEGENERATIVE DISORDERS

CO-GRAFTS IN DOPAMINE-DEPLETED PRIMATES: PRELIMINARY RESULTS AND THEORETICAL ISSUES RELATED TO HUMAN APPLICATIONS FOR PARKINSON'S DISEASE

J.R. Sladek, Jr.,[1] J.D. Elsworth,[3] R.H. Roth,[3] B.C. Blanchard,[1] J.R. Taylor,[3] T.J. Collier,[2] and D.E. Redmond, Jr.[3]

[1]Department of Neuroscience, Chicago Medical School, Chicago, Il., 60064, [2]Department of Neurological Sciences, Rush College of Medicine, Chicago, IL, 60612 and [3]Departments of Pharmacology and Psychiatry, Yale School of Medicine, New Haven, CT, 06510, U.S.A.

INTRODUCTION

Transplantation of embryonic mesencephalon has been attempted in several hundred Parkinson's disease patients worldwide. Initial reports support the role of these dopamine-producing grafts in ameliorating some of the signs and symptoms of the disease (Lindvall et al., 1989; Lindvall et al., 1990; Freed et al., 1992; Spencer et al., 1992; Widner et al., 1992). In particular, many patients 1) require less daily levodopa after grafting, 2) have a greater percentage of "on" time, and 3) show reduced dyskinesias. Nevertheless, considerable variation exists in the methodological approach followed by each group of clinical investigators, in part because it is difficult to perform experiments with human neural tissue that would provide unequivocal answers to technical questions about optimal survival parameters for the grafted tissue. For example, a study of the survival rate of grafted human dopamine neurons utilized the rat as a host (Brundin et al, 1988). This investigation found poor (i.e. 2-5%) survival of dopamine neurons, which may be influenced by several factors including the relatively small size of the rat brain. Perhaps, there is insufficient room to permit survival of great numbers of these proportionally larger human neurons in the confines of the rat striatum. Moreover, the role, if any, played by immunosuppression is unclear. As a consequence of this low survival, some clinical experiments have utilized mesencephalic tissue from several embryos in attempts to increase the total number of viable dopamine neurons, but this results in a considerably larger number of grafted non-dopaminergic neurons and the role of these presumably "silent"

Neural Cell Specification: Molecular Mechanisms and Neurotherapeutic Implications
Edited by Juurlink *et al.*, Plenum Press, New York, 1995

219

neurons is unknown. Since these grafts contain substantial numbers of neuroblasts and progenitor cells of the ventral mesencephalon, they presumably could be replete with neuronal and glial representatives of the mesencephalic reticular formation, the red nucleus, the ventral tegmentum, the oculomotor and trochlear nuclei, the proprioceptive nucleus of the trigeminal complex, serotonergic neurons of the dorsal raphe and many others for example. Since we are unaware of the influence that these neurons may exert on the host brain, it might be prudent to limit the potential contribution of this non-dopaminergic constituent of ventral mesencephalic grafts. This suggests that attention should be paid to mechanisms for enhancing the yield of grafted dopamine neurons.

Trophic and tropic effects are known to exist for striatal tissue on dopaminergic mesencephalic neurons. Enhanced survival and neurite outgrowth have been demonstrated in vitro (Prochiantz et al., 1979; Hemmindinger et al., 1981) and preliminary tests demonstrated similar phenomena after double-grafting intracranially (De Beaurepaire and Freed, 1987) or in the anterior chamber of the eye (Olson et al., 1987). An earlier experiment from our laboratory demonstrated a profound influence of striatal grafts on mesencephalic graft-induced amelioration of turning behavior in a rat model of nigrostriatal cell loss (Yurek et al., 1990). We have performed similar experiments in African green monkeys to test the effect of a presumed target-derived factor on the survival and growth of neurites from grafted dopamine neurons of mesencephalic origin. Our initial results suggest that more extensive neurite outgrowth can be achieved with co-grafts and that dopamine neuron survival can be considerably greater in a higher order non-human primate than that predicted from human xenografts.

GRAFTS IN PRIMATES: UNRESOLVED ISSUES

Such studies have been performed with the presumption that success in non-human primates would provide a predictive value for human clinical trials because of the close relationship and species similarities in the anatomy and physiology of the involved neural circuits. Since 1985, several laboratories have published results of fetal cell grafting in a variety of primate species and all support the concept that neuroblasts, and even developing neurons of mesencephalic origin, can survive grafting, grow into the host brain, release dopamine and cause an improvement in the conditions associated with experimental parkinsonism (Bakay et al., 1985; Sladek et al., 1986; 1987; 1988; Bankiewicz et al., 1990; 1991; Annett et al., 1993) In a recent analysis of biochemical data from 26 different primate graft recipients (Elsworth et al., 1995), we found that when experimental parkinsonism was improved (Taylor et al., 1991; 1994) dopamine levels were elevated by a mean of 500% (Fig. 1). Dopamine neuron survival in these same animals (Sladek et al., 1994) averaged over 12,000 (Fig. 2) from each donor when tissue was collected from embryos during the period of neurogenesis (i.e. E36-43 in the old world macaque) .

Survival of a maximal number of neurons from each donor mesencephalon is important for two reasons. First, optimal survival possibly might negate the need for the collection of tissue from several fetuses for each human recipient, as performed by others (Widner et al., 1992), and help to standardize the clinical procedure. More importantly, it would provide the best chance for maintaining improved function in light of the progressive nature of Parkinson's disease. Simply supplying enough new dopamine to elevate the system above a critical threshold would be insufficient to maintain function as the disease process continues. Although there is no evidence that the disease process also will not attack the new cells, grafts might be somewhat protected if the disease is due to a gradual accumulation of toxins, for example.

Another obstacle to overcome is the relatively large size of the human striatum in relation to the degree to which a single graft may "innervate" its surrounding striatal neuropil. Specifically, solid grafts in the monkey striatum can achieve a dimension of approximately 1.5 mm in width and 6-8 mm in height, and can occupy up to 15% of the area of the host caudate nucleus (Sladek et al., 1993). Fiber outgrowth has been seen to extend 3-5 mm into the host, but even these large grafts do not fully re-supply the host with dopaminergic fibers. Assuming that a more complete innervation may be necessary to restore motor control as well as mitigate against further decline as the disease progresses, it is conceivable that a human striatum could require dozens of implants to achieve this goal. Since a large total number of implants might not be feasible with available technology, an alternative would be the enhancement of fiber outgrowth from a smaller number of grafts to supply the most widespread distribution possible.

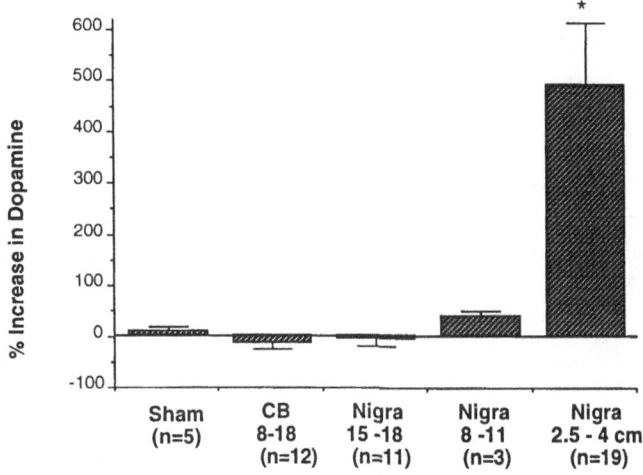

Figure 1. This graph shows the effects of different types of grafts, as well as sham surgery, on dopamine concentration in the caudate nucleus of MPTP-treated monkeys. The percent change was determined by dividing dopamine levels in micropunches of the caudate within 2 mm of either a graft or the site of sham surgery by the dopamine values in micropunches that are distant from the graft (or surgery) in the caudate of the same animal. The percent increase is shown for five groups of 26 monkeys. Differences were not seen with sham surgery or the implantation of non dopaminergic tissue (cerebellum: CB). Mesencephalic tissue taken from embryos or fetuses (i.e. 8-11cm crown-rump length (crl), 15-18 cm crl) other than the time of nigral neurogenesis, likewise were not effective in elevating dopamine content. Only embryonic tissue from an early stage (bar to the right) showed significantly elevated dopamine content in proximity to grafts.

There would appear to be at least two stategies to achieve greater fiber outgrowth. First, one could co-apply a tropic factor specific to neurite extension from mesencephalic neurons. This might involve simultaneous application of cells that release specific factors (e.g. BDNF, GDNF) or a striatal-derived embryonic factor (Sladek, et al., 1993a). Second, it theoretically is possible to achieve greater long distance outgrowth by implantation of xenografts of comparatively larger cells (e.g. human) to smaller animals as reported by Wictorin and colleagues, who performed a series of human xenografts in rats and discovered exceptional outgrowth and connectivity of axons (Wictorin et al., 1990). Thus, it might be possible to achieve maximal innervation from grafted neurons by utilizing cross-species transplants from animals with neurons proportionally larger than human neurons. One confound might be the identification of a larger cell type for xenografting into humans and attendant questions of supply. Nevertheless, bovine mesencephalic neurons reportedly have axons substantially longer than those in humans, and might be able to be secured in consistent quantity.

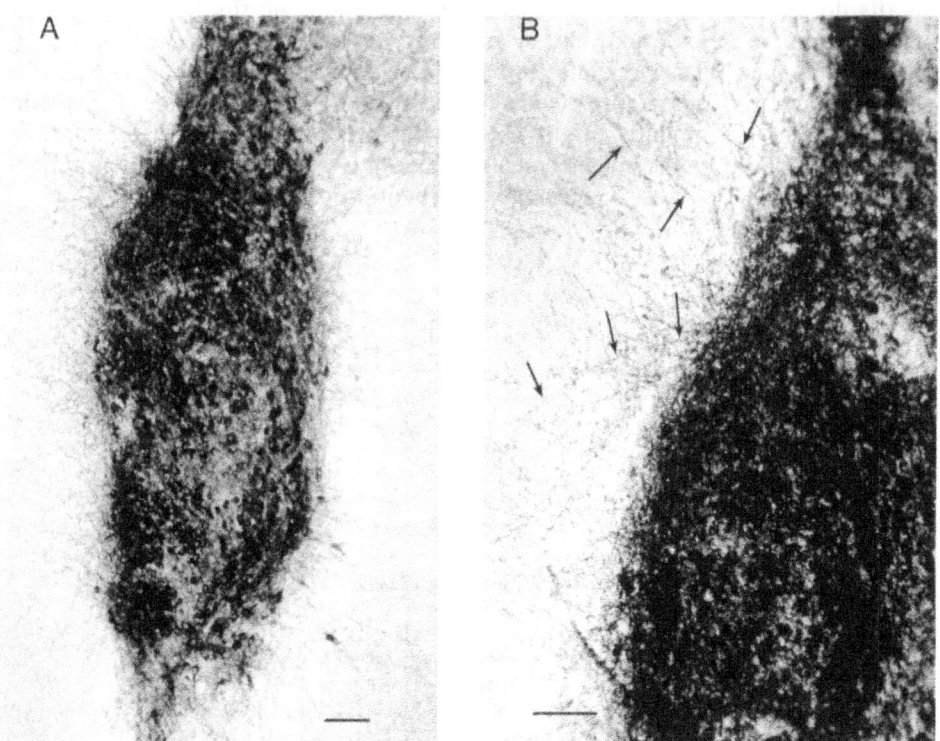

Figure 2. A and B are low and medium power magnifications of a mesencephalic graft located in the head of the caudate nucleus. The section was stained immunohistochemically for tyrosine hydroxylase. This graft is representative of those seen in animals that contained as many as 28,000 total surviving dopaminergic neurons derived from a single donor. Individual grafts in such animals often contained 5000 to 6000 tyrosine hydroxylase positive neurons and some had as many as 11,000. The neurons are densely packed and neuritic outgrowth extended into the host neuropil of the surrounding caudate nucleus. The neuritic extensions (arrows) are prominent in B. Bar = 100 μM.

A third consideration is the degree to which circuits may be "re-established" following grafting. Presumably, complete incorporation of grafted neurons could lead to better nigrostriatal control of dopamine release, for example, and this would be expected to support receptor stability through a regulated dopamine release and reuptake. Achievement of this will require placement of mesencephalic dopamine neurons into the mesencephalon instead of the striatum, which in turn would require the dopamine axons to grow over long distances, successfully find a path to the striatum and synaptically connect with nigrostriatal and striatonigral loops. For this to occur, it might be necessary to provide tropic factor "guideposts" along the route of the nigrostriatal system, perhaps in the form of factor-producing cells. Although this was attempted in rodent brain (Dunnett et al., 1989), it deserves further study in primates.

CO-GRAFT ENHANCEMENT OF MESENCEPHALIC GRAFTS

Earlier studies from our laboratory examined the effect of embryonic grafts of striatum on mesencephalic dopamine neurons that were grafted simultaneously into the striatum in 6-OHDA lesioned rats (Yurek et al., 1990). In comparison to solitary mesencephalic grafts, co-grafts resulted in more rapid reversal of apomorphine-induced rotations. Grafted dopaminergic neurons appeared larger and showed greater neuritic outgrowth into host striatum. We have extended this approach to MPTP-treated African green monkeys and have found similar, although more robust, terminal patterns in the striatal portion of co-grafts (Figs. 3-4). These were characterized also by the appearance of a patch-matrix configuration reminiscent of the striosomes that are a benchmark of nigrostriatal development. Some grafts displayed more homogenous patterns that also were of maximal density.

When the striatal and mesencephalic grafts were implanted into separate locations at a distance of 2-3 μm, fiber outgrowth appeared to favor the direction of the striatal graft (Sladek et al., 1993a). Such extensions also appeared to gain access to the corpus callosum and extended across the midline when grafts were located in contralateral striata (Fig. 5). The fiber growth from solo mesencephalic grafts, in contrast, appeared to extend uniformly in all directions from the center of the grafts (Fig. 6).

Thus, the polarity of fiber outgrowth suggests that the presence of embryonic striatum can exert a powerful influence on the extent and direction of neuritic growth. This could prove to be a useful tool for maximizing fiber outgrowth and consequently, the potential for "reinnervation" of striatal targets if a method can be developed to deliver the putative growth factor during critical periods of graft development. One possibility for delivery is through the implantation of striatal neuroblasts as described above, but this raises the question of whether the dopamine fibers will be attracted preferentially to grafted striatal neurons instead of host neurons. Even if this were to occur, it might not be counterproductive to restoration of function, since Wictorin and colleagues have demonstrated that grafted striatal neurons readily incorporate into striatal circuits to interconnect with cerebral cortex, thalamus and substantia nigra, for example (Wictorin et al., 1990).

Finally, we have observed that the presence of striatal grafts stimulates a regenerative or sprouting response by the host tyrosine-hydroxylase fibers as discussed in more detail below. A possible therapeutic strategy could include the combined effectiveness of striatal-nigral co-grafts to enhance dopamine production from the host as well as from the grafted mesencephalic tissue.

FUTURE STRATEGIES

Co-Grafts

The co-transplantation of mesencephalic and striatal neurons might provide a useful approach for human investigation in Parkinson's disease. Dopamine neuron replacement through the use of mesencephalic grafts is well established as

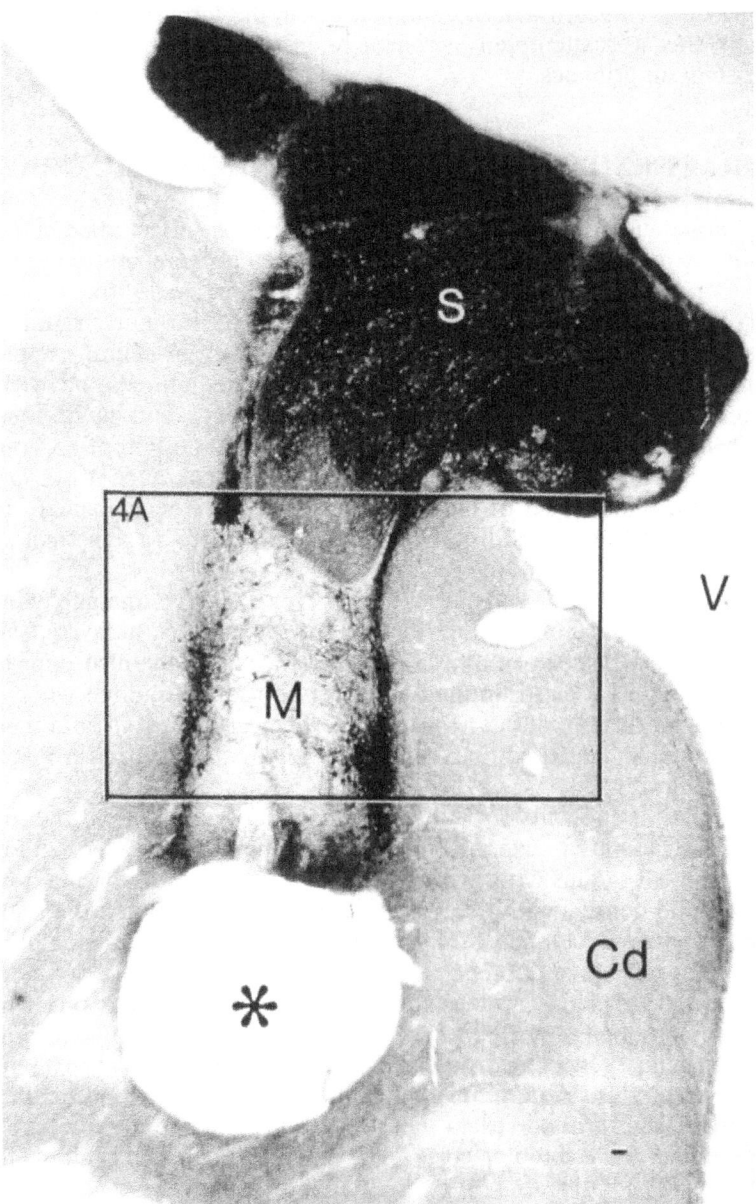

Figure 3. This scanning view depicts the extent of a co-graft of embryonic mesencephalon (M) and striatum (S); the individual components of the graft were placed juxtaposed in the head of the caudate nucleus (Cd). A portion of the graft was micropunched (*) for biochemical determinations. The lateral ventricle (V) serves as a landmark for orientation. Bar = 100 μ M

a means of replenishing lost dopamine in animal models of Parkinson's disease, and has been shown to be of sufficient utility in human experiments to warrant further application in controlled clinical trials. Unfortunately, current animal models are not capable of mimicking the progression of neuronal degeneration that characterizes Parkinson's disease, and other neurodegenerative disorders.

Consequently, the effectiveness of dopamine-producing grafts may be compromised by the continued degeneration of the residual population of neurons

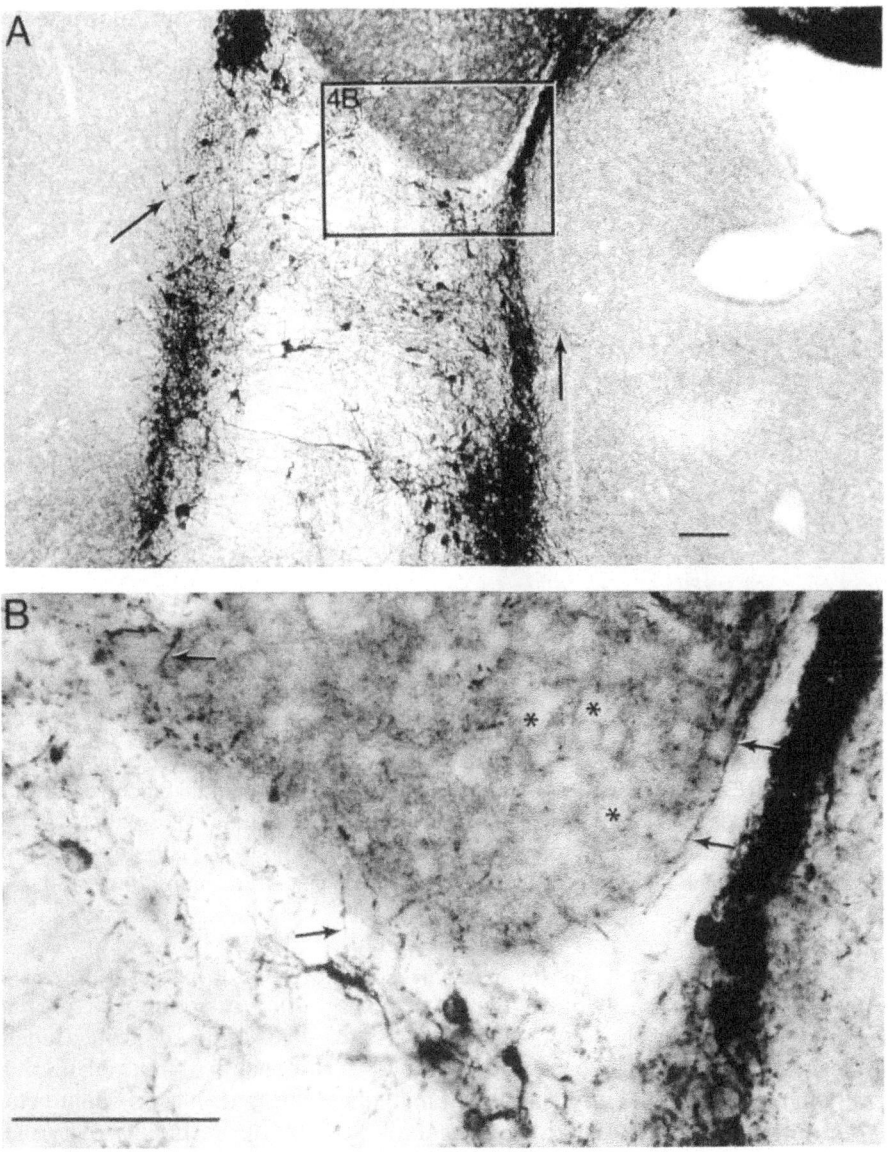

Figure 4. In **A** some tyrosine hydroxylase positive fibers appear to project into the host brain from the co-graft (arrows), but the majority of these fibers extend dorsally to terminate in a dense dopaminergic plexus in the striatal portion of the co-graft. Here, the fibers appeared to form pericellular arrays around striatal target neurons (*), seen to advantage in **B**. Some individual fibers appear to invade the striatal portions of the grafts (arrows) from the more ventral mesencephalic component of the co-graft. Bar = 100 μM

in the substantia nigra. This might be seen as a leveling-off of graft-induced improvements followed by a decline to pre-graft conditions, and possibly further decline as degeneration proceeds. Thus, a therapeutic strategy that incorporates either repair or protection of the host system could be of profound benefit when coupled with replacement of the degenerated dopamine neurons. The addition of embryonic striatal neurons and glia to grafts of ventral mesencephalon appears to be one method for providing this dual therapeutic advantage and should be tested further in non-human primate models to determine the extent of protection achieved.

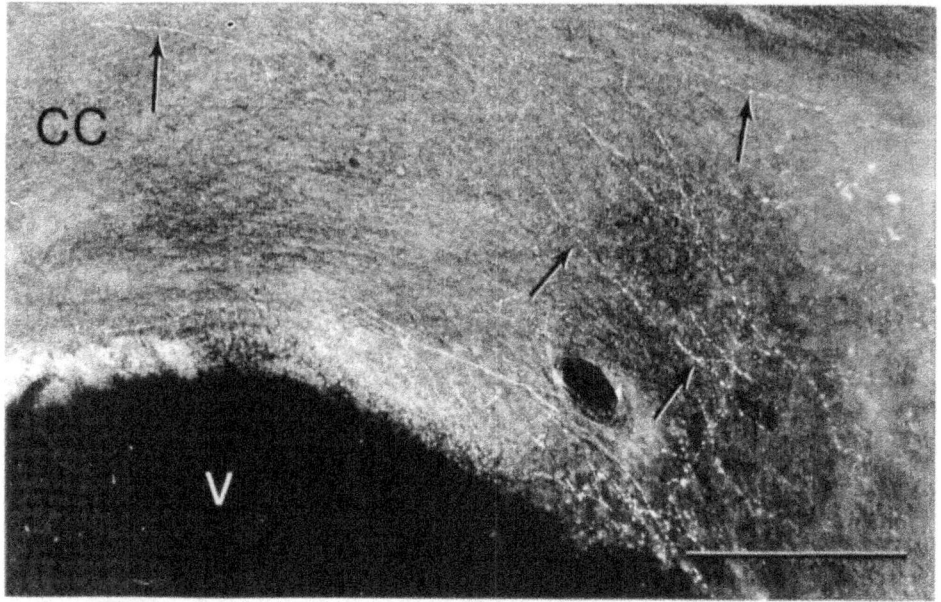

Figure 5. A number of tyrosine hydroxylase positive fibers (arrows) are seen within the corpus callosum (CC) overlying the lateral ventricle (V) in an animal that received a striatal graft in the left caudate and a mesencephalic graft in the right caudate. These fibers could be traced from the dopaminergic neurons of the mesencephalic graft, dorsally into the corpus callosum through which they extended to the contralateral side of the brain. Bar = 100 µM.

Considerable interest has been focused on the possible preservation of the nigrostriatal system in lesioned animals through the introduction, through transplantation, of growth factor-producing tissue. This type of protection has been seen in the form of a regenerative sprouting response by fibers of the host, generally in response to the implantation of medullary chromaffin cells (Bohn, et al., 1987) or surgical trauma (Bankiewicz et al., 1990). A dense plexus of tyrosine hydroxylase positive fibers near the implant or region of trauma was reported in these investigations. It was not possible to trace these fibers to their perikarya of origin; thus it remains unclear whether these catecholamine fibers arose from the nigrostriatal pathway or mesolimbic system, or alternately represent postganglionic sympathetic fibers that manufacture norepinephrine and are known to sprout in the CNS in response to lesions. While the molecule responsible for this type of response by the host brain is unknown, the ability to sprout new axons in the region of dopamine depletion generally has been viewed as supporting the concept that new growth is a type of restoration or preservation of dopaminergic input to the host striatum. Some investigators have reported that this presumptive sprouting response is responsible for improved function of the striatum following dopamine depletion caused by MPTP.

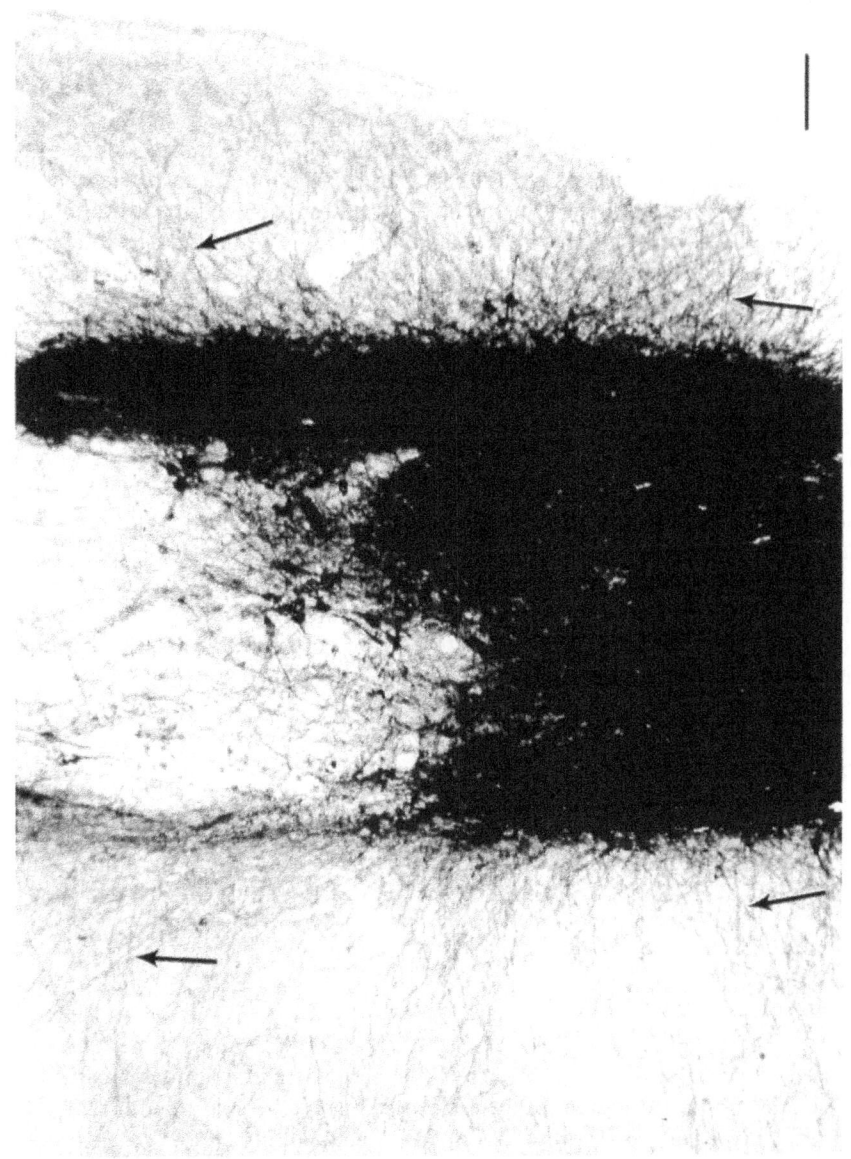

Figure 6. This densely packed graft of dopaminergic neurons shows extensive neuritic outgrowth into the host caudate nucleus. This growth appeared uniformly distributed around the graft in contrast to the directional preference exhibited by the dopaminergic components of co-grafts as seen in Figs. 3 and 4. Bar = 100 μM.

Utilization of a protection strategy clinically would require knowledge of a specific trophic influence since the creation of cavities for the purpose of stimulating growth would appear to be inappropriate for human tests. Thus, we have attempted to identify a naturally occurring source of trophic/tropic influence that would be suitable for direct intracerebral grafting. To this purpose, we have demonstrated that the embryonic striatum can exert a powerful influence on fiber outgrowth in the primate striatum. Specifically, we observed a robust growth pattern of tyrosine hydroxylase fibers that was exceptionally dense within the co-grafted embryonic striatum. These fibers appeared to produce pericellular arrays around small striatal neurons of the grafts that also stained positive for striatal transmitters such as acetylcholine (Wechsler et al., 1994). We also have seen dense tyrosine hydroxylase positive fibers in solo striatal grafts placed into the caudate nucleus of partially-lesioned monkeys (Wechsler et al., 1993). The importance of this finding with respect to potential clinical applications appears to be two-fold. First, the animals exhibited a partially-lesioned nigrostriatal system in response to MPTP; dopamine content in the striatum was reduced to about 30-50 % of normal (Elsworth et al., 1989) and dopamine cell numbers in the substantia nigra were about 40-60 % of normal. This is somewhat representative of an early stage of Parkinson's degeneration with depleted dopamine and minimal dysfunction. Thus, if striatal grafts can stimulate dopamine fiber growth in the partially-depleted animal model, then the application of tropic factors at such an early stage in a progressive disease might provide a strategy for the prevention of further degeneration, or might alter the progression of the disease by the delivery of increased amounts of dopamine as a result of the generation of new nerve terminals in target regions of the striatum. Although MPTP-induced degeneration in the primate brain has not been reported to be progressive, the apparent sprouting response observed in monkeys to the implantation of embryonic striatum suggests the capability of a partially damaged dopamine system to regenerate. Identification and isolation of the protein responsible for this growth would appear to be an important next step in the consideration of the applicability of this approach.

Acknowledgements

This work was supported in part by PO1 NS 24032 and KO1 MH 00643 (to DER). The authors are grateful for the skilled technical support by the staff of the St. Kitts Biomedical Research Institute.

REFERENCES

Annett LE, Martel FL, Rogers DC, Ridley RM, Baker HF and Dunnett SB (1993): Behavioural assessment of the effects of embryonic nigral grafts in marmosets with unilateral 6-OHDA lesions of the nigrostriatal pathway. Exp Neurol 125: 228-246.

Bakay RAE, Fiandaca MS, Barrow DL, Schiff A and Collins DC (1985): Preliminary report on the use of fetal tissue transplantation to correct MPTP-induced parkinsonian-like syndrome in primates. Appl Neurophys 48: 358-361.

Bankiewicz KS, Plunkett RJ, Jacobowitz DM, Kopin IJ and Oldfield EH (1991): Fetal nondopaminergic neural implants in parkinsonian primates. J Neurosurg 74: 97-104.

Bankiewicz KS, Plunkett RJ, Jacobowitz DM, Porrino L, Di Porzio U, London WT, Kopin IJ and Oldfield EH (1990): The effect of fetal mesencephalon implants on primate MPTP-induced parkinsonism. J Neurosurg 72: 231-244.

Bohn MC, Cupit L, Marciano F and Gash DM (1987): Adrenal medulla grafts enhance recovery of striatal dopaminergic fibers. Science 237: 913-916.

Brundin P, Strecker RE, Widner H, Clarke DJ, Nilsson OG, Astedt B, Lindvall O and Björklund A (1988): Human fetal dopamine neurons grafted in a rat model of

Parkinson's disease: immunological aspects, spontaneous and drug-induced behaviour, and dopamine release. Exp Brain Res 70: 192-208.

De Beaurepaire R and Freed WJ (1987): Embryonic substantia nigra grafts innervate embryonic striatal co-grafts in preference to mature host striatum. Exper Neurol 95: 448-454.

Dunnett SB, Rogers DC and Richards SJ (1989): Nigrostriatal reconstruction after 6-OHDA lesions in rats: Combination of dopamine-rich nigral grafts and nigrostriatal "bridge" grafts. Exper Brain Res 75: 523-535.

Elsworth JD, Deutch AY, Redmond DE Jr, Taylor JR, Sladek JR Jr and Roth RH (1989): Symptomatic and asymptomatic 1-methyl-4-phenyl-1,2,3,6-tetrahydropyridine-treated primates:bichemical changes in striatal regions. Neuroscience 33: 323-331

Elsworth JD, Sladek Jr, JR, Redmond Jr, DE, Taylor JR, Collier TJ and Roth RH (1994): Early gestational mesencephalon grafts, but not later gestational mesencephalon, cerebellum, or sham grafts, increase dopamine in caudate nucleus of MPTP-treated monkeys. Neurosci (submitted).

Freed CR, Breeze RE, Rosenberg NL, Schneck SA, Kriek E, Qi JX, Lone T, Zhang YB, Snyder JA, Wells TH, Ramig LO, Thompson L, Mazziotta JC, Huang SC, Grafton ST, Brooks D, Sawle G, Schroter G and Ansari AA (1992): Survival of implanted fetal dopamine cells and neurologic improvement 12 to 46 months after transplantation for Parkinson's disease. New Eng J Med 327: 1549-55.

Hemmendinger LM, Garber BB, Hoffman PC and Heller A (1981): Target neuron-specific process formation by embryonic mesencephalic dopamine neurons in vitro. Proc Natl Acad Sci USA 78: 1264-1268.

Lindvall O, Rehncrona S, Brundin P, Gustavii B, Astedt B, Widner H, Lindholm T, Björklund A, Leender K, Rothwell JC, Frackowiak R, Marsden CD, Johnels B, Steg G, Freedman R, Hoffer BJ, Seiger Å, Bygdeman M, Strömberg I and Olson L (1989): Human fetal dopamine neurons grafted into the striatum in two patients with severe parkinson's disease. Arch Neurol 46: 615-631.

Lindvall O, Brundin P, Widner H, Rehncrona S, Gustavii B, Frackowiak R, Leenders KL, Sawle G, Rothwell JC, Marsden CD and Björklund A (1990): Grafts of fetal dopamine neurons survive and improve motor function in parkinson's disease. Science 247: 574-577.

Olson L, Strömberg I, Bygdeman M, Granholm A-Ch, Hoffer B, Freedman R and Seiger Å (1987): Human fetal tissues grafted to rodent hosts: structural and functional observations of brain, adrenal and heart tissues in oculo. Exp Brain Res 67: 163-178.

Prochiantz A, Di Porzio I, Dato A, Berger B and Glowinski J (1979): In vitro maturation of mesencephalic dopaminergic neurons from mouse embryos is enhanced in presence of their striatal target cells. Proc Natl Acad Sci USA 76: 5387-5391.

Sladek JR Jr, Collier TJ, Haber SN, Roth RH and Redmond DE Jr (1986): Survival and growth of fetal catecholamine neurons tranplanted into primate brain. Brain Res Bull 17: 809-818.

Sladek JR Jr, Redmond DE Jr, Collier TJ, Haber SN, Elsworth JD, Deutsch AY and Roth RH (1987): Transplantation of fetal dopamine neurons into primate brain reverses MPTP-induced parkinsonism. Prog Brain Res 71: 309-322

Sladek JR Jr, Redmond DE Jr, Collier TJ, Blount JP, Elsworth JD, Taylor JR and Roth RH (1988): Fetal dopamine neural grafts: extended reversal of methylphenyltetrahydropyridine-induced parkinsonism in monkeys. Prog Brain Res 78: 497-506.

Sladek JR Jr, Collier TJ, Elsworth JD, Taylor JR, Roth RH and Redmond DE Jr (1993a): Can graft-derived neurotrophic activity be used to direct axonal outgrowth of grafted dopamine neurons for circuit reconstruction in primates? Exp Neurol 124: 134-139.

Sladek JR Jr, Elsworth JD, Roth RH, Evans LE, Collier TJ, Cooper SJ, Taylor JR and Redmond DE Jr (1993b): Fetal dopamine cell survival after transplantation is dramatically improved at a critical donor gestational age in non-human primates. Exp Neurol 122: 16-27.

Sladek JR Jr, Blanchard BC, Collier TC, Elsworth JD, Taylor JR, Roth RH and Redmond DE Jr (1994): Optimal survival of grafted mesencephalic dopamine (DA) neurons is attained from tissue taken during neurogenesis of the substantia nigra in non-human primate. Soc Neurosci Abstr 20: 1328.

Spencer DD, Robbins RJ, Naftolin F, Marek KL, Vollmer T, Leranth C, Roth RH, Price LH, Gjedde A, Bunney BS, Sass KJ, Elsworth JD, Kier LE, Makuch R, Hoffer PB and Redmond DE Jr (1992): Unilateral transplantation of human fetal mesencephalic tissue into the caudate nucleus of Parkinsonian patients: functional effects for 18 months. New Eng J Med 327: 1541-1548.

Taylor JR, Elsworth JD, Roth RH, Sladek JR Jr, Collier TJ and Redmond DE Jr (1991) Fetal grafting of substantia nigra to striatum reverses behavioral deficits induced by MPTP in primates: A comparison with other types of grafts as controls. Exp Brain Res 85: 335-348.

Taylor JR, Elsworth JD, Sladek JR Jr, Collier TJ, Roth RH and Redmond DE Jr (1994): Sham surgery does not ameliorate MPTP-induced behavioral deficits in monkeys. Cell Transplant, in press.

Wechsler RT, Sladek JR Jr, Blanchard BC, Collier TJ, Elsworth JD, Taylor JR, Roth RH and Redmond DE Jr (1993): Sprouting of host dopamine fibers in response to grafts of embryonic striatum in adult African green monkeys. Soc Neurosci Abstr 19: 1509.

Wechsler RT, Blanchard BC, Collier TJ, Elsworth JD, Taylor JR, Roth RH, Redmond DE Jr and Sladek JR Jr (1994): Host striatum projects fibers into transplanted fetal mesencephalon in a parkinsonian primate model. Soc Neurosci Abstr 20: 1328.

Wictorin K, Brundin P, Gustavii B, Lindvall O and Björklund A (1990): Reformation of long axon pathways in adult rat central nervous system by human forebrain neuroblasts. Nature 347: 556-558.

Widner H, Tetrud J, Rehncrona S, Snow B, Brundin P, Gustavii B, Björklund A, Lindvall O and Langston JW (1992): Bilateral fetal mesencephalic grafting in two patients with parkinsonism induced by 1-methyl-4-phenyl-1,2,3,6-tetrahydropyridine (MPTP). New Eng J Med 327: 1557-1563.

Yurek DM, Collier TJ and Sladek JR Jr (1990): Embryonic mesencephalic and striatal co-grafts: Development of grafted dopamine neurons and functional recovery. Exp Neurol 109: 191-199.

USING FETAL MESENCEPHALIC GRAFTS TO TREAT MPTP-INDUCED PARKINSONISM

Håkan Widner*

Restorative Neurology Unit Department of Neurology,
University Hospital, S-221 85 Lund, SWEDEN and The
Parkinson's Institute, 1170 Morse Avenue, Sunnyvale,
CA 94089-1605 U.S.A.

INTRODUCTION

Transplantation of immature neuronal tissue is a novel approach with the potential to reverse functional defects in neurological disorders. The development of this technique has come the furthest in animal models of dopamine depletion, extending into early clinical attempts to restore the dopamine deficiency in Parkinson's disease. Although the degree of functional recovery induced by immature tissue rich in dopaminergic neurons has been impressive in rodent and monkey models, clinical neural tissue transplants are still to be regarded as experimental and under refinement. In this review the development of the intracerebral transplantation approach is briefly outlined. In particular, the role of the selective dopaminergic neurotoxin methyl-phenyl-tetrahydropyridine (MPTP)-induced human and non-human primate models of parkinsonism in this development will be addressed. Other aspects of intracerebral transplantation technique are reviewed elsewhere in this volume.

BACKGROUND

*Data and material appearing in this text have been accumulated through on-going collaborations among the following institutions, departments and persons:

Lund University, Sweden: *Department of Neurology,* Olle Lindvall, Patrik Brundin; *Department of Neurosurgery*, S. Rehncrona; *Department of Obstetrics and Gynecology*, Björn Gustavi; *Department of Medical Cell Research*, Anders Björklund; **The Parkinson's Institute, Sunnyvale, CA, U.S.A.**: James W. Tetrud, J. William Langston; **University of British Columbia, Vancouver, BC, Canada**: *Department of Neurology*, Barry Snow.

Neural Cell Specification: Molecular Mechanisms and Neurotherapeutic Implications
Edited by Juurlink *et al.*, Plenum Press, New York, 1995

231

Parkinson's Disease

Parkinson's disease (PD) is a progressive neurodegenerative disorder, of unknown cause. The prevalence is about 40 / 100,000 at the age of 40, but 1,600 / 100,000 at age 70. The dopaminergic neurons are slowly degenerating leading to a progressive dopamine loss within the striatum. The degeneration is more pronounced in the putamen but there is dopamine loss also in the caudate nucleus. When the initial compensatory mechanisms are exhausted, which occurs usually when more than 70- 80% of the normal dopamine cells and dopamine content have been lost, symptoms of parkinsonism become obvious (Agid et al. 1987). The dopamine deficiency leads to an imbalance between two of the efferent pathways, one direct and one indirect, which is thought to lead to double inhibition of movements, explaining the symptoms of bradykinesia and hypokinesia (Alexander and Crutcher 1990).

Grafting in Rodent Models

The first reliable survival of immature neural tissue grafts was demonstrated for catecholaminergic neurons in 1976 in rodents (Stenevi et al. 1976). The first examples that dopaminergic neural tissue grafts exert important functional effects in rats with a selective dopamine depletion was demonstrated in 1979 (Björklund and Stenevi 1979; Perlow et al. 1979). In this rodent model the neurotoxin 6-hydroxydopamine is injected unilaterally into the medial forebrain bundle, causing >95% depletion of the dopamine content in the striatum. This imbalance between the intact and lesioned sides results in a tendency for rotational behavior, which can be quantified and augmented by drugs affecting the dopamine transmission (Ungerstedt and Arbuthnott 1970). Immature dopaminergic neurons can be found within the ventral mesencephalic region. The single most important factor for graft survival is the donor age. Embryonic cells have the unique capacity to withstand ischemia and the neurons can be isolated without suffering extensive damage since they lack extensive neurite outgrowths. The fraction of dopaminergic neurons that survive the grafting process is still only about 1/5, or between 5-20% of the number of dopaminergic neurons in the donor tissue (Sauer and Brundin 1991). The surviving neurons integrate into the host brain, reinnervate the striatum, reverse the abnormal post-synaptic receptor changes and restore dopamine synthesis to near normal . The presence of a viable graft is essential, since at least in the rodent model destruction of the graft leads to a rapid return of the behavioral defect. The mechanisms of action of the neural tissue grafts have been reviewed recently (Björklund et al. 1987; Freed 1992).

Early Clinical Trials in Parkinson's Disease

Human transplantation was first tried using adrenal medullary tissue in order to avoid any immunological problems as well as to avoid ethical constraints for using fetal tissue. The first attempts, from Backlund et al. (1985) and Lindvall et al. (1987) showed only very short lasting effects, with only few clinically important improvements observed. In 1985, when the results of the adrenal tissue grafts were at hand, a conscious approach to attempt grafting fetal neural tissue into humans with Parkinson's disease was made at Lund University. One important step in performing early clinical trials with fetal tissue was to try to determine if there was a disease process in Parkinson's disease that might affect the grafted cells. If that is the case, the transplantation technique may not be a fruitful avenue to follow. It was also of interest to assess if grafted tissue at all could affect parkinsonism in humans. The Swedish Medical Association has formulated ethical guidelines, which have subsequently been revised, and will become law in 1995. In 1987, tissue from three donors was stereotaxically grafted into the putamen and the caudate of two patients (Lindvall et al 1989). In summary, the antiparkinsonian effects were of only modest clinical importance, but could be detected and measured. One patient, who prior to the implant usually could not walk in her "off condition" improved so as to being able to do so some months after the implantation. However, compared to the dramatic effects behavior recovery observed in animal experiments of fetal neural grafts, the effects were disappointing. There were several possible explanations for the lack of effects. Lack of graft survival because of either immunological or neurobiological reasons; not enough graft survival to reach a critical threshold for symptomatic relief; or a disease process attacking the grafted cells. In order to address the latter point, we sought a human condition, exhibiting parkinsonism and an essentially non-progressive disorder.

MPTP-induced Parkinsonism

In 1982 a group of heroin addicts turned up at various Northern California hospitals, after having suddenly become completely immobile, unable to talk, communicate or care for themselves. The features they presented was like advanced parkinsonism, with muscular rigidity, tremor, extreme bradykinesia and poor balance. Their initial response to L-dopa was excellent, with complete disappearance of all parkinsonian features. It was soon unraveled that the causative agent was a chemical substance produced by accident by a designer-drug chemist, trying to make synthetic heroin. The substance was found to be 1-methyl-4-phenyl-1,2,3,6-tetrahydropyridine (MPTP) (Langston et al. 1983; Ballard et al. 1985) and it was shown that the toxin lead to a selective dopaminergic loss. Non-human primates develop a condition with striking similarities to humans with both MPTP-induced parkinsonism and the Idiopathic Parkinson's Disease (Burns et al. 1983; Langston et al. 1984). Transplants of embryonic and fetal neural tissue has been attempted in approximately 50 animals with MPTP-induced parkinsonism. The results are varied, but graft survival has been demonstrated and in some instances the behavior recovery has been reported to be dramatic (Annett et al. 1994; Bakay et al. 1987; Bankiewicz et al. 1990; Dunnett and Annett 1991; Elsworth et al. 1989; Fine et al. 1988; ; Freed et al. 1988; Taylor et al. 1991; Sladek et al. 1994; Widner et al. in preparation).

GRAFTING INTO NON-HUMAN MPTP-INDUCED PARKINSONIAN PRIMATES: THE LUND-SYNNYVALE EXPERIENCE

Eleven adult squirrel monkeys (saimiri sciureus) were included and nine animals were given one or more subcutaneous injections of MPTP (2.0 mg/kg). The symptoms of parkinsonism were assessed with an automated movement monitoring device, and a rating scale. All animals demonstrated stable symptoms of parkinsonian for a period over 60 days using these criteria. Seven animals received implants of immature ventral mesencephalic tissue. Implants were made into the putamen bilaterally with 20 µL donor tissue on each side. Histological graft survival was found in four animals and in these animals there were improvements in their behavior, both in terms of parkinsonism ratings and automated general activity measurements (Fig. 1). There were no signs of spontaneous recovery in sham grafted animals (n = 2), nor were there any signs of behavioral improvement in the animals without any surviving graft tissue (n = 3). The lack of surviving tissue is likely to be due to immunological rejection or neurobiological reasons such as inadequate donor age.

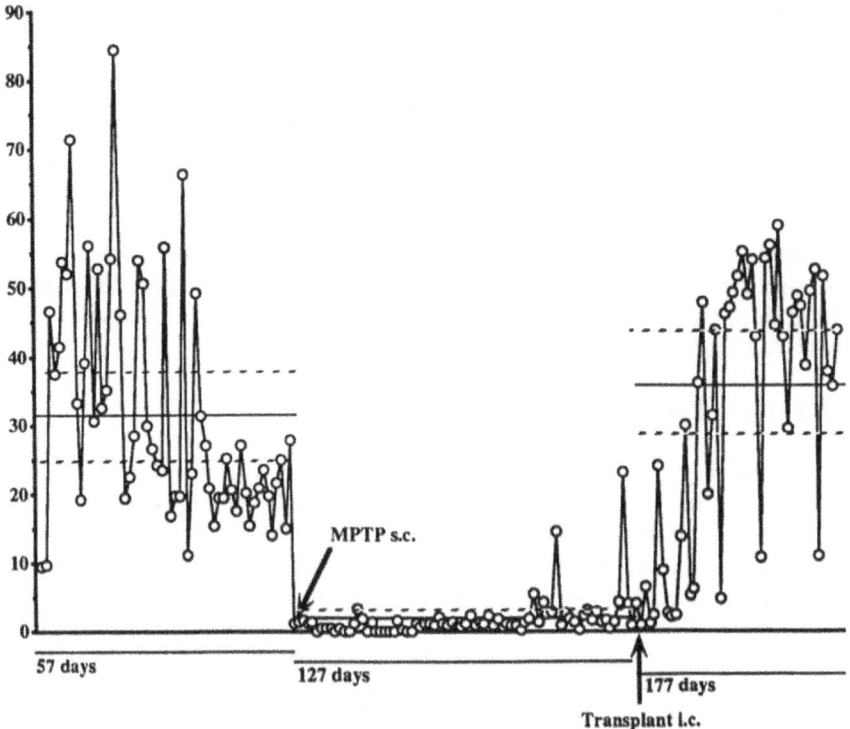

Figure 1. Time pattern of the spontaneous movements of a squirrel monkey prior to being injected with MPTP, after toxin exposure and after intracerebral transplantation with embryonic dopaminergic neurons. Horizontal lines indicate the mean value of movements/min at that period ± 99% confidence interval.

GRAFTING OF SEVERELY MPTP-INDUCED PARKINSONIAN PATIENTS

Two patients of the original seven index cases (Langston et al. 1983 and Ballard et al. 1985) were selected as transplantation candidates. During a preoperative assessment period of 15 - 18 months, repeated functional tests were performed, including [18F]-6L-fluorodopa positron emission tomography (PET) scanning, largely following a protocol summarized in the CAPIT program (Langston et al. 1992). The patients were on stable antiparkinsonian medication during this whole period. The procedure and results have been summarized earlier (Widner et al. 1992).

Patient #1 had in "defined off", ie. the condition 12 - 15 hours after the last L-dopa dose, an average total score in the Unified Parkinson's Disease Rating Scale (UPDRS) of 110±19 (Fig. 2), and pronounced hypokinesia and muscular rigidity in all limbs. He also had early fatigue when walking, frequent freezing and shuffling gait without any arm swings. He had frequent falls. The regular medication was 50 mg L-dopa combined with 5 mg carbidopa 5 times per day. On each dose, he experienced peak-dose visual hallucinations and generalized dyskinesias. During a single dose of L-dopa the amount of dyskinesias could be rated using an Abnormal Involuntary Movement Scale (AIMS, after Guy 1976). The integrated amount of dyskinesias were found to be 7.7±1.2 arbitrary units (AU) pre-operatively. The "best on" function, defined as the best condition experienced when taking antiparkinsonian medication, was rated on the UPDRS scale to 76±21 and his gait improved to be smooth and with a good stride and length of step, but he fatigued easily, after some 50 m. A pre-operative PET scan showed a net influx constant for fluorodopa of 0.220 ml/min for the striatum.

The equivalent pre-operative data for the second patient were: "Defined off" UPDRS total score of 93±19, improving to 51±16 in the "best on" condition. Her standard medication was 100 mg levodopa with 25 mg carbidopa 5 times a day. The integrated AIMS score were 11.9±3.1 AU. The main clinical problem was an accumulation of dyskinesias. Her net influx value in the PET was 0.210 ml/min for the striatum.

The transplantations were made in two surgical sessions with a two week interval. The first patient received for implantation in each hemisphere the tissue from 3 - 4 donors. Three implants, each of 20 µL, were made into each putamen, and 1 implant into each caudate nucleus. The second patient received a graft in the right caudate only. Donor ages were 6 - 8 weeks post-conception. Both patients were immunosuppressed with cyclosporin, azathioprine and prednisolone.

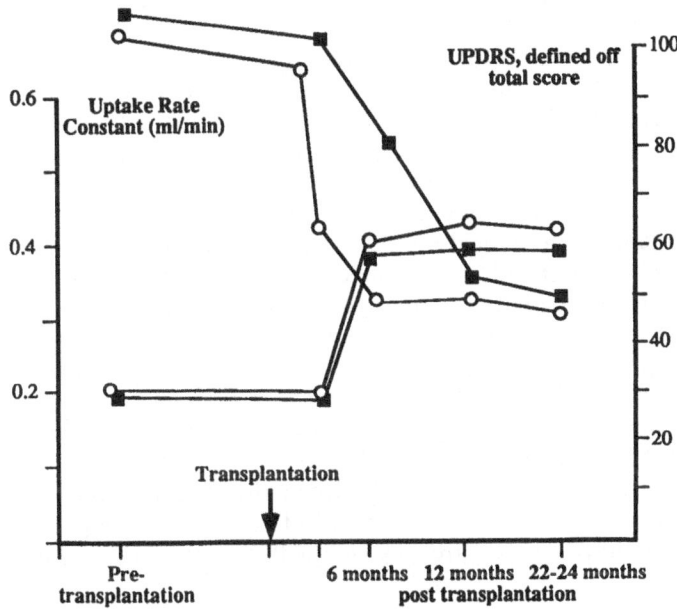

Figure 2. The relationship between total UPDRS score and net influx rate of fluorodopa. Solid squares indicate patient #1 and open circles patient #2.

The post-operative course was slow and protracted. The earliest signs of decreased muscular rigidity appeared 3 - 6 months after grafting in patient #2 . The same effect was observed only 6 - 9 months later in the first patient. At 22 - 24 months post- operatively both had improved significantly. In the single L-dopa-dose test, neither patient could be seen to fluctuate between the pre-L-dopa "off-condition" to the "on-condition". Both patients also experienced lesser degree of

L-dopa-induced dyskinesias. There was a correlation between the PET influx values and the UPDRS ratings.

Patient #1 improved on unchanged medication in "defined off" to 58±3 in the total UPDRS scale, mainly due to improvement in the motor scoring, and in "best on" to a value of 23. The AIMS integrated score on a 50 mg L-dopa/5 mg carbidopa dose decreased by 95% to 0.5 AU. The PET influx was increased 160% at 12 months after grafting and remained increased at 24 months. On a global scale he managed better in all aspects, in particular because of a general ability to initiate movements better and by the increase in beneficial time of his medication with fewer side-effects. For instance, he was able to walk several hundred meters, and the tendency of freezing-up when walking was decreased.

The second patient was similarly improved with "defined off" total UPDRS values of 44 at 22 months and the "best on" value of 30. The antiparkinsonian medication was reduced in two steps, to a final 1/3 of the starting dose. In spite of the difficulties in evaluating patients when the medication has been changed, these reductions were necessary since the duration of the individual dosages were prolonged so as to cause symptoms of over- medication. When stabilized on 50 mg L-dopa/12.5 mg carbidopa, the integrated AIMS test were reduced to 75% of the pre-operative value. The patient now manages to live completely on her own completing all chores of daily living without any assistance. She is attending adult education classes.

GRAFTING IN IDIOPATHIC PARKINSON'S DISEASE

The long-term follow-up of two patients with Idiopathic Parkinson's Disease, that were grafted in parallel with these two MPTP-patients has been published recently (Lindvall et al. 1994). These patients were grafted with an identical technique, but unilaterally and in the putamen only. There is evidence of an on-going degeneration of the intrinsic dopaminergic system (Sawle et al. 1992; Lindvall et al. 1992; Lindvall et al. 1994) and in spite of this, there are graft effects that are evident for prolonged periods. One patient in this report is the first graft recipient who has been able to completely stop all L-dopa medication. There seem to be no indications, when comparing the net effects between the MPTP- cases and the Parkinson's disease cases with intracerebral fetal neural grafts, that there is any disease process attacking the immature grafted cells. However, the net effect of the implantation may be obscured in patients with an active, rapid degeneration of their dopaminergic pathways in Parkinson's disease. Other centers are reporting graft survival with some clinical benefits eg. Freed et al. (1992) and Spencer et al. (1992). Quantitative comparison between the grafts effects are difficult, but appears to be approximately 1/4 of the effects when compared to the studies accounted for above. This correlates well with the amount tissue implanted.

CONCLUSIONS

Intracerebral grafting of immature dopaminergic neurons is a novel technique with potential to reinnervate and reverse neurological deficits, not only in Parkinson's disease. The early clinical trials have indicated that graft survival is possible, and that the effects can be of major clinical importance. However, the technique is still investigational, and the optimal techniques for donor tissue procurement, method of implantation and graft placement sites remain to be determined. Also the clinical value of the methods should be evaluated against other treatment methods. Although animal experimental data are very promising and clinical trials have given encouraging results in terms of symptomatic relief there exists, at present, no treatment for PD based on intracerebral transplantation. It is important that patients and relatives are informed that this research is still at an experimental stage.

ACKNOWLEDGEMENTS

The studies were supported by various grants. In Sweden: the Riksbanken Tricentennial Fund, the Swedish Medical Research Council (projects 14X-8666 and K93/16P 10135-02B), the Swedish Society for Medicine; Swedish Society for Medical Research; the Åke Wiberg, Elsa Schmitz, Rut and Erik Hardebo, Westerström and Thorsten and Elsa Segerfalk Foundations, the County of Malmöhus. In California and Canada: the California Parkinson's Foundation, the Parkinson's Disease Foundation, the Parkinson's Institute, the Dystonia Research Foundation of Canada, and the Valley Foundation.

REFERENCES

Agid Y, Javoy-Agid F and Ruberg M (1987): Biochemistry of neurotransmitters in Parkinson's disease. In: Marsden CD and Fahn S (eds): "Movement Disorders 2." London, Butterworth, pp 166-230.

Alexander GE and Crutcher MD (1990): Functional architecture of basal ganglia circuits: neural substrates of parallel processing. Trend Neurosci 13: 266-271.

Annett LE Martel FL Rogers DC Ridley RM Baker HF and Dunnett SB (1994): Behavioral assessment of the effects of embryonic nigral grafts in marmosets with unilateral 6-OHDA lesions of the nigrostriatal pathway. Exp Neurol 125: 228-246.

Backlund E-O, Granberg PO, Hamberger B, Knutsson E, Mårtensson A, Sedvall G, Seiger Å and Olson L (1985): Transplantation of adrenal medullary tissue to the striatum in parkinsonism. First clinical trials. J Neurosurg 62: 169-173.

Bakay RAE, Barrow DL, Fiandaca MS, Ivuone PM, Schiff A and Collins DC (1987): Biochemical and behavioral correction of MPTP Parkinson-like syndrome by fetal cell transplantation. Ann NY Acad Sci 495: 623-640.

Ballard PA, Tetrud JW and Langston JW (1985): Permanent parkinsonism in humans due to 1- methyl-4-phenyl-1,2,3,6-tetrahydropyridine (MPTP): Seven cases. Neurol 35: 949-956.

Bankiewicz KS, Plunkett RJ, Jacobowitz DM, Porrino L, di Porzio U, London WT, Kopin IJ andOldfield EH (1990): The effect of fetal mesencephalon implants on primate MPTP- induced parkinsonism. J Neurosurg 72: 231-244.

Björklund A and Stenevi (1979): Reconstruction of the nigrostriatal dopamine pathway by intracerebral nigral transplants. Brain Res 177: 555-560.

Björklund A, Lindvall O, Isacson O, Brundin P, Wiktorin K, Strecker RE, Clarke DJ and Dunnett SB (1987): Mechanisms of action of intracerebral neural implants: studies on nigral and striatal grafts to the lesioned striatum. Trend Neurosci 10: 509-516.

Burns S, Chiueh CC, Markey SP, Ebert MH, Jacobowitz DM and Kopin IJ (1983): A primate model of parkinsonism: selective destruction of dopaminergic neurons in pars compacta of the substantia nigra by N-methyl-4-phenyl-1,2,3,6-tetrahydropyridine. Proc Natl Acad Sci USA. 80: 4546-4550.

Dunnett SB and Annett LE (1991): Nigral transplants in primate models of parkinsonism. In: Lindvall O, Björklund A and Widner H (eds): "Intracerebral Transplantation In Movement Disorders." Restorative Neurology 4, Amsterdam, Elsevier, pp 27-51.

Elsworth JD, Redmond DE, Sladek JR, Deutsch AY, Collier TJ and Roth RH (1990): Reversal of MPTP-induced parkinsonism in primate by fetal dopamine cell transplants. In: Franks AJ, Ironside JW, Mindham RHS, Smith RJ, Spokes EGS and Winlow W (Eds): "Function And Dysfunction Of The Basal Ganglia". Manchester Univ Press, pp 161-180.

Fine A, Hunt SP, Oertel WH, Nomoto M, Chong PN, Bond A, Waters C, Temlett JA, Annett LE, Dunnett SB, Jenner P and Marsden CD (1988): Transplantation of embryonic dopaminergic neurons to the corpus striatum of marmosets rendered parkinsonian by 1-methyl-4-phenyl-1,2,5,6,-tetra-hydropyridine. Prog Brain Res 78: 479-489.

Freed CR, Breeze RE, Rosenberg NL, Schneck SA, Kriek E, Qi J-X, Lome T, Zhang Y-B, Snyder JA, Wells TH, Olson Ramig L, Thompson L, Mazziotta JC, Huang SC, Grafton ST, Brooks D, Sawle G, Schroter G and Ansari AA (1992): Survival of implanted fetal dopamine cells and neurologic improvement 12 to 46 months after transplantation for Parkinson's disease. N Engl J Med 327: 1549-1555.

Freed CR, Richards JB, Sabol KE and Reite ML (1988): Fetal substantia nigra transplants lead to dopamine cell replacement and behavioral improvement in bonnet monkeys with MPTP induced parkinsonism. In: Beart PM, Woodruff GN and Jackson D (eds): "Pharmacology and function of dopaminergic neurons." London, Macmillan, pp 353- 360.

Freed WJ (1991): Substantia nigra grafts and Parkinson's disease: from animal experimetns to human therapeutic trials. Restor Neurol Neurosci 3: 109- 134.

Guy W (1976): Abnormal involuntary movement scale (AIMS) In Guy (ed): "ECDEU Assessment Manual For Psychopharmacology, Revised Edition". US Dept Health, Education and Welfare, HDHEW, Rockville MD, pp 534-537.

Langston JW, Ballard P, Tetrud JW and Irwin I (1983): Chronic parkinsonism in humans due to a product of meperidine analog synthesis. Science 219: 979-980.

Langston JW, Forno LS, Rebert CS and Irwin I (1984): Selective nigral toxicity after systemic administration of 1-methyl-4-phenyl-1,2,5,6-tetrahydropyridine (MPTP) in the squirrel monkey. Brain Res 292: 390-394.

Langston JW, H Widner, D Brooks, S Fahn, T Freeman, C Goetz and R Watts (1992): Core Assessment Program for Intracerebral Transplantations. CAPIT. Mov Disord 7: 1- 13.

Lindvall O, Backlund E-O, Farde L, Sedvall G, Freedman R, Hoffer B, Nobin A, Seiger Å and Olson L (1987): Transplantation in Parkinson's disease: two cases of adrenal medullary grafts to the putamen. Ann Neurol 22: 457-468.

Lindvall O, Rehncrona S, Brundin P, Gustavii B, Åstedt B, Widner H, Lindholm T, Björklund A, Leenders KL, Rothwell JC, Frackowiak R, Marsden CD, Johnels B, Steg G, Freedman R, Hoffer BJ, Seiger Å, Bygdeman M, Strömberg I and Olson L (1989): Human fetal dopamine neurons grafted into the striatum in two patients with severe Parkinson's disease: A detailed account of methodology and a 6-month follow- up. Arch Neurol 46: 615-631.

Lindvall O, Widner H, Rehncrona S, Brundin P, Odin P, Gustavii B, Frackowiak R, Leenders KL, Sawle G, Rothwell JC, Björklund A and Marsden CD (1992): Transplantation of fetal dopamine neurons in Parkinson's disease: 1-year clinical and neurophysiological observations in two patients with putaminal implants. Ann Neurol 31: 155 - 165.

Lindvall O, Sawle G, Widner H, Rothwell JC, Björklund A, Brooks D, Brundin P, Frackowiak R, Marsden CD, Odin P and Rehncrona S (1994): Evidence for long term survival and function of dopaminergic grafts in progressive Parkinson's disease. Ann Neurol 35: 172-180.

Perlow MJ, Freed WJ, Hoffer BJ, Seiger , Olson L and Wyatt RJ (1979): Brain grafts reduce motor abnormalities produced by destruction of nigrostriatal dopamine system. Science 204: 643-647.

Sauer HJ and Brundin P (1991): Effects of cool storage on survival and function of intrastriatal ventral mesencephalic grafts. Rest Neurol Neurosci 2: 123-135.

Sawle G, Bloomfield PM, Björklund A, Brooks DJ, Brundin P, Leenders KL, Lindvall O, Marsden CD, Rehncrona S, Widner H and Frackowiak SJ (1992): Transplantation of fetal dopamine neurons in Parkinson's disease: positron emission tomography [18F]-6-L- fluorodopa studies in two patients with putaminal implants. Ann Neurol 31: 166 - 173.

Sladek Jr JR, Elsworth JD, Roth RH, Evans LE, Collier TJ, Cooper SJ, Taylor JR and Redmond Jr, DE (1993): Fetal dopamine cell survival after transplantation is dramatically improved at a critical donor gestational age in nonhuman primates. Exp Neurol 122: 16-27.

Spencer DD, Robbins RJ, Naftolin F, Marek KL, Vollmer T, Leranth C, Roth RH, Price LH, Gjedde A, Bunney BS, Saas KJ, Elsworth JD, Kier EL, Makuch R, Hoffer PB and Redmond Jr, ER (1992): Unilateral transplantation of human fetal mesencephalic tissue into the caudate nucleus of patients with Parkinson's disease. N Engl J Med 327: 1541-1548.

Stenevi U, Björklund A and Svendgaard NA (1976): Transplantation of central and peripheral monoaminergic neurons to the adult rat brain: techniques and conditions for survival. Brain Res 114:1-20.

Taylor JR, Elsworth JD, Roth RH, Sladek JR, Collier TJ and Redmond DE (1991): Grafting of fetal substantia nigra to striatum reverses behavioral deficits induced by MPTP in primates: a comparison with other types of grafts as controls. Exp Brain Res 85: 335- 348.

Ungerstedt U and Arbuthnott GW (1970): Quantitative recording of rotational behavior in rats after 6-hydroxydopamine lesions of the nigrostriatal dopamine system. Brain Res 24: 485-493.

Widner H Tetrud J Rehncrona S Snow B Brundin P Gustavii B Björklund A Lindvall and Langston JW (1992): Bilateral fetal mesencephalic grafting in two patients with parkinsonism induced by 1-methyl-4-phenyl-1,2,3,6-tetrahydropyridine (MPTP). N Engl J Med 327: 1556-1563.

INCORPORATION OF CELLS INTO ADULT FIBRE TRACTS

Geoffrey Raisman

Norman and Sadie Lee Research Centre, Laboratory of
Neurobiology, National Institute for Medical Research,
Mill Hill, London NW7 1AA, U.K.

INTRODUCTION

Neurons in the adult brain are capable of receiving and giving new synapses after injury (Raisman, 1969) or transplantation (Zhou et al., 1990). These events only occur in grey matter. When nerve fibres are cut in white matter tracts, they fail to regenerate. However, cut fibres can be induced to regenerate all the way to form functional contacts with their targets by the use of peripheral nerve grafts (Vidal-Sanz et al., 1987).

Peripheral nerve grafts, or implantation of cultured Schwann cells (Neuberger et al., 1992) may provide a way of bridging a lost connection between a localised area of neurons and a circumscribed, accessible target, but they cannot provide repair in systems (such as the long ascending and descending fibre tracts of the spinal cord) where axons are recruited from many different, dispersed areas, and distributed to similarly dispersed targets. For repair to occur in such a situation the cut adult nerve fibres would have to reenter the adult fibre tracts, and regenerate along them.

In a series of experiments from this laboratory we have investigated the following questions:
(1) What is the structure of adult central fibre tracts?
(2) Are axons able to grow along adult central fibre tracts?
(3) Can Schwann cells be incorporated into normally organised adult central fibre tracts?
(4) Can Schwann cells induce axon growth in adult central fibre tracts?

We find that the astrocytes and oligodendrocytes are arranged in a regular meshwork in adult central fibre tracts. This closely knit cellular framework of adult myelinated fibre tracts is permissive to the rapid growth of axons from transplanted embryonic neurons. These axons elongate for considerable distances, and can reach terminal areas of neuropil. The routes taken by the axons of these donor embryonic neurons through the tracts are aligned with the arrangement of the longitudinal astrocytic processes. However, the requirements for axon growth in tracts do not show the same specificity as observed in normal development: axons

Neural Cell Specification: Molecular Mechanisms and Neurotherapeutic Implications
Edited by Juurlink et al., Plenum Press, New York, 1995

from donor neurons taken from completely foreign areas of brain tissue grow as rapidly, and exuberantly, and with the same distribution as those taken from the correct areas of brain.

It has been known for some time that cut central tract fibres can be induced to regenerate all the way to form functional contacts with their targets by the use of peripheral nerve grafts. This property of peripheral nerve grafts depends on the presence of viable Schwann cells (Berry et al., 1988). We find that when injected as a suspension, cultured Schwann cells can be incorporated into the meshwork of adult host central fibre tracts, and their presence induces sprouting of the host axons.

These studies show that the structure and arrangement of adult fibre tracts are important in axon growth, and provide preliminary evidence that modification of this structure by the incorporation of Schwann cells may be capable of inducing regeneration of cut axons.

STRUCTURE OF ADULT CENTRAL FIBRE TRACTS

The glial structure of adult central fibre tracts is complex and regular (Suzuki and Raisman, 1992). The basis of the arrangement is a series of elongated, unicellular columns which pass along the longitudinal axis of the tract, in parallel to each other, and regularly spaced in the transverse plane. In the fimbria, these cell columns consist of about 5 to 10 contiguous oligodendrocytes separated by single, solitary astrocytes (Fig. 1). Each oligodendrocyte gives rise to 30 to 40 very fine processes which traverse the tract in the radial dimension and give rise to the 100 to 200 µm long myelinating processes. The astrocytic cell bodies are prolonged in the radial plane into tapering transverse processes (which, unlike their predecessor radial glia, no longer traverse the full ventriculopial width of the adult fimbria). The characteristic of the adult tract astrocytes is that they give rise to a very large number of fine, uniform, untapering longitudinal processes which are aligned along the long axis of the tract (see also Butt and Ransom, 1993; Li and Raisman, 1994).

ARE AXONS ABLE TO GROW ALONG ADULT CENTRAL FIBRE TRACTS?

Despite this complexity, the closely knit cellular framework of adult myelinated fibre tracts is permissive to the rapid growth of axons from transplanted embryonic neurons. When suspensions of embryonic neurons are injected into the fimbria (Davies et al., 1993), corpus callosum (Davies et al., 1994), or spinal dorsal columns (Li and Raisman, 1993), they give rise to profuse axons which grow rapidly into the host tract for distances considerably in excess of 1 cm. The axons travel at about 1 mm per day, initially in a mass, which later breaks up into individual axons interspersed singly among the myelinated host tract axons (Fig. 2). The main branching pattern appears to be the formation of collaterals given off at right angles to the main axonal stems (cf O'Leary et al., 1990). The fibres are able to reach terminal fields and arborise in them.

The growth of embryonic fibres along adult tracts does not require the same specificity of matching as occurs in normal development (Davies et al., 1994). Thus donor cells from hippocampus, neocortex, or superior colliculus all produce axons which grow equally rapidly, and with the same topographical distribution in the fimbria (which normally has hippocampal axons) or the corpus callosum (which normally has neocortical axons). Embryonic hippocampal donor cells also grow long axons in the long tracts of the adult spinal dorsal columns.

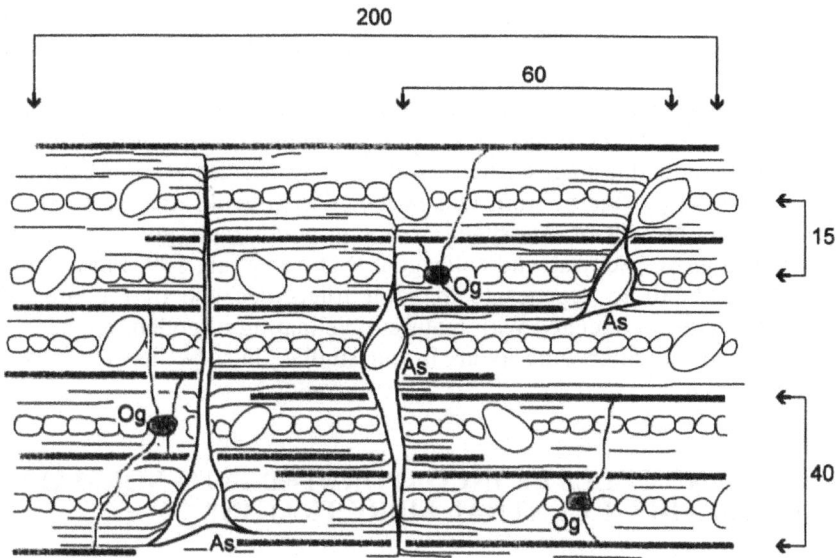

Figure 1. Regular arrangement of astrocytes and oligodendrocytes in an adult myelinated fibre tract (the fimbria). The cell bodies of the solitary astrocytes (As) lie singly in interfascicular rows of cell bodies, separated by stretches of 5 - 10 contiguous oligodendrocytes (Og). The tapering radial processes of the astrocytes give rise to large numbers of untapering, threadlike longitudinal processes. The oligodendrocytes give rise to fine radial processes which support the longitudinally oriented myelinating internodal segments. Sample spacing and lengths given in microns. (Adapted from Suzuki and Raisman, 1992).

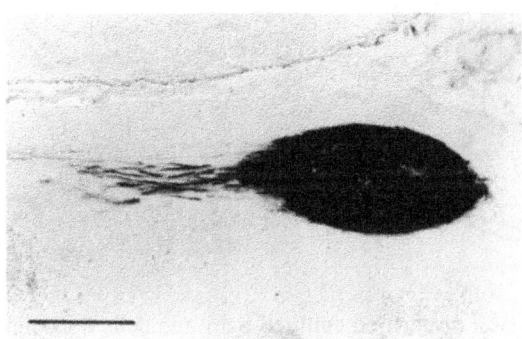

Figure 2. A suspension of E14 rat hippocampal cells transplanted into an adult mouse fimbria adjacent to the lateral ventricle gives rise to a mass of axons projecting along the myelinated fibre bundles of the host tract. Thy-1.2 immunohistochemistry. Survival, 37 days after transplantation. Scale bar = 500 μm. (Adapted from Davies et al., 1993).

Whereas the donor axons do not have to be matched in cell type to the tract in which they grow, they do show a strict adherence to the orientation of the glial framework of the host tracts (Davies et al., 1994). Within a tract such as the fimbria, the donor axons remain confined to the longitudinal band which is in continuity with the position into which the donor cells are placed. In situations where the host axons and the longitudinal processes of the host fibre tract are inflected, the donor axons faithfully follow the same inflections. Thus the directional cues for the growth of the donor axons reside in the orientation of the structures in the adult host tract.

These observations on the growth of embryonic axons in adult tracts do not constitute a reconstruction of the normal developmental pattern of growth, since at the early developmental stages when axons first grow, the glial structure of the tracts is completely different from that in the adult (Suzuki and Raisman, 1994). Neither does the growth of embryonic axons in adult tracts guarantee that adult axons will also grow. Schwab and colleagues (Schwab et al., 1993) have suggested that adult axons fail to regenerate in adult tracts because of the presence of inhibitory molecules associated with mature, myelinating oligodendrocytes. If that is the case, then the embryonic axons in our studies must be insensitive to these factors.

CAN SCHWANN CELLS BE INCORPORATED INTO NORMALLY ORGANIZED ADULT CENTRAL FIBRE TRACTS?

Transplants of peripheral nerve have two problems - the damage caused at the interface, and the impossibility of placing nerve grafts that will correctly repair severed central tracts in which many fibre components are recruited from different areas and distributed to spatially dispersed targets. These might be surmounted if cultured purified Schwann cells could be transplanted (Neuberger et al., 1992).

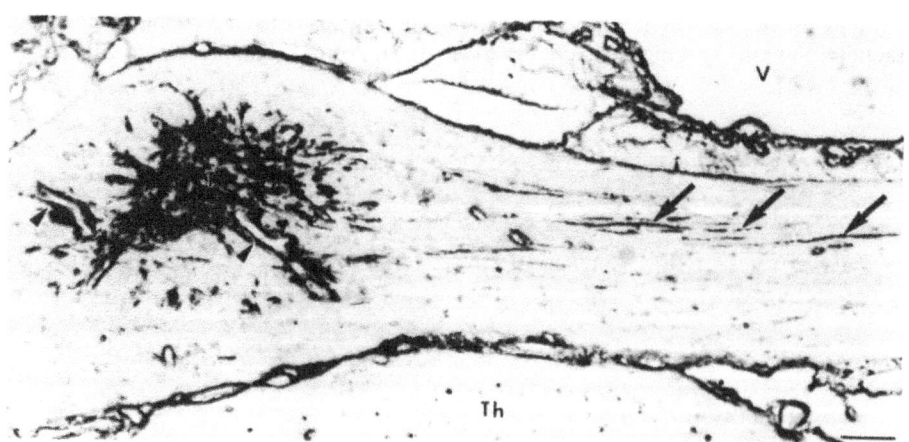

Figure 3. A suspension of purified cultured Schwann cells transplanted (T) into the adult rat fimbria shows migration of Schwann cells along blood vessels (arrowheads) and also transformation of Schwann cells into elongated, thread-like cells which migrate (arrows) along the myelinated fibre bundles of the host tract. Th, thalamus; V, lateral ventricle. Schwann cells identified by immunohistochemistry for low affinity nerve growth factor receptor. Survival, 6 days after transplantation. Scale bar = 200 μm. (Adapted from Brook et al., 1993).

We have found that suspensions of purified Schwann cells cultured from peripheral nerve can survive transplantation into adult tracts, where they migrate initially along blood vessels, and later become transformed into threadlike cells interspersed singly along the longitudinal axis of the host fibre tract (Fig. 3; Brook et al., 1993).

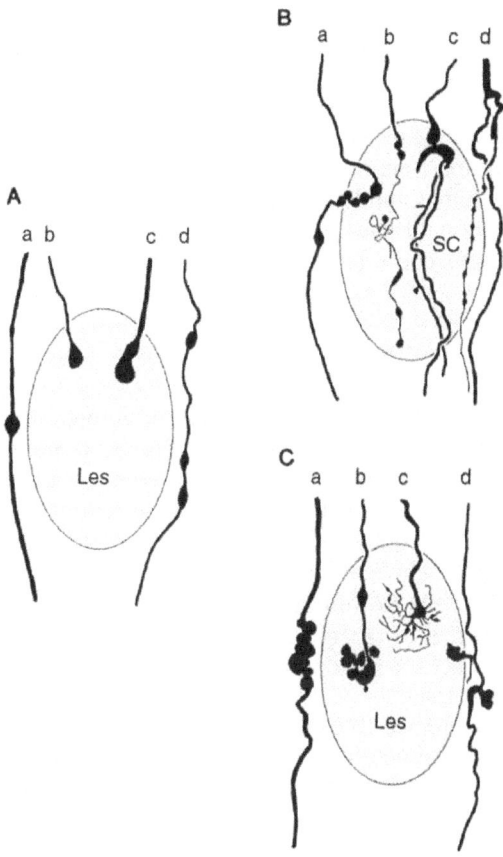

Figure 4. Effect of a small, circumscribed lesion (Les) in the upper cervical region of the corticospinal tract in the dorsal horn of the adult rat spinal cord. A, acute lesion induces varicosities on both cut axons (b,c) and uncut axons (a,d). C, long term lesion causes increased varicosities and profuse local sprouting. B, Transplantation of Schwann cells (SC), causes elongated sprouting of both the cut and uncut axons. The corticospinal axons were labelled by orthograde transport of HRP after an injection into the medullary pyramid. (Adapted from Li and Raisman, 1994).

If the Schwann cells are extruded continuously during withdrawal of the micropipette, the cells form a long column. In a few days the individual Schwann cells become highly elongated along the axis of withdrawal. These Schwann cell columns induce the ingrowth of host axons which become aligned along the long axis of the column. Such columns can convey axons out of the brain at the dorsal surface of the thalamus, across the choroid fissure, and back into the brain on the ventral surface of the hippocampus (Brook et al., 1994).

In other experiments, Schwann cells were injected into the dorsal spinal funiculi (Li and Raisman, 1994). The cells were readily incorporated into the surrounding tract structure. Either the descending motor (corticospinal) axons or the ascending sensory axons were labelled by orthograde transport. Both cut and uncut axons emitted sprouts either at the cut ends, or *en passant*. Under the influence of the injected Schwann cells the axonal sprouts elongated along the axis of the tract, and branched (Fig. 4). The branches had a strong tendency to fasciculate with each other or with other axons. We do not yet know the final length of these Schwann cell induced branches, nor their ultimate fate.

It was striking that, without injection of Schwann cells, the surgical procedure alone severed a number of axons. In this situation too, both the cut ends of the axons and the axonal stems also emitted terminal and *en passant* branches, but these were short, profuse and localised (Fig. 4). From the point of view of possible future repair of long term clinical spinal cord lesions, it is significant that appreciable numbers of cut axons remain for long periods (without retraction) in the region of the lesion, where they are presumably still able to respond if Schwann cells could later be introduced.

CONCLUSIONS

These observations indicate that the glial structure of adult tracts is highly complex and regularly arranged. What appear to be largely intact, undisturbed adult myelinated tracts permit the rapid and profuse growth of axons from embryonic neurons. These axons intermingle singly with the host tract axons and they follow the tract architecture closely but appear to be nonspecific with regard to the cells of origin. Schwann cells can survive, migrate and incorporate as interspersed threadlike cells in minimally disturbed adult host central tracts. The effect of Schwann cells is to change the nature of the sprouts formed by cut and uncut axons, and to direct the resultant new branches into fascicles which become interspersed in the host tract.

Future work is designed to examine the degree to which the incorporation of Schwann cells is able to induce long regeneration of cut axons along their original tracts and in such a way as to be able to reestablish correct functional patterns of connections.

REFERENCES

Berry M, Rees L, Hall S, Yiu P, Sievers J (1988): Optic axons regenerate into sciatic nerve isografts only in the presence of Schwann cells. Brain Res Bull 20: 223-231.

Brook GA, Lawrence JM, Raisman G (1993): Morphology and migration of cultured Schwann cells transplanted into the fimbria and hippocampus in adult rats. Glia 9: 292-304.

Brook GA, Lawrence JM, Shah B, Raisman G (1994): Extrusion transplantation of Schwann cells into the adult rat thalamus induces directional host axon growth. Exp Neurol 125: 1-13.

Butt AM, Ransom BR (1993): Morphology of astrocytes and oligodendrocytes during development in the intact rat optic nerve. J Comp Neurol 338: 141-158.

Davies SJA, Field PM, Raisman G (1993): Long fibre growth by axons of embryonic mouse hippocampal neurons micro-transplanted into the adult rat fimbria. Eur J Neurosci 5: 95-106.

Davies SJA, Field PM, Raisman G (1994): Long interfascicular axon growth from embryonic neurons transplanted into adult myelinated tracts. J Neurosci 14: 1596-1612. .

Li Y, Raisman G (1993): Long interfascicular axon growth from embryonic mouse hippocampal neurons transplanted into the myelinated corticospinal tracts and dorsal columns of immunosuppressed adult rat hosts. Brain Res 629: 115-127.

Li Y, Raisman G (1994): Schwann cells induce sprouting in motor and sensory axons in the adult rat spinal cord. J Neurosci 14: 4050-4063.

Neuberger TJ, Cornbrooks CJ, Kromer LF (1992): Effects of delayed transplantation of cultured Schwann cells on axonal regeneration from central nervous system cholinergic neurons. J Comp Neurol 315: 16-33.

O'Leary DDM, Bicknese AR, De Carlos JA, Heffner CD, Koester SI, Kutka LJ, Terashima T (1990): Target selection by cortical axons: alternative mechanisms to establish axonal connections in the developing brain. Cold Spring Harbor Symp Quant 55: 453-468.

Raisman G (1969): Neuronal plasticity in the septal nuclei of the adult rat. Brain Res 14: 25-48.

Schwab ME, Kapfhammer JP, Bandtlow CE (1993): Inhibitors of neurite growth. Annu Rev Neurosci 16: 565-595.

Suzuki M, Raisman G (1992): The glial framework of central white matter tracts: Segmented rows of contiguous interfascicular oligodendrocytes and solitary astrocytes give rise to a continuous meshwork of transverse and longitudinal processes in the adult rat fimbria. Glia 6: 222-235.

Suzuki M, Raisman G (1994): Multifocal pattern of postnatal development of the macroglial framework of the rat fimbria. Glia (in press).

Vidal-Sanz M, Bray GM, Villegas-Pérez MP, Thanos S, Aguayo AJ (1987): Axonal regeneration and synapse formation in the superior colliculus by retinal ganglion cells in the adult rat. J Neurosci 7: 2894-2909.

Zhou CF, Li Y, Morris RJ, Raisman G (1990): Accurate reconstruction of three complementary laminar afferents to the adult hippocampus by embryonic neural grafts. Neurosci Res Suppl 13: S43-S53.

CHARACTERISTICS OF INTRASPINAL GRAFTS AND LOCOMOTOR FUNCTION AFTER SPINAL CORD INJURY

Douglas K. Anderson[1,2], Dena R. Howland[1] and Paul J. Reier[1]

[1]Departments of Neuroscience and Neurological Surgery, University of Florida College of Medicine, Gainesville, Florida, 32610, U.S.A. [2]Gainesville Veterans Affairs Medical Center, Gainesville, Florida, 32608, U.S.A.

INTRODUCTION

It is well established that fetal neural tissue placed into acute lesions in both the newborn and adult rat spinal cord undergoes extensive differentiation and survives for extended periods (Bregman and Reier, 1986; Das, 1983a; Gelderd and Quarles, 1990; Himes et al., 1994; Jakeman and Reier, 1989; Nornes et al., 1983; Nothias and Pechanski, 1990; Nygren et al., 1977; Patel and Bernstein, 1983; Privat et al., 1986, 1989; Reier et al., 1986; Sieradzan and Vrbova, 1989). Projections from long–tract and segmental host systems, as well as neurons in the fetal graft, grow across the host–graft interface (Bernstein–Goral and Bregman, 1993; Björklund et al., 1986; Bregman, 1987; Foster et al., 1985; Itoh et al., 1993; Itoh and Tessler, 1989, 1990; Jakeman and Reier, 1987a,b, 1989, 1991; Nothias and Pechanski, 1990; Privat et al., 1986; Reier et al., 1986; Tessler et al., 1988). In adult animals, long–tract host systems that grow into the graft for short distances (Reier et al., 1986) in combination with axons from the graft that project into host motoneuron pools (Jakeman and Reier, 1991) may provide the anatomical basis for a functional relay between the host and graft. The active participation of the graft in relaying neuronal information is only one of several potential mechanisms by which fetal grafts may affect function. Grafts may also enhance function by supplying depleted transmitters (Björklund et al., 1986; Buchanan and Nornes, 1986; Foster et al., 1989; Privat et al., 1989), by providing trophic support (Arenas and Persson, 1994; Bregman, 1988; Bregman and Reier, 1986; Himes et al., 1994), by promoting sprouting (Schnell et al., 1994) or by myelinating demyelinated axons (Rosenbluth et al., 1990).

Our findings reviewed in this chapter, in combination with studies by others, show that intraspinal fetal grafts can promote the recovery of segmental reflexes (Buchanan and Nornes, 1986; Moorman et al., 1990; Privat et al., 1986, 1988) and

Neural Cell Specification: Molecular Mechanisms and Neurotherapeutic Implications
Edited by Juurlink *et al.*, Plenum Press, New York, 1995

249

locomotor function (Bregman et al., 1993; Howland et al., 1995; Kunkel–Bagden and Bregman, 1990) following spinal cord injury. Taken together, these studies suggest that intraspinal transplantation may have therapeutic benefits. Consequently, experimental models that permit the rigorous study of both structure and function (Goldberger et al., 1990; Kunkel–Bagden et al., 1993) and approximate human spinal cord injury are imperative to assess the potential therapeutic benefits of fetal grafts. The contusion/compression models of spinal cord injury are currently thought to be a good approximation of a common form of human injury (Young, 1989). For the analysis of locomotion, the cat provides an appropriate model for correlating structure and function because the neural mechanisms which underlie locomotion in the cat have been studied extensively (for reviews see Armstrong, 1988; Grillner, 1975; Shik and Orlovsky, 1976; Wetzel and Stuart, 1976). Additionally, the cat has been used effectively to study the pathophysiology and pharmacology of acute, traumatic spinal cord injury (Anderson et al., 1976, 1980, 1985, 1988a,b; Faden and Jacobs, 1985; Hall and Braughler, 1982; Means et al., 1981; Young and Flamm, 1982). However, there are few studies that have adapted the technical and biological principles of intraspinal transplantation that were developed in the rat to other species (Dunnett and Richards, 1990; Gash and Sladek, 1988; Howland et al., 1995).

To determine the effects of fetal intraspinal grafts on the recovery of locomotor function, we are transplanting fetal CNS tissue into the chronically injured spinal cords of adult cats. In addition, we have also begun to address a number of basic biological and practical issues including optimal donor age, donor tissue origin, method of tissue presentation into the lesion site, graft survival, maturity, and integration as well as host– graft histocompatibility. Our results show that solid and suspension grafts of fetal tissue from several areas of the neuraxis survive intraspinal transplantation, mature and integrate with the contused adult host spinal cord. In addition, our findings suggest that the ability of these grafts to promote recovery of locomotor function depends, at least in part, upon the origin of the donor tissue, the time interval between injury and transplantation, and whether or not the lesion cavity was debrided prior to grafting. Fetal spinal cord and brainstem grafts introduced within 3 months of injury with minimal or no debridement of the lesion site seemed to improve locomotor function whereas debridement of and transplantation into more chronic injuries appeared to adversely affect locomotion.

This review briefly summarizes ongoing studies from our laboratories. More detailed reports of our earlier findings have been published elsewhere (Anderson et al., 1991; Wirth et al., 1992).

METHODS

Thirty–five adult female cats received high lumbar spinal cord compression injuries. In 29 of these animals, fetal CNS tissue was grafted into the resulting chronic injury cavities 2 to 30 weeks post–injury. Grafts consisted of either fetal spinal cord (FSC) from embryonic day 21 to embryonic day 24 (E21 – E24), fetal brainstem (FBSt) from E21 – E38, fetal neocortex (FNCx) from E38, a mixture of spinal cord and brainstem, or fetal nasal epithelium (FNE) from E41 – E44. The use of FNE was initiated in collaboration with S.K. Pixley in an attempt to identify a potential source of non–CNS neuronal stem cells that could be useful for intraspinal grafts. Six cats did not receive a fetal graft. Locomotor function was evaluated pre–injury, post–injury and post–grafting (Fig. 1). Injury and transplantation methods have been described in detail previously (Anderson et al.,

1976, 1980, 1985, 1988a,b, 1991; Means et al., 1981) and, therefore, are described below only briefly.

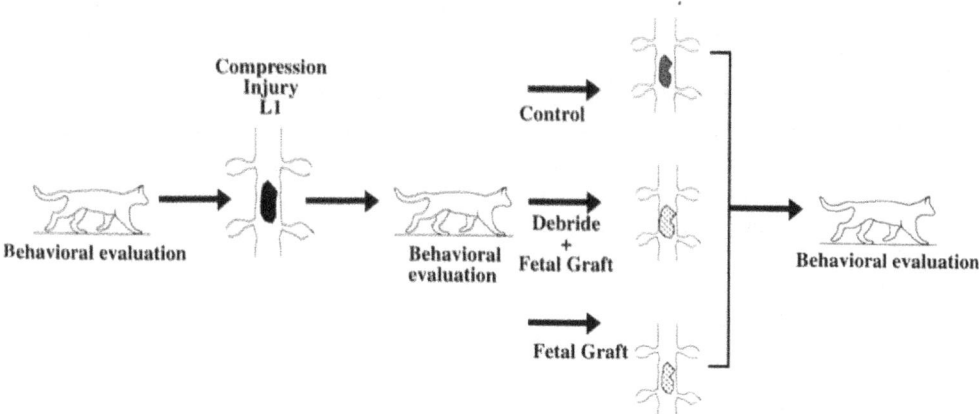

Figure 1. Illustration of the lesion paradigm used in these experiments. The locomotor function of all animals was evaluated pre–injury. In each animal, a contusion spinal cord injury was made at vertebral level L1, causing the formation of a central cyst. The locomotor function of these animals was evaluated for up to 7 months post–injury. The spinal cord was then re–exposed at the level of the injury. In control animals, the lesion cavity was filled with gelfoam, saline or a piece of peripheral nerve . Animals that received fetal tissue grafts fell into two groups. In the first group, the scar tissue was removed from the lesion cavity (debridement) immediately prior to filling the cavity with fetal tissue. In the second group there was little or no debridement of the lesion cavity prior to grafting. The locomotor function of all animals was evaluated for 2–8 months post–grafting.

Surgical Procedures

Static Load Compression Injury. Cats were anesthetized with intraperitoneal injections of sodium pentobarbital (30 mg/kg), immobilized in a stereotaxic frame, and intubated to maintain a patent airway. Under sterile conditions a full laminectomy was made at vertebral level L1. The 7–mm tip of a stainless steel rod weighted to 180 g was placed extradurally on the exposed spinal cord for 5 minutes. The incision was sutured in layers and cats were placed in veterinary intensive care units (Thermo Care) for 48 hours. All cats received an antibiotic (Amoxicillin) for 7 days.

Transplantation Procedures. Two to 31 weeks after compression injury, cats were again anesthetized with sodium pentobarbital, intubated and the injured spinal cord segment re-exposed. Cats received either solid pieces or suspension grafts of fetal tissue. All cats received Amoxicillin for 7 days.

Solid grafts. The dura and pia-arachnoid were slit. Gentle aspiration was used to debride the lesion cavity. Solid pieces of fetal tissue were drawn into a tapered glass pipette and gently expelled into the lesion cavity. The dura and pia–arachnoid were sutured to hold the graft within the cavity. A piece of Durafilm (Codman and Shurtliff, Inc.) was placed over the dural stitches. The muscle and skin incisions were sutured in layers.

Suspension grafts, Fetal tissue was minced into small fragments in Hank's Balanced Salt Solution and then dissociated in 0.6% glucose–saline by trituration with pipettes of progressively smaller tip diameters. The resulting cell suspensions were sampled for viable cell density using Acridine Orange–Ethidium Bromide (Brundin et al., 1985). The original injury site was then re–exposed and the dura slit, leaving the pia intact. In some cats, translucent, fluid–filled, pulsating cavities were readily apparent. In other cats more subtle lesions with varying amounts of spared tissue were seen. Dependent upon the size of the lesion cavity, between 50 and 250 μL of cell suspension (density ~1.5 x 10^5 to 2 x 10^6 cells/cc) were injected into the lesion site with a Hamilton syringe fitted with a blunted 25 gauge needle.

All cats were immunosuppressed with cyclosporine A (Sandimmune, Sandoz Pharmaceutical Company, East Hanover, N.J.). 10 mg/kg were given orally beginning the day prior to transplantation and continuing daily thereafter until sacrifice. Cyclosporine A was discontinued in one cat 10–12 weeks prior to sacrifice.

Evaluation of Locomotor Function

Two methods of locomotor evaluation were used. In initial studies, locomotion was assessed using a categorical rating scale (Anderson et al., 1985, 1988a,b; Means et al., 1981). Each cat's locomotor ability was evaluated in three categories: the ability to i) walk, ii) change gait and iii) climb stairs. A score was assigned for each performance category, the sum of which served as an index of generalized locomotor function. Eleven was the highest score and indicated normal or pre–injury function while zero was the lowest and indicated complete paraplegia. Locomotion was studied weekly after injury prior to grafting and for up to 30 weeks after grafting.

More recently, conditioned overground locomotion was evaluated. Cats were trained to cross horizontal runways and ladders for a food reward. The runways varied in width from 2 to 12 inches and the flat 1 inch rungs of the ladder were spaced 6 inches apart. First it was determined whether an animal could cross a runway or ladder using its hindlimbs. If the cat was successful in crossing, the time to cross and the number of errors in hindlimb foot placement were determined from slow motion analysis of videotapes. Hindlimbs foot fall errors occur when placement of the foot causes it to miss or slip off the runway or ladder surface.

Morphology and Immunocytochemistry

Following the completion of behavioral testing, cats were overdosed with intraperitoneal injections of sodium pentobarbital and perfused intracardially with 0.9% saline followed by a fixative for electron microscopy (5% glutaraldehyde, 4% paraformaldehyde in 0.1 M Sorenson's phosphate buffer) or immunohistochemistry (4% paraformaldehyde in 0.1% Sorenson's phosphate buffer). Tissue blocks prepared for immunocytochemistry were cut into 40 μm sections. Peroxidase–antiperoxidase (PAP) immunocytochemical techniques (Sternberger, 1976) were used to identify projections between the host and graft using antibodies to serotonin (5HT, Pel–freez) and Calcitonin Gene Related Peptide (CGRP, Pennisula). The PAP process was also used with an antibody to Glial Fibillary Acidic Protein (GFAP, Sigma) to identify any apparent glial scar formation. The Biotin–Avidin immunoperoxidase process (Vector Laboratories) was used to identify projections using an antibody to the phosphorylated M_r 200,000 subunit of neurofilament protein (RT97, Boehringer Mannheim). Mature neurons were identified in grafts of fetal nasal epithelium using the Biotin–Avidin immunoperoxidase process and an antibody against Olfactory Marker Protein (OMP, gift from F.Margolis). Tissue blocks prepared for electron microscopy were embedded in epon, sectioned at 2μm, stained with toluidine blue, and first examined by light microscopy.

RESULTS

Graft Survival and Characteristics

Fetal spinal cord, FBSt, FNCx and FNE all survived transplantation and were well–vascularized. Viable grafts were found 6–30 weeks post–grafting in 25 of 29 animals that received FSC, FBSt or FNCx tissue. The basic histological characteristics of these grafts into chronic compression lesions of the cat spinal cord were similar to characteristics described for grafts into contusion lesions of the rat spinal cord (Reier et al., 1992a,b; Stokes and Reier, 1992). Both neuronal and glial elements were seen. For example, grafts were filled with neurons of various sizes and were extensively myelinated (Fig. 2). Tissue from 2 of the 4 animals without viable grafts showed active rejection of the transplant. These grafts showed significant lymphocytic infiltration and spongiform degeneration. Additionally, the rejection process appeared to be restricted to the grafts. Cyclosporine A had been discontinued in one of these animals two and one–half months prior to sacrifice (4 months post–grafting).

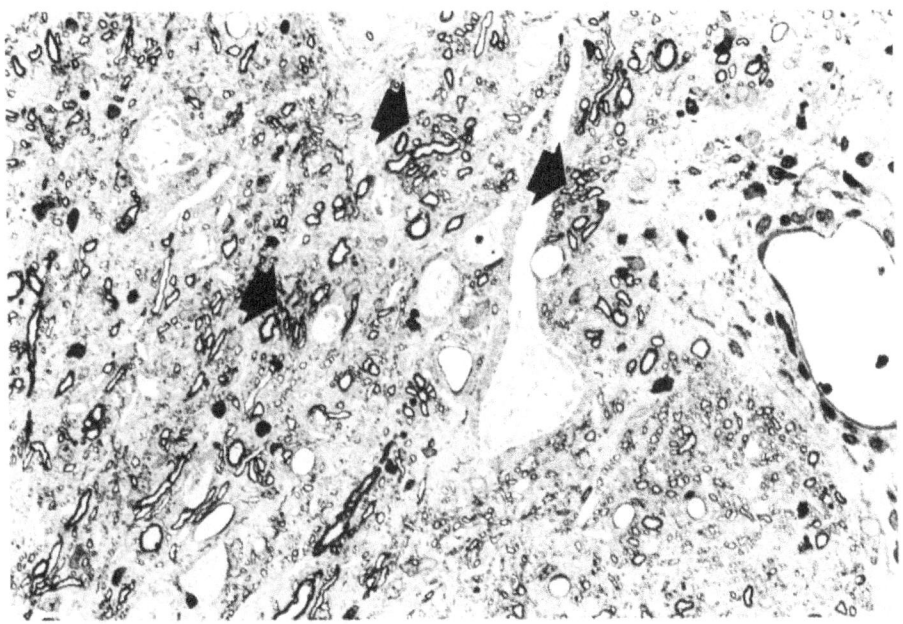

Figure 2. Toludine blue stained plastic section of a fetal spinal cord (FSC) graft at 20 weeks post– transplantation, showing a mature graft with a high degree of vascularization, myelination, and various sized neurons. In some areas, fascicles of myelinated axons are seen (arrows).

Both contusion injury of the rat spinal cord and compression injury of the cat spinal cord result in a single, sometimes highly partitioned, central cyst surrounded peripherally by white matter. Some solid and suspension grafts of fetal tissue filled the cysts, merging rostrally, laterally and caudally with the host spinal cords. Minimal cavitation or glial scarring was seen at most host–graft interfaces. In some grafts, fetal tissue merged with spared subpial axons apparently by filling large extracellular spaces (Fig. 3).

Viable grafts were found 12–22 weeks post–grafting in 3 animals that received FNE. These grafts showed grossly different histological characteristics than the other graft types. Grafts of FNE tissue were disorganized, containing primarily cartilage, bone and epithelial–lined cavities (Pixley et al., 1994). Some OMP immunoreactive cells were seen indicating these grafts also contained mature olfactory–specific neurons. These neurons, however, were a minor component of the grafts.

Graft Integration

Immunocytochemical methods were used to determine if projections were established between the host and graft in animals that received grafts of FSC, FBSt or a mix of FSC and FBSt. Calcitonin Gene Related Peptide–immunoreactive (CGRP–IR) fibers identified one host afferent segmental system that grew into the graft. CGRP–IR axons could be seen crossing the host–graft interfaces and arborizing throughout the grafts. Collaterals emerged from some of the axons that originated from both rostral and caudal host spinal sources. Serotinergic–immunoreactive (5HT–IR) axons also grew across the host–graft interfaces. These axons arborized only in the most rostral areas of the grafts and identified a descending host fiber system. Growth of the 5HT–IR axons was less extensive than the growth of the CGRP–IR axons in density as well as distance.

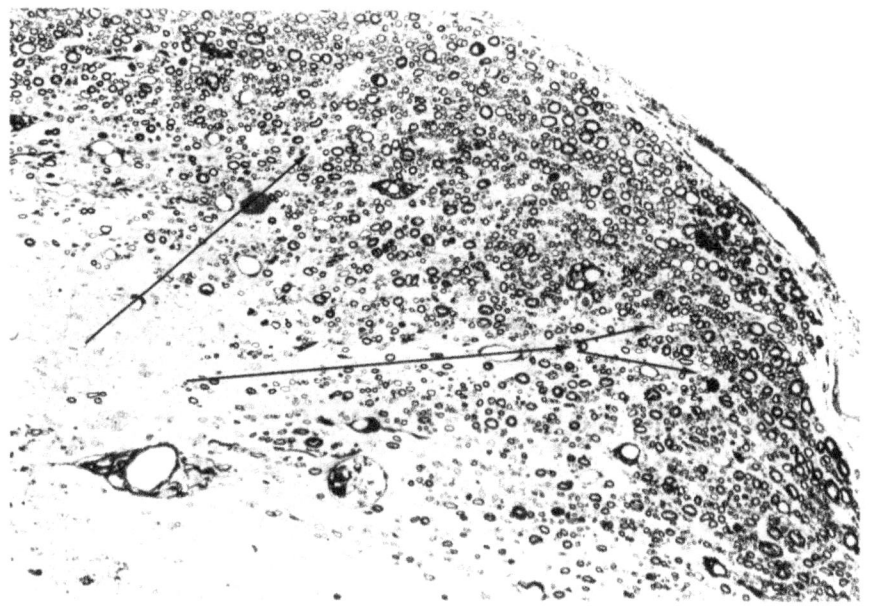

Figure 3. Toludine blue stained plastic section of a caudal fetal brainstem (FBSt) graft at 29 weeks post–transplantation, showing the radial expansion of graft tissue that fills areas of degenerated host white matter (arrows).

Immunocytochemical staining for the phosphorylated heavy subunit of neurofilament protein showed a more dense pattern of axonal growth than was seen using antibodies against CGRP and 5HT. Complex axonal networks crossing the rostral, lateral and caudal host–graft interfaces were identified. Fasciculated axon bundles, some oriented longitudinally, were apparent in the grafts. Although this population of axons included the CGRP and 5HT fibers, the greater areal extent and density of staining indicated that other fiber types also were entering and exiting the grafts.

Table 1. Effects of grafts on locomotion*.

Spinal Cord and Brainstem Grafts Enhance Locomotion

Animal	Functional change	Time interval	Lesion debridged	Graft
1	no change	9 weeks	no	solid FNCx
2	loss	7 weeks	no	solid FNCx
3	no change	7 weeks	no	solid FNCx
4	improvement	5 weeks	no	solid FBSt
5	no change	9 weeks	no	solid FBSt
6	improvement	7 weeks	no	suspension FBST
7	no change	4 weeks	no	solid FSC
8	improvement	8 weeks	no	solid FSC
9	improvement	8 weeks	no	solid FSC
10	improvement	2 weeks	no	suspension FSC
11	improvement	6 weeks	no	suspension FSC+FBSt
12	no change	5 weeks	no	solid FNE
13	no change	11 weeks	no	solid FNE
14	no change	20 weeks	no	solid FNE

*All cats represented in this table had healthy, surviving grafts at the time of sacrifice. Only cats receiving fetal spinal (FSC) or fetal brainstem (FBSt) grafts showed an improvement in locomotor function post–grafting. Eight of the fourteen cats evaluated using the categorical behavioral scale received FSC and/or FBSt grafts. The locomotor score of six cats receiving fetal neocortex (FNCx) or fetal nasal epithelium (FNE) did not change after grafting, even though they received the same type of spinal injury and the time interval between injury and grafting was similar.

Recovery of Function

Recovery of locomotor function was seen in some cats that received grafts of FSC or FBSt. No functional improvement was seen in the cats that received grafts of FNCx or FNE (Fig. 4 and Table 1). The following descriptions are of animals that received grafts of FSC, FBSt, or both.

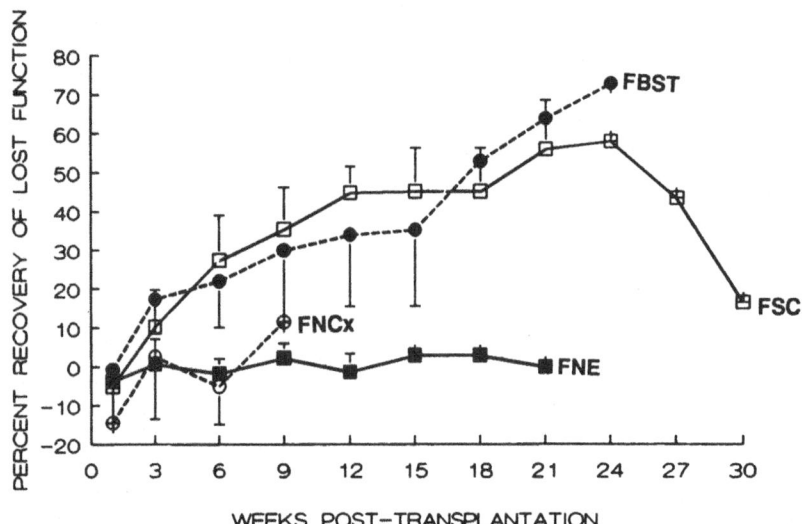

Figure 4. Mean percent recovery of lost function (± SEM) based upon the results of the categorical rating scale for cats receiving either fetal spinal cord (FSC), brainstem (FBSt), neocortical (FNCx), or nasal epithelial (FNE) tissue grafts. Percent recovery of lost function was calculated for each animal according to the equation: $100(x-y)/(n-y)$, where x = post–graft score; y = post–injury, pre–graft score; n = normal or pre–injury score. Recovery is plotted as a function of weeks post–transplantation. The animal number in each group decreased over time because cats were sacrificed at various time points for histological study. Lack of a SEM indicates that a time point value was calculated from the scores of less than three animals. The mean percent recovery for animals with FSC or FBSt grafts increased over time. The decline in percent recovery seen between 24 and 30 weeks post–grafting in animals with FSC grafts was due to the loss of locomotor function in one cat whose transplant was being rejected. Reprinted with permission from Reier et al. (1994): Neural cell grafting: anatomical and functional repair of the spinal cord. In Salzman SK and Faden AI (eds): "The Neurobiology of Central Nervous System Trauma." Oxford University Press.

Ten cats that received grafts 2–9 weeks post–injury were evaluated using the categorical rating scale described above. Of the 8 cats that had surviving grafts at the time of sacrifice, locomotor function had permanently improved in 6 (Table 1 and Anderson et al., 1991). Deterioration of locomotor performance during the last month of survival was seen in 2 cats that initially had shown an improvement. This decline in locomotor function was correlated with graft rejection in both cats. In one of these animals, Cyclosporine A had been discontinued (see Graft Survival section). Improvement in locomotor performance suggests that grafts of FSC and/or FBSt introduced 2–9 weeks post–injury can promote partial recovery of hindlimb locomotor function. Moreover, the correlation between deterioration of locomotor function and graft rejection suggests that permanent graft survival may be necessary to maintain any transplant–enhanced characteristics of locomotion.

Recently, 13 cats that received grafts 8–31 weeks post–injury were evaluated using the horizontal runways and ladders. After transplantation, locomotor function was permanently improved in 3 cats whereas 4 cats showed no improvement, and 6 showed an immediate and permanent loss of locomotor function. The animals' post–grafting function appeared to be dependent upon at least two factors, the time interval between injury and grafting and whether or not the lesion site was debrided at the time of grafting (Table 2).

The 3 animals that showed an improvement in locomotor function were grafted 8–15 weeks following injury with minimal or no debridement of the lesion cavity. Of these 3 animals the 2 cats that showed the most dramatic improvement in locomotor function had the shortest injury to grafting interval (8 and 11 weeks). Prior to grafting, both animals could cross a 12" wide runway. After transplantation, animal Op was able to cross narrower runways as well as a ladder and animal Ba was able to cross a 4" wide runway. The post–graft locomotor performances required greater balance, narrower bases of support and more accurate foot placement than was seen pre– grafting. The third cat who received its graft at 15 weeks post–injury showed a less dramatic improvement. Before grafting, this animal (Pe) could walk in an open arena but could not cross any runways. Several weeks after grafting, Pe was able to cross a 12" wide runway which indicated that this cat's balance had also improved although not to the level of the other two animals (Fig. 5B). The animals that showed no change in their post–graft locomotor performance received grafts 11–23 weeks after injury with no debridement of the lesion cavities (Fig. 5C). The cats that lost locomotor function immediately after grafting received grafts 10–31 weeks after injury. The loss of function was dramatic in that the hindlimbs no longer contributed to overground locomotion (Fig. 5D). In all of the cats with immediate loss of function, the lesion cavities were extensively debrided prior to the placement of the fetal tissue.

Four control animals received sham transplants with no debridement of the lesion sites 14–29 weeks post–injury (Fig. 5A and Table 2). These animals showed no post–graft change in locomotor performance. Taken together, the results using both types of behavioral analyses suggest that the time between injury and grafting is critical. Intraspinal grafts of FSC or FBSt transplanted within 3 months of injury appear capable of enhancing locomotor function in adult cats with compression spinal cord injuries. Grafting at intervals greater than 3 months after injury may not result in functional improvement. Debridement of the lesion site appears to be generally detrimental and, even in conjunction with grafts of fetal CNS tissue, frequently results in loss of function.

CONCLUSIONS

Our results indicate that fetal CNS tissue can survive grafting into chronic compression lesions of the adult spinal cord. These grafts not only survive, but appear to differentiate and integrate with the host spinal cord and to support the ingrowth of the host's descending supraspinal and segmental systems.

The locomotor performance of these cats provides evidence that under certain conditions, fetal neural grafts can enhance recovery of post–injury locomotor function. Although tissue from different regions of the fetal neuraxis (FSC, FBSt, FNCx, and FNE) survived when grafted into the injured spinal cord, only those animals that received homotypic transplants (FSC and FBSt) showed an improvement in locomotor performance. This finding may be associated with the

Table 2. Effects of grafts on locomotion*.

Animals with Fetal Spinal Cord and Brainstem Grafts and Fetal Graft Controls

Experimental Group

Animal	Functional change	Time interval	Lesion debridged	Graph
Op	dramatic improvement	8 weeks	minimal	solid BST
Ba	dramatic improvement	11 weeks	no	suspension FSC
Pe	minimal improvement	15 weeks	minimal	solid FSC
MM	no change	11 weeks	no	suspension BST & FSC
Bu	no change	14 weeks	no	solid FSC
Tw	no change	15 weeks	no	solid BST
To	no change	23 weeks	no	solid BST
Du	severe loss	10 weeks	yes	solid BST & FSC
Cl	severe loss	15 weeks	yes	solid BST & FSC
Re	severe loss	22 weeks	yes	solid FSC
Do	severe loss	23 weeks	yes	solid FSC
Bo	severe loss	30 weeks	yes	solid FSC
Ge	severe loss	31 weeks	yes	solid BST & FSC

Control Group

Animal	Functional change	Time interval	Lesion debridged	Graph
Toa	no change	14 weeks	no	sciatic nerve
Wi	no change	20 weeks	no	gel foam
Di	no change	20 weeks	no	gel foam
Be	no change	29 weeks	no	saline

*Three of the fourteen cats receiving fetal spinal cord (FSC) and/or fetal brainstem (FBSt) grafts improved their performance on the runways post–grafting. Four of the animals, like four controls, did not change their performance post–grafting and the performance of six animals was immediately and permanently diminished. The lesion site had been debrided in all animals showing a loss of function. In general, the best functional results were obtained with the early graft introduction and no/or minimal debridement.

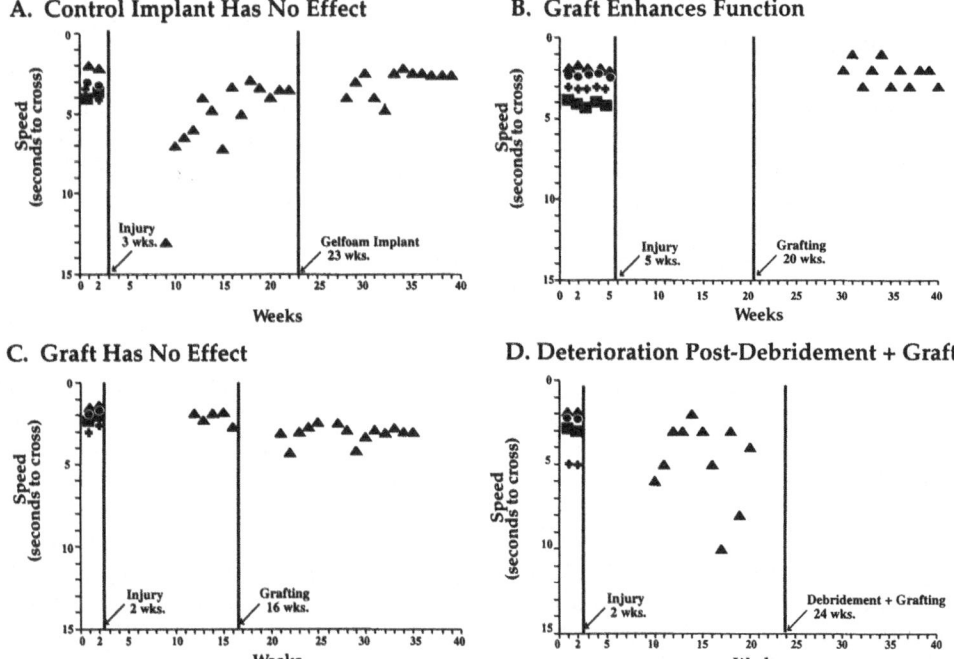

Figure 5. Locomotor function: effects of compression injury, fetal spinal cord (FSC) grafts, and debridement on hindlimb locomotor function. The time in seconds to cross different horizontal runways and a ladder is plotted as a function of weeks. No symbol on the graph indicated that the cat was unable to perform the task. (A) Function does not change in control cats (animals without lesion debridement or fetal grafts). This graph represents the data from a cat (Wi) that received a gelfoam implant, without lesion debridement, 20 weeks after a compression spinal cord injury. The gelfoam implant did not alter the cat's locomotor performance. (B) Some grafts appeared to enhance locomotor function. This graph represents the data from a cat (Pe) that received a compression spinal cord injury after five weeks of locomotor evaluation. Fifteen weeks post–injury, the cat received a FSC graft. Prior to grafting, this cat was unable to cross any of the runways. Approximately ten weeks post–grafting, this animal began to cross the 12" runway which suggests that this improvement in locomotor performance may have been graft mediated. (C) Some grafts did not appear to affect locomotion. This graph represents the data from a cat (Bu), that received a compression spinal cord injury after two weeks of locomotor evaluation. Fourteen weeks post–injury, the cat received a FSC graft. This animal's pre– and post–grafting performances were virtually the same. The graft, therefore, did not change the cat's performance. (D) Extensive debridement prior to grafting appeared to cause a deterioration in locomotion. This graph represents the data from a cat (Re) that received a compression spinal cord injury after two weeks of locomotor evaluation. Twenty–two weeks post–injury, the chronic lesion cavity was debrided and a FSC graft was placed into the lesion cavity. Prior to grafting, this cat could cross a 12" runway. Immediately post–grafting, this animal became permanently paraplegic. This type of loss appears to be due to the surgical manipulation (debridement) that occurred prior to grafting which may have damaged spared host tissue.

capacity of neurons within the homotypic graft to project to the appropriate terminal fields in the host spinal cord (Privat et al., 1989) and for homotypic transplants to provide a better terrain for axonal growth (Itoh and Tessler, 1990). Moreover, homotypic grafts also form greater numbers of permanent synapses between host axons and grafted neurons than is seen with heterotypic transplants (Bregman and Kunkel–Bagden, 1988; Itoh and Tessler, 1990). These findings coupled with our anatomical observations in the rat (Reier, 1986; Jakeman and Reier, 1991) and cat (Anderson et al., 1991) suggest that grafts might establish a polysynaptic relay between the rostral and caudal ends of the injured spinal cord. It has also been theoretically suggested that a graft may serve as a tissue bridge through which axons could regenerate in the adult animal from the rostral and caudal spinal regions (Nornes, et al., 1984; Reier, 1985).

A consistent finding throughout our studies has been that cats showing some improvement in post–graft locomotor function had, on average, the shortest injury to graft interval whereas those animals whose function was either unchanged or declined after transplantation had the longest periods between injury and grafting. This suggests that there may be a restricted time frame following injury during which FSC or FBSt grafts may enhance locomotion. Following injury the spinal cord attempts to repair itself by mechanisms such as sprouting (Liu and Chambers, 1958; Murray and Goldberger, 1986; Zhang et al., 1993) and the unmasking of latent functions in existing pathways (Alstermark et al., 1987). During the acute and subchronic post–injury phases a graft may magnify the plastic capabilities of the host spinal cord. It is also possible that grafts may rescue fibers or circuits that would otherwise undergo progressive degeneration (Reier et al., 1992a). Increasing the delay between injury and transplantation appears to contribute to a decrease in the functional efficacy of homotypic fetal grafts. This may be due to the relative dormant nature of the chronic spinal lesion in that neuron death and fiber degeneration have occurred and re–modeling in the spared systems has stabilized.

For the contusion/compression injury, graft–mediated enhancement of locomotor function may occur by mechanism(s) other than or in addition to the rescue of host neurons, axonal bridging, or polysynaptic relays. Since these lesions generally have some surviving though apparently dysfunctional pathways, grafts may exploit this anatomical substrate to promote recovery of function by re–establishing transmission in some of these intact but non– or poorly conducting fibers. Perhaps the role of embryonic graft tissue involves modification of the physiological environment of the lesion site (Barres et al., 1990; Stokes and Reier, 1991), potassium channel blockade in the spared fibers (Blight, 1989; Blight, et al., 1991; Hansebout et al., 1993), remyelination (Blight 1983a,b; Rosenbluth et al., 1990), or the transfer of sodium channels from the glial population of the graft to the spared axons of the host (Waxman, 1992). Finally, it is known that the monoaminergic system is important to spinal locomotor processes and that stepping movements and walking on a treadmill can be enhanced in spinalized animals by administration of a variety of monoamines, monoamine precursors, and pharmacological monoamine agonists (Grillner, 1973; Barbeau and Rossignol, 1991; Edgerton et al, 1992). A similar effect has been achieved by transplantation of fetal noradrenergic neurons below the level of a spinal transection in the rat (Yakovleff et al., 1989). Thus, it is conceivable that grafts can augment motoneuron excitability caudal to a lesion by replacing neurotransmitters lost from supraspinal sources.

One factor that may reduce or prevent any graft–mediated improved locomotion is injury to the host spinal cord at the time of grafting. The immediate and permanent loss of function that was seen following transplantation in only those animals whose lesion cavity was debrided prior to grafting suggests that it was the surgical procedure and not the graft that damaged the host spinal cord.

Any loss of function due to overgrowth of the grafts would be expected to occur over several weeks, not immediately.

It is likely that removal of the scar tissue that has formed at the lesion site can damage adjacent spared host systems (Fig. 6). Scar tissue anastomoses tightly with the host neuropil (Barrett et al., 1984) and host fibers are seen within this scar tissue (Reier et al., 1983). Thus, removal of the scar would damage host fibers in and immediately adjacent to the lesion. In addition, it is possible that the same secondary pathophysiological events that contributed to the destruction of the spinal cord following the original injury (Means and Anderson, 1987; Anderson and Thomas, 1994) may be reactivated by the debridement procedure.

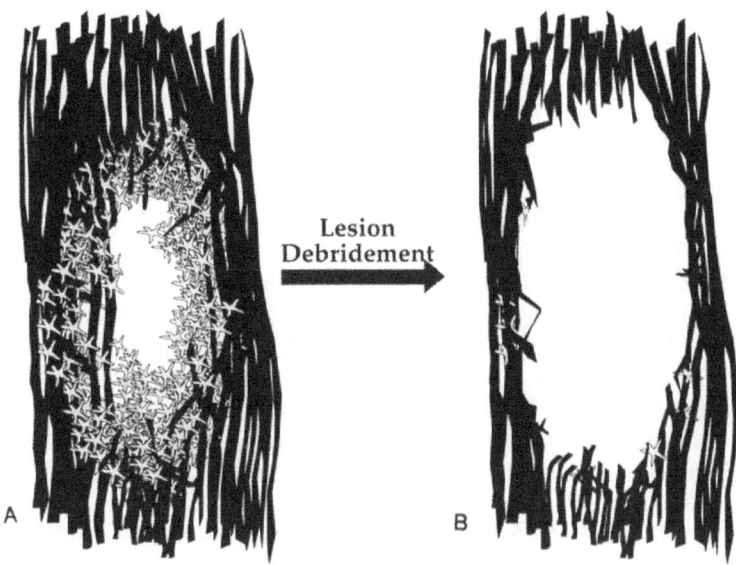

Lesion
Debridement

A   B

Figure 6. Lesion preparation for grafting. Schematic representing an established spinal lesion and the possible effects of debridement. (A) shows the lesion cavity of a chronic spinal cord injury characterized by an astroglial scar that is intermeshed with the host border of the cyst. (B) shows a lesion cavity in which the scar tissue has been removed (debrided) and the host fibers spared by the original injury are damaged. Fibers of passage, collateral sprouts, and regenerative sprouts that are in contact with the scar tissue may be damaged or removed upon even gentle aspiration of the scar tissue. Damaged fibers are represented in (B) by the darker fiber profiles.

Regardless of the mechanism(s), the benefits of removing the scar tissue in order to place the graft in contact with healthy neuropil appears to be negated by the damage this procedure causes to remodeled systems in the host spinal cord. Thus, it is likely that in the cats showing either no change or an immediate loss in function following transplantation, any graft–mediated effects might have been masked by a loss in residual function caused by debridement of the lesion area prior to grafting. This may be particularily true for the more established lesions which appear to be the most susceptible to surgical manipulation. Consequently, a permanent loss of spared function due to debridement plus grafting may argue for the exclusive use of less invasive approaches for implanting donor tissue into lesions of the spinal cord such as injection of dissociated CNS tissue. Injections of cell suspensions cause relatively little damage to the spared host structures they penetrate in contrast to the surgical incisions that must be made in the tissue when placing a solid graft.

ACKNOWLEDGEMENTS

This work was supported in part by the Department of Veterans Affairs (D.K.A.), the American Paralysis Association (D.K.A.), the Spinal Cord Research Foundation of the Paralyzed Veterans of America (D.K.A.), the National Institutes of Health (P.J.R., NIH PO1 NS27511), and the C.M. and K.E. Overstreet (D.K.A.) and Mark F. Overstreet (P.J.R.) Endowments for Spinal Cord Injury Research. Special gratitude is extended to M. A. Smith for her commitment to the large volume of histological preparation involved in these studies.

REFERENCES

Alstermark B, Lundberg A, Pettersson LG, Tantisira B and Walkowska M (1987): Motor recovery after serial spinal cord lesions of defined descending pathways in cats. Neurosci Res 5: 68–73.

Anderson DK, Braughler JM, Hall ED, Waters TR, McCall JM and Means ED (1988a): Effects of treatment with U-74006F on neurological outcome following experimental spinal cord compression injury. J Neurosurg 69: 562-567.

Anderson DK and Thomas CE (1994): Mechanisms and role of oxygen free radicals in CNS pathology. In: "Recent Advances in the Treatment of Neurodegenerative Disorders and Cognitive Dysfunction" Proceedings of the International Academy for Biomedical and Drug Research CINP President's Workshop. S Karger AG, Basal, vol. 7, pp 119–124.

Anderson DK, Waters TR and Means ED (1988b): Pretreatment with alpha tocopherol enhances neurological recovery after experimental spinal cord compression injury. J Neurotrauma 5: 61-68.

Anderson DK, Means ED, Waters TR and Spears CJ (1980): Spinal cord energy metabolism following compression trauma to the feline spinal cord. J. Neurosurg 53: 375-380.

Anderson DK, Prockop LD, Means ED and Hartley LE (1976): Cerebrospinal fluid lactate and electrolyte levels following experimental spinal cord injury. J Neurosurg 44: 715-722.

Anderson DK, Reier PJ, Wirth III ED, Theele DP, Mareci T and Brown SA (1991): Delayed grafting of fetal CNS tissue into chronic compression lesions of the adult spinal cord. Restor Neurol Neurosci 2: 309–325.

Anderson DK, Saunders RD, Demediuk P, Dugan LL, Braughler JM, Hall ED, Means ED and Horrocks LA (1985): Lipid hydrolysis and peroxidation in injured spinal cord. Partial protection with methylprednisolone and vitamin E. CNS Trauma 2: 257-267.

Arenas E and Persson H (1994): Neurotrophin–3 prevents the death of adult central noradrenergic neurons in vivo. Nature 367: 368–371.

Armstrong DM (1988): The supraspinal control of mammalian locomotion. J. Physiol 405: 1–37.

Barbeau H and Rossignol S (1991): Initiation and modulation of the locomotor pattern in the adult chronic spinal cat by nonadrenergic, serotonergic, and dopaminergic drugs. Brain Res 546: 250–260.

Barrett CP, Donati EJ and Guth, L (1984): Differences between adult and neonatal rats in their astroglial response to spinal injury. Exp Neurol 84: 374–385.

Barres BA, Loroshetz WJ, Chun LL and Corey DP (1990): Ion channel expression by white matter glia: the type–1 astrocyte. Neuron 5: 527–544.

Bernstein–Goral H and Bregman BS (1993): Spinal cord transplants support the regeneration of axotomized neurons after spinal cord lesions at birth: A quanitative double labelling study. Exp Neurol 23: 118–132.

Björklund A, Nornes H and Gage FH (1986): Cell suspension grafts of noradrenergic locus coeruleus neurons in rat hippocampus and spinal cord: Reinnervation and transmitter turnover. Neuroscience 18: 685-698.

Blight AR (1983a): Cellular morphology of chronic spinal cord injury in the cat: Analysis of myelinated axons by line–sampling. Neuroscience 10: 521–543.

Blight AR (1983b): Axonal physiology of chronic spinal cord injury in the cat: Intracellular recording *in vitro*. Neuroscience 10: 1471–1486.

Blight AR (1989): Effect of 4–aminopyridine on axonal conduction–block in chronic spinal cord injury. Brain Res Bull 22: 47–52.

Blight AR, Toombs JP, Bauer MS and Widmer WR (1991): The effects of 4–aminopyridine on neurological deficits in chronic cases of traumatic spinal cord injury in dogs: A phase I clinical trial. J Neurotrauma 8: 103–118.

Bregman BS (1987): Spinal cord transplants permit growth of serotonergic axons across the site of neonatal spinal cord transections. Dev Brain Res 34: 265–279.

Bregman BS (1988): Requirements of immature axotomized CNS neurons for survival and axonal elongation after injury. In PJ Reier, RP Bunge and FJ Seil (eds): "Neurology and Neurobiology: Current Issues in Neural Regeneration Research." Alan R. Liss: New York, pp. 75–87.

Bregman BS and Kunkel–Bagden E (1988): Effect of target and non–target transplants on neuronal survival and axonal elongation after injury to the developing spinal cord. In DM Gash and JR Sladek, Jr (eds.) 'Progress in Brain Res' Elsevier: Amsterdam, pp. 205–211.

Bregman BS, Kunkel–Bagden E, Reier PJ, Dai HN, McAtee M and Gao D (1993): Recovery of function after spinal cord injury: Mechanisms underlying transplant–mediated recovery of function differ after spinal cord injury in newborn and adult rats. Exp Neurol 123: 3–16.

Bregman BS and Reier PJ (1986): Neural tissue transplants rescue axotomized rubrospinal cells from retrograde death. J Comp Neurol 244: 86-95.

Brundin P, Isacson O and Björklund A (1985): Monitoring of cell viability in suspensions of embryonic CNS tissue and its use as a criterion for intracerebral graft survival. Brain Res 331: 251–259.

Buchanan JT and Nornes HO (1986): Transplants of embryonic brainstem containing the locus coeruleus into spinal cord enhances the hindlimb flexion reflex in adult rats. Brain Res 381: 225-236.

Das GD (1983): Neural transplantation in the spinal cord of adult rats. J Neurol Sci 62: 191–210.

Dunnett SB and Richards SJ (1990): Neural Transplantation: From Molecular Basis To Clinical Applications. Prog Brain Res Volume 82: Amsterdam: Elsevier.

Edgerton VR, Ray RR, Hodgson JA, Prober RJ, Camille P, de Guzman CP and de Leon R (1992): Potential of adult mammalian lumbosacral spinal cord to execute and acquire improved locomotion in the absence of supraspinal input, In Jane JA, Anderson DK, Torner JC and Young W (eds): "Central Nervous System Trauma Status Report". J Neurotrauma 9 (Supp 1): S119–S128.

Faden AI and Jacobs TP (1985): Effect of TRH analogs on neurologic recovery after experimental spinal trauma. Neurol 35: 1331–1334.

Foster GA, Schultzberg M, Gage FH, Björklund A, Hokfelt T, Nornes H, Cuello AC, Verhofstad AAJ and Visser T J (1985): Transmitter expression and morphological development of embryonic medullary and mesencephalic raphe neurones after transplantation to the adult rat central nervous system. Exp Brain Res 60: 427-444.

Foster GM, Roberts MHT, Wilkinson LS, Björklund A, Gage FH, Hokfelt R, Schultzberg M and Sharp T (1989): Structural and functional analysis of raphe neurone implants into denervated rat spinal cord. Brain Res Bull 22: 131–137.

Gash DM and Sladek Jr JR (1988): Transplantation into the mammalian CNS. Prog. in Brain Res. Vol. 78, Amsterdam: Elsevier.

Gelderd JB and Quarles JE (1990): A preliminary study of homotopic fetal cortical and spinal cotransplants in adult rats. Brain Res Bull 25: 35-48.

Goldberger ME Bregman BS Vierck, Jr CK and Brown M (1990): Criteria for assessing recovery of function after spinal cord injury: Behavioral methods. Exp Neurol 107: 113–117.

Grillner S (1973): Locomotion in the spinal cat, In Stein RB, Pearson KG, Smith RS and Redford JB (eds): "Control of Posture and Locomotion," Advances in Behavorial Biology, Vol 7. Plenum Press: New York, pp 515–535.

Grillner S (1975): Locomotion in vertebrates: central mechanisms and reflex interaction. Physiol Rev 55: 247–304.

Hall ED and Braughler MJ (1982): Effects of intravenous methylprednisolone on spinal cord lipid peroxidation and $(Na^+ + K^+)$–ATPase activity. J Neurosurg 57: 247–253.

Hansebout RR, Blight AR, Fawcett S and Reddy K (1993): 4–Aminopyridine in chronic spinal cord injury: A controlled, double–blind, crossover study in eight patients. J Neurotrauma 10: 1–18.

Himes BT, Goldberger ME and Tessler A (1994): Grafts of fetal central nervous system tissue rescue axotomized Clarke's nucleus neurons in adult and neonatal operates. J Comp Neurol 339: 117–131.

Howland DR, Bregman BS, Tessler A and Goldberger ME (1995): Transplants enhance locomotion in neonatal kittens whose spinal cords are transected. Expt. neurol. in press.

Itoh Y and Tessler A (1989): Ultrastructural organization of regenerated adult dorsal root axons within transplants of fetal spinal cord. J Comp Neurol 292: 396-411.

Itoh Y and Tessler A (1990): Regeneration of adult dorsal root axons into transplants of fetal spinal cord and brain - a comparison of growth and synapse formation in appropriate and inappropriate targets. J Comp Neurol 302: 272-293.

Itoh Y, Sugaware T, Kowda M and Tessler A (1993): Time course of dorsal root axon regeneration into transplants of fetal spinal cord: An electron microscopic study. Exp Neurol 123:,133–146.

Jakeman LB and Reier PJ (1987a): Interactions between corticospinal tract axons and fetal CNS transplants in the adult rat. Schmitt Neurol Symposium, University of Rochester, Rochester, New York.

Jakeman L and Reier PJ (1987b): The response of corticospinal tract fibers following injury and transplantation in the adult rat spinal cord. Soc Neurosci Abst 13: 750.

Jakeman LB and Reier PJ (1989): Regeneration or sprouting of corticospinal tract axons into fetal spinal cord transplants in the adult rat. Soc Neurosci Abst 15: 1242.

Jakeman LB and Reier PJ (1991): Axonal projections between fetal spinal cord transplants and the adult rat spinal cord: A neuroanatomical tracing study of local interactions. J Comp Neurol 307: 311-334.

Kunkel–Bagden E and Bregman BS (1990): Spinal cord transplants enhance the recovery of locomotor function after spinal cord injury at birth. Exp Brain Res 81: 25-34.

Kunkel–Bagden E, Dai HN and Bregman BS (1993): Methods to assess the development and recovery of locomotor function after spinal cord injury in rats. Exp Neurol 119: 153–164.

Liu CN and Chambers WW (1958): Intraspinal sprouting of dorsal root axons. Arch Neurol Psychiat 49: 46–61.

Means ED and Anderson DK (1987): The pathophysiology of acute spinal cord injury. In Davidoff RA (ed): "Handbook of The Spinal Cord" Vols. 4 and 5, Marcel Dekker Publishers. Chap. 2, pp 19-61.

Means ED, Anderson DK, Waters TR and Kalaf L (1981): Effect of methylprednisolone in compression trauma to the feline spinal cord. J Neurosurg 55: 200-208.

Moorman SJ, Whalen LR and Nornes HO (1990): A neurotransmitter specific functional recovery mediated by fetal implants in the lesioned spinal cord of the rat. Brain Res 508: 194–198.

Murray M and Goldberger ME (1986): Replacement of synaptic terminals in lamina II and Clarke's nucleus after unilateral lumbosacral dorsal rhizotomy in adult cats. J. Neurosci 6: 3205–3217.

Nornes H, Björklund A and Stenevi U (1983): Reinnervation of the denervated adult spinal cord of rats by intraspinal transplants of embryonic brain stem neurons. Cell Tiss Res 230: 15-35.

Nornes H, Björklund A and Stenevi U (1984): Transplantation strategies in spinal cord regeneration, In Sladek Jr. JR, Gash DM (eds): "Neural Transplants-Development and Function." Plenum Press: New York.

Nothias F and Pechanski M (1990): Homotypic fetal transplants into an experimental model of spinal cord neurodegeneration. J Comp Neurol 301: 520–534. .

Nygren LG, Olson L and Seiger A (1977): Monoaminergic reinnervation of the transected spinal cord by homologous fetal brain grafts. Brain Res 129: 227–235.

Patel U and Bernstein JJ (1983): Growth, differentiation, and viability of fetal rat cortical and spinal cord implants into adult rat spinal cord. J Neurosci Res 9: 303–310.

Pixley SK, Anderson DK and Reier PJ (1994): Transplantation of embryonic cat nasal tissue into adult cat spinal cord. Manuscript in preparation.

Privat A, Mansour H and Geffard M (1988): Transplantation of fetal serotonin neurons into the transected spinal cord of adult rats: Morphological development and functional influence. In Gash DM and Sladek JR (eds): "Transplantation into the Mammalian CNS." Prog Brain Res, Vol 78: 155–167.

Privat A, Mansour H Pavy A Geffard M and Sandillon F (1986): Transplantation of dissociated foetal serotonin neurons into the transected spinal cord of adult rats. Neurosci Lett 66: 61-66.

Privat A, Mansour H, Rajaofetra N and Geffard M (1989): Intraspinal transplants of serotonergic neurons in the adult rat. Brain Res Bull 22: 123-129.

Reier PJ (1985): Neural tissue grafts and repair of the injured spinal cord. Neuropath Appl Neurobiol 11: 81-104.

Reier PJ, Anderson DK and Stokes BT (1992a): Neural tissue transplantation and CNS trauma: Anatomical and functional repair of the injured spinal cord. In Jane JA, Anderson DK, Torner JC and Young W (eds): "Central Nervous System Status Report" J. Neurotrauma 9(Suppl 1): S223–S248.

Reier PJ, Bregman BS and Wujek JR (1986): Intraspinal transplantation of embryonic spinal cord tissue in neonatal and adult rats. J Comp Neurol 247: 275-296.

Reier PJ, Stensaas LJ and Guth L (1983): The astrocytic scar as an impediment to regeneration in the central nervous system. In Kao CC Bunge RP and Reier PJ (eds): "Spinal Cord Reconstruction." Raven Press, New York pp 163–196.

Reier PJ, Stokes BT, Thompson FJ and Anderson DK (1992b): Fetal cell grafts into resection and contusion/compression injuries of the rat and cat spinal cord. Exp Neurol 115:,177–188.

Rosenbluth J, Hasegawa M, Shirasaki N, Rosen CL and Liu Z (1990): Myelin formation following transplantation of normal fetal glia into myelin–deficient rat spinal cord. J Neurocytol 19(5): 718–730.

Schnell L, Schneider R, Kolbeck R, Barde Y–A and Schwab ME (1994): Neurotrophin–3 enhances sprouting of corticospinal tract during development and after adult spinal cord lesion. Nature 367: 170–173.

Shik ML and Orlovsky GN (1976): Neurophysiology of locomotor automatism. Physiol Rev 58: 465–501.

Sieradzan K and Vrbova G (1989): Replacement of missing motoneurons by embryonic grafts in the rat spinal cord. Neurosci. 31: 115-130.

Sternberger LA (1976): "Immunocytochemistry" 2nd Ed, Wiley, New York.

Stokes BT and Reier PJ (1991): Oxygen transport in intraspinal fetal grafts: graft–host relations. Exp Neurol 111: 312–323.

Stokes BT and Reier PJ (1992): Fetal grafts alter chronic behavioral outcome after contusion damage to the adult rat spinal cord. Exp Neurol 116: 1–12.

Tessler A, Himes BT, Houle J and Reier PJ (1988): Regeneration of adult dorsal root axons into transplants of embryonic spinal cord. J Comp Neurol 270: 537–548.

Waxman SG (1992): Demyelination in spinal cord injury and multiple sclerosis: what can we do to enhance functional recovery? In Jane JA, Anderson DK, Torner JC, Young W (eds): "Central Nervous System Trauma Status Report". J Neurotrauma 9(Supp 1): S105–S117.

Wetzel MC and Stuart DG (1976): Ensemble characteristics of cat locomotion and its neural control. Prog Neurobiol 7: 1–98.

Wirth ED, Theele DP, Mareci TH, Anderson DK, Brown SA and Reier PJ (1992): *In vivo* magnetic resonance imaging of fetal cat neural tissue transplants in the adult cat spinal cord. J Neurosurg 76: 261–274.

Yakovleff A, Roby–Brami A, Guezard B, Mansour A, Brussel B and Privat A (1989): Locomotion in rats transplanted with noradrenergic neurons. Brain Res Bull 22: 115–121.

Young W (1989): Recovery mechanisms in spinal cord injury: implications for regenerative therapy. In: FJ Seil (ed): 'Neural Regeneration and Transplantation'. Alan R. Liss, Inc., New York, pp 157–169.

Young W and Flamm ES (1982): Effects of high dose corticosteroid therapy on blood flow, evoked potentials, and extracellular calcium in experimental spinal injury. J Neurosurg 57: 667–673.

Zhang B, Goldberger ME and Murray M (1993): Proliferation of SP– and 5HT–containing terminals in lamina II of rat spinal cord following dorsal rhizotomy: quantitative EM–immunocytochemical studies. Exp Neurol 123: 51–63.

ROLE OF TARGET-DERIVED NEUROTROPHINS IN THE MAINTENANCE OF DEVELOPING AND ADULT BASAL FOREBRAIN NEURONS

M.A. Burke[1], E.J. Mufson[2,3], and J.H. Kordower[2,3]

[1]Department of Anatomy and Cell Biology, University
of Illinois School of Medicine, Chicago, IL 60612;
[2]Department of Neurological Sciences and [3]Rush
Alzheimer's Disease Center, Rush Presbyterian/ St.
Lukes Medical Center, Chicago, IL 60612, U.S.A.

INTRODUCTION

Nerve growth factor (NGF) is the prototype neurotrophic molecule initially characterized by Levi-Montalcini and colleagues as a protein which supports the viability and phenotypic differentiation of catecholamine-containing peripheral ganglia (see reviews: Hefti et al., 1989; Levi-Montalcini and Angeletti, 1968; Thoenen and Barde, 1980; Thoenen and Edgar, 1985; Thoenen et al., 1987). Other members of the neurotrophin superfamily include brain-derived neurotrophic factor (BDNF), neurotrophin-3 (NT-3), and neurotrophin 4/5 (NT4/5), each of which display structural homology with NGF and are also thought to provide critical support for the development, survival, and maintenance of specific populations of neurons. Throughout development and adulthood, the highest levels of the both protein and mRNA of these neurotrophins are produced within cells of the cerebral cortex, hippocampus and olfactory bulb (Maisonpierre et al., 1990a; Maisonpierre et al., 1990b; Whittemore et al., 1986) which are the major projection sites of cholinergic basal forebrain neurons (e.g. Mesulam et al., 1983). Autoradiographic studies demonstrate that neurons which bind radiolabeled NGF codistribute with cholinergic basal forebrain neurons in the rat (Raivich and Kreutzberg, 1987; Richardson et al., 1986). Furthermore, intraventricular injections of NGF in newborn rats (Gnahn et al., 1983) or administration of NGF or BDNF to cultured basal forebrain neurons (Hefti et al., 1985; Knusel et al., 1991) increases cholinergic enzyme levels. Choline acetyltransferase (ChAT) activity has also been shown to increase in the neonatal striatum following NGF administration (Mobley et al., 1985). In the intact animal, the medial septal (MS) and vertical limb of the diagonal band (VLDB) neurons project to the hippocampus via the fimbria-fornix (ffx) and ventral pathway (Milner and Amaral, 1984) where axons bind and transport both

NGF and BDNF produced within this target region (Conner et al.,1992; Conner and Varon, 1992; DiStefano et al., 1992). The basal forebrain neurons, in turn, express the mRNA for trkA and trkB (Kokaia et al., 1993), the high affinity receptors for NGF and BDNF, respectively.

These studies and others prompted many investigators to study the effects of neurotrophins, principally NGF and BDNF, in lesion models of neurodegeneration. Fimbria-fornix transection results in a marked reduction in the size and number of cholinergic and noncholinergic MS/VLDB neurons in both rats and monkeys (Diatz and Powell, 1954; Gage et al., 1986; Hefti, 1986; Kordower and Fiandaca, 1990; Tuszynski et al., 1990); an effect believed to be due in part to disconnecting the septal neurons from their endogenous source of trophic factors (see below for further discussion). In vivo, intraventricular infusions of NGF reverses atrophy of cholinergic basal forebrain neurons and spatial memory impairment associated with aging (Fischer et al., 1987) and prevents the degeneration of the septal/diagonal band neurons following fimbria-fornix transection in rats (Gage et al., 1988; Hefti et al., 1989; Kromer, 1987; Williams et al., 1986) and nonhuman primates (Tuszynski et al., 1990; Emerich et al., 1994; Kordower et al., 1994). More recently, it has been demonstrated that BDNF infusions following fimbria transection in rodents significantly reduced cholinergic basal forebrain cell loss associated with such lesions (Knusel et al., 1992). These data and others have led to the hypothesis that NGF and BDNF provide neurotrophic support for intact adult cholinergic basal forebrain neurons.

Recent observations by Sofroniew and colleagues (Sofroniew et al., 1990, 1993), as well as our group (Kordower et al., 1992) have questioned the role of endogenous NGF and BDNF in the maintenance of adult cholinergic basal forebrain neurons. Excitotoxic lesions of the hippocampus resulted in the destruction of neurons that normally synthesize NGF and BDNF, thereby drastically reducing trophic factor support to the cholinergic basal forebrain. Up to 500 days postlesion, the hippocampal-projecting cholinergic septal/diagonal band neurons had not degenerated (Fig. 1A; Kordower et al., 1992; Sofroniew et al., 1990); yet analysis of the hippocampal remnant revealed as much as a 97% reduction in NGF and BDNF mRNA levels (Sofroniew et al., 1993). The sustained viability of septohippocampal neurons was not specific to cholinergic cells since, the number of septal/diagonal band neurons immunoreactive for parvalbumin, a putative marker for GABAergic neurons within the MS/VLDB, were also unchanged (Kordower et al., 1992). This viability of both cholinergic and GABAergic neurons following excitotoxic removal of their target region is in contrast to the response of these neurons following fornix transection. Both cholinergic and GABAergic neurons (Batchelor et al., 1989; Koliatsos et al., 1989; Kromer, 1987; Peterson et al., 1987; Tuszynski et al., 1990; Williams et al., 1986) become nonviable following axotomy, a lesion model which injures the cell by physically disconnecting it from its hippocampal target region. Additional data indicate that basal forebrain neurons carry on normal function following loss of their target cells. Following hippocampal lesion, the staining patterns of phosphorylated neurofilaments within the septal/diagonal band complex 2-24 weeks following the lesion are unchanged (Kordower et al., 1992). Again this is in contrast to that seen following fimbria-fornix transection. Koliatsos and colleagues (1989) demonstrated the transient synthesis of nonphosphorylated neurofilament proteins within the septal perikarya 4 weeks following axotomy. Thus septohippocampal neurons synthesize new neurofilaments in response to dynamic degenerative processes following axotomy but not following lesion of their neurotrophin synthesizing trophic factor secreting cells.

The viability of cholinergic septal/diagonal band neurons in the absence of their normal source of neurotrophins suggests that these cells may sprout to a novel source in search of target derived support. The septohippocampal pathway is not highly collateralized (Amaral and Kurz, 1985), and thus it is unlikely that intact projections to nonlesioned structures maintain these neurons. Alternatively, axons that normally project to the lesioned hippocampus could have sprouted to the contralateral hippocampus, possibly via the hippocampal commisure, where high levels of NGF and BDNF remain following this type of lesion. Injection of two different retrograde tracers into both the lesioned and nonlesioned hippocampus resulted principally in transport of each tracer to the ipsilateral medial septum and VLDB (Fig. 1B; Kordower et al., 1992). Rather than sprouting to the intact side, the septal/diagonal band neurons maintain their ipsilateral projection at least 12 weeks following hippocampal lesions. Therefore, the viability of adult septal/diagonal band neurons following loss of their target is not due to collateral sprouting towards a novel source of trophic factors. The failure of cholinergic septal/diagonal band neurons to degenerate under these conditions suggests that (a) NGF and BDNF are not required for the maintenance of basal forebrain neurons in adult animals, (b) NGF and BDNF are synthesized from extrahippocampal sites following the lesion, and/or (c) other trophic molecules that are synthesized within the hippocampus or septohippocampal pathway under normal or pathological conditions are sufficient to maintain the viability of these adult neurons.

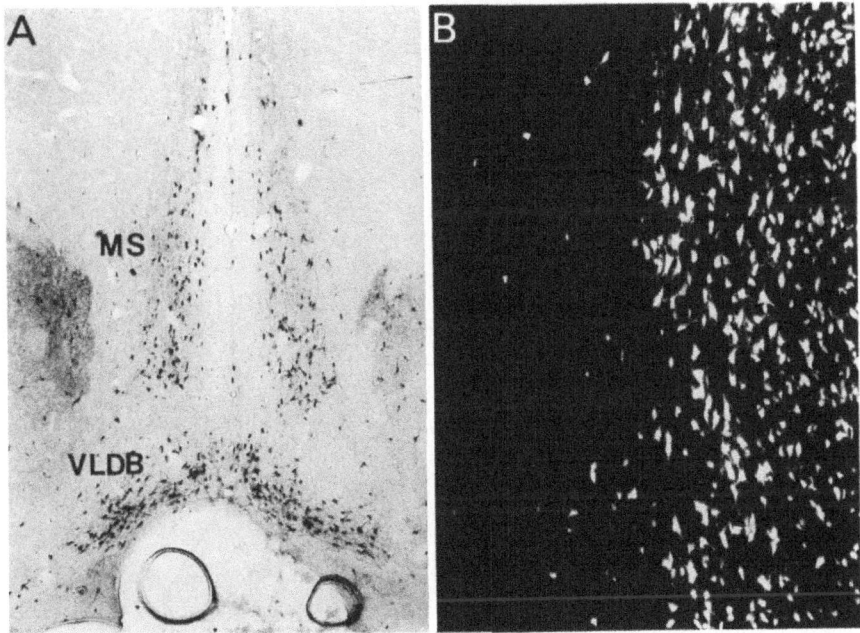

Figure 1. (A) The number of ChAT-ir septal neurons is unchanged ipsilateral to a excitotoxic hippocampal lesion (right) relative to the contralateral side (left). (B) Injections of fluorogold into the small aneuronal hippocampal remnant resulted in extensive retrograde labeling within the ipsilateral septal/diagonal band region.

In contrast to normal adult basal forebrain neurons which appear not require at least high levels of neurotrophins for their viability, developing cholinergic septal/diagonal band neurons may be more vulnerable to loss of target-derived neurotrophic support. We have recently carried out a study which created excitotoxic lesions of the hippocampus in developing rat pups between postnatal days 7-28. These lesions result in a significant loss of both ChAT and the low affinity nerve growth factor receptor (p75 NGFr)-immunoreactive neurons in the MS/VLDB (Fig. 2A; Burke et al., 1994). Although the reduction in the number of cholinergic neurons was significant, the degree of immunoreactive cell loss was modest (<30%), with the majority of these neurons maintaining their viability up to 4 weeks post-lesion (Burke et al., in 1994). It is interesting that the magnitude of cell loss became progressively attenuated when the lesions were made at older developmental time points (Fig. 2B). The vulnerability of developing cholinergic septodiagonal band neurons appears to inversely correlate with the developmental pattern of cholinergic hippocampal innervation which displays an adult pattern acetylcholinesterase fiber staining by postnatal day 14 (Milner et al., 1983). Hippocampal lesions created on postnatal days 7, 10, and 14 resulted in significant reductions in the number of cholinergic septal diagonal band neurons ipsilateral to the lesion. However, there was not a significant loss of ChAT-immunoreactive medial septal neurons in animals receiving hippocampal lesions on postnatal day 28 (Burke et al., 1994).

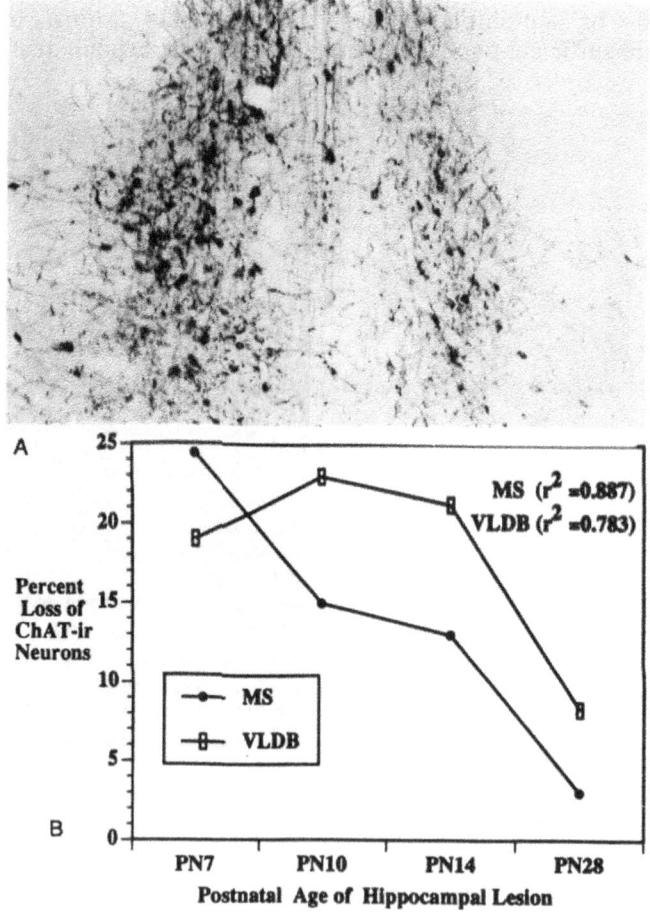

Figure 2. (A) In contrast to adult rats, excitotoxic lesions of the hippocampus on postnatal day 10 result in a significant loss of cholinergic septal neurons. (B) The magnitude of the loss of cholinergic septal/diagonal band neurons was inversely correlated with the age at which the lesion was made.

As described previously, intraventricular infusions of NGF or BDNF following fimbria-fornix transection prevent the degeneration of the rat cholinergic basal forebrain neurons (Gage et al., 1988; Hefti et al., 1989; Knusel et al., 1992; Kromer, 1987; and Williams et al., 1986). Therefore, it could be reasoned that intraventricular infusions of either NGF or BDNF following neonatal excitotoxic hippocampal lesions would prevent the loss of cholinergic neurons within the septal/diagonal band observed in developing animals. Preliminary data suggest that intraventricular infusions of NGF rescue the cholinergic septal/diagonal band neurons which degenerate following hippocampal lesions in the developing rat (Burke et al., 1993). In contrast, infusions of BDNF following excitotoxic lesions of the developing hippocampus do not result in significant recovery of the cholinergic basal forebrain neurons. The inability of BDNF to rescue MS/VLDB may be due to poor penetration of the neurotrophin to the degenerating neurons as intraventricularly injected BDNF principally binds to the truncated trk B receptor located upon the ependymal cells lining the lateral ventricle (Kordower et al., 1993). However, it is interesting that both NGF and BDNF induced a dramatic cholinergic sprouting response within the small hippocampal remnant ipsilateral to the infusion. These data suggest that BDNF is penetrating sufficiently to influence cholinergic basal forebrain neurons in a tropic, but not trophic, fashion. Perhaps higher concentrations of BDNF would prevent the degeneration of developing cholinergic septal/diagonal band neurons in a manner presently seen with NGF.

The present series of experiments illustrates that the viability of developing, but not adult, cholinergic basal forebrain neurons are dependent upon target derived neurotrophins. The lack of degeneration in adult cholinergic basal forebrain neurons following almost complete loss of target derived neurotrophic support is in direct contrast to that seen following axotomy. A plausible explanation is that only small levels of NGF and BDNF are required to maintain these neurons under normal adult conditions where plasticity of the septohippocampal system is at a minimum. This concept is supported by the fact that residual NGF and BDNF levels observed within the lesioned hippocampus (Kordower et al., 1992; Sofroniew et al., 1990, 1993) are far less than is required for the rescue of axotomized septohippocampal neurons. However, during development when synapse formation and stabilization within the septohippocampal projection is dynamic, the loss of target derived neurotrophic support significantly influences the viability of cholinergic basal forebrain neurons in an age-dependent manner; an effect which can be completely prevented by intraventricular infusions of NGF. This suggests that trophic factors derived from target neurons play a crucial role in basal forebrain neuron development. However, once this system has matured, target-derived neurotrophic support molecules may promote other homeostatic functions serving primarily phenotypic and neurochemical differentiation of these neurons.

ACKNOWLEDGEMENTS

We would like to thank Janice Apter and Dr. Amir Umar for expert technical assistance. Supported by AG094416, NS25655, NS29585, and The Washington Square Foundation.

REFERENCES

Amaral DG and Kurz J (1985): An analysis of the origins of the cholinergic and noncholinergic septal projections to the hippocampal formation of the rat. J Comp Neurol 240: 37-59.

Batchelor PE, Armstrong DM, Blaker SN and Gage FH (1989): Nerve growth factor receptor and choline acetyltransferase colocalization in neurons within the rat forebrain: response to fimbria-fornix transection. J Comp Neurol 284: 187-204.

Burke MA, Apter JR, Wainer BH, Mufson EJ and Kordower JH (1994): Age-related vulnerability of developing cholinergic basal forebrain neurons following excitotoxic lesions of the hippocampus. Exp. Neurol (in press).

Burke MA, Mobley WC, Mufson EJ and Kordower JH (1993): NGF prevents the loss of developing cholinergic basal forebrain neurons following excitotoxic lesions of the hippocampus. Soc Neurosci Abstr 23: 279

Conner JM, Muir D, Varon S, Hagg T and Manthorpe M (1992): The localization of nerve growth factor-like immunoreactivity in the adult rat basal forebrain and hippocampal formation. J Comp Neurol 319: 454-462.

Conner JM and Varon S (1992): Distribution of nerve growth factor-like immunoreactive neurons in the adult rat brain following colchicine treatment. J Comp Neurol 326: 347-362.

Diatz HM and Powell TPS (1954): Studies on the connections of the fornix system. J Neurol Neurosurg Psychiatry 17: 75-82.

DiStefano PS, Friedman B, Radziejieski AC, Boland P, Schick CM, Lindsay RM and Wiegand SJ (1992): The neurotrophins BDNF, NT-3, and NGF display distinct patterns of retrograde axonal transport in peripheral and central neurons. Neuron 8: 983-993.

Emerich DF, Winn SR, Harper J, Hammang JP, Baetge EE and Krodower JH (1994): Implants of polymer encapsulated human NGF-secreting cells in the non-human primate: rescue and sprouting of degenerating cholinergic basal forebrain neurons. J Comp Neurol 349: 148-164.

Fischer W, Wictorin K, Björklund A, Williams LR, Varon S and Gage FH (1987): Amelioration of cholinergic neuron atrophy and spatial memory impairment in aged rats by nerve growth factor. Nature (London) 329: 65-68.

Gage FH, Wictorin K, Fischer W, Williams LR, Varon S and Björklund A (1986): Retrograde cell changes in medial septum and diagonal band following fimbria-fornix transection: quantitative temporal analysis. Neuroscience 19: 241-255.

Gage FH, Armstrong DM, Williams LR and Varon S (1988): Morphological response of axotomized septal neurons to nerve growth factor. J Comp Neurol 269: 147-155.

Gnahn H, Hefti F, Heumann R, Schwab ME and Thoenen H (1983): NGF-mediated increase of choline acetyltransferase (ChAT) in the neonatal forebrain: evidence for a physiological role of NGF in the brain? Dev Brain Res 9: 45-52.

Hefti F (1986): Nerve growth factor promotes survival of septal cholinergic neurons after fimbrail transections. J Neurosci 6: 2155-2162.

Hefti F, Hartikka J, Eckenstein F, Gnahn H, Heumann R and Schwab M (1985): Nerve growth factor increases choline acetyltransferase but not survival or fiber outgrowth of cultured fetal septal cholinergic neurons. Neuroscience 14: 55-68.

Hefti F, Hartikka J and Knusel B (1989): Function of neurotrophic factors in the adult and aging brain and their possible use in the treatment of neurodegenerative diseases. Neurobiol Aging 10: 515-533.

Knusel B, Beck KD, Winslow JW, Rosenthal A, Burton LE, Widmer HR, Nikolics K and Hefti F (1992): Brain-derived neurotrophic factor administration protects basal forebrain cholinergic but not nigral dopaminergic neurons from degenerative changes after axotomy in the adult rat brain. J Neurosci 12: 4391-4402.

Knusel B, Winslow JW, Rosenthal A, Burton LE, Seid DP, Nikolics K and Hefti F (1991): Promotion of central cholinergic and dopaminergic neuron differentiation by brain-derived neurotrophic factor but not neurotrophin 3. Proc Natl Acad Sci USA 88: 961-965.

Kokaia Z, Bengzon J, Metsis M, Kokaia M, Persson H and Lindvall O (1993): Coexpression of neurotrophins and their receptors in neurons of the central nervous system. Proc Natl Acad Sci USA 90: 6711-6715.

Koliatsos VE, Applegate MD, Kitt CA, Walker LC, DeLong MR and Price DL (1989): Aberrant pohophorylation of neuronfilaments accompanies transmitter-related changes in the rat septal neurons following transection of the fimbria-fornix. Brain Res 482: 205-218.

Kordower JH, Burke-Watson MA, Roback JD and Wainer BH (1992): Stability of septohippocampal neurons following excitotoxic lesions of the rat hippocampus. Exp Neurol 117: 1-16.

Kordower JH and Fiandaca MS (1990): Response of the monkey septohippocampal system to fornix transection: a histochemical and cytochemical analysis. J Comp Neurol 298: 443-457.

Kordower JH, Mufson EJ, Granholm A-C, Hoffer B and Friden PM (1993) Delivery of trophic factors to the primate brain. Exp Neurol 124: 21-30.

Kordower JH, Winn SR, Liu Y-T, Mufson EJ, Sladek JR Jr, Hammang JP, Baetge EE and Emerich DF (1994): The aged monkey basal forebrain: rescue and sprouting of degenerating cholinergic neurons following grafts of polymer encapsulated NGF secreting cells. Proc Natl Acad Sci USA (in press).

Kromer LF (1987): Nerve growth factor treatment after brain injury prevents neuronal death. Science 235: 214-216.

Levi-Montalcini R and Angeletti PU (1968): Nerve growth factor. Physiol Rev 8: 534-569.

Maisonpierre PC, Belluscio L, Friedman B, Alderson RF, Wiegand SJ, Furth ME, Lindsay RM and Yancopoulos GD (1990a): NT-3, BDNF, and NGF in the developing rat nervous system: parallel as well as reciprocal patterns of expression. Neuron 5: 501-509.

Maisonpierre PC, Belluscio L, Squinto S, Ip NY, Furth ME, Lindsay RM and Yancopoulos GD (1990b): Neurotrophin-3: a neurotrophic factor related to NGF and BDNF. Science 247: 1446-1451.

Mesulam M-M, Mufson EJ, Wainer BH and Levey AI (1983): Central cholinergic pathways in the rat: an overview based upon an alternative nomenclature. Neuroscience 10: 1185-1201.

Milner TA and Amaral DG (1984): Evidence of a ventral septal projection to the hippocampal formation of the rat. Exp Brain Res 55: 579-585.

Milner TA, Loy R and Amaral DG (1983): An anatomical study of the development of the septo-hippocampal projection in the rat. Dev Brain Res 8: 343-371.

Mobley WC, Rutkowski JL, Tennekoon GI, Buchanan K and Johnston MV (1985): Choline acetyltransferase activity in striatum of neonatal rats increased by nerve growth factor. Science 229: 284-287.

Peterson GM, Williams LR, Varon S and Gage FH (1987): Loss of GABAergic neurons in medial septum after fimbria-fornix transection. Neurosci Lett 76: 140-144.

Raivich G and Kreutzberg GW (1987): The localization and distribution of high affinity BNGF binding sites in the central nervous system in the adult rat: a light microscopic autoradiographic study using (125-I) BNGF. Neuroscience 20: 23-36.

Richardson PM, Verge Issa VMK and Riopelle RJ (1986): Distribution of neuronal receptor for nerve growth factor in the rat. J Neurosci 6: 2312-2321.

Sofroniew MV, Cooper JD, Svendsen CN, Crossman P, Ip NY, Lindsay RM, Zafra F and Lindholm D (1993): Atrophy but not death of adult septal cholinergic neurons after ablation of target capacity to produce mRNAs for NGF, BDNF, and NT3. J Neurosci 13: 5263-5276.

Sofroniew MV, Galletly NP, Isacson O and Svendsen CN (1990): Survival of adult basal forebrain cholinergic neurons after loss of target neurons. Science 247: 338-342.

Thoenen H and Barde DYA (1980): Physiology of nerve growth factor. Physiol Rev 60: 1284-1335.

Thoenen H and Edgar D (1985): Neurotrophic factors. Science 229: 238-242.

Thoenen H Bandtlow C and Heumann R (1987): The physiological function of nerve growth factor in the central nervous system: comparison with the periphery. Rev Phys Biochem Pharmacol 109: 145-178.

Tuszynski MH, Armstrong DM and Gage FH (1990): Basal forebrain cell loss following fimbria-fornix transection. Brain Res 508: 45-52.

Tuszynski MH, Sang U, Amaral DG and Gage FH (1990): Nerve growth factor infusion in the primate brain reduces lesion-induced cholinergic neuronal degeneration. J Neurosci 10: 3604-3614.

Whittemore SR, Ebendal T, Larkfort L, Olson L, Seiger A, Stromberg I and Persson H (1986): Developmental and regional expression of B nerve growth factor messenger RNA and protein in the rat central nervous system. Proc Natl Acad Sci USA 83: 817-821.

Williams LR, Varon S, Peterson GM, Wictorian K, Fischer W, Björklund A and Gage FH (1986): Continuous infusion of nerve growth factor prevents basal forebrain neuronal cell death after fimbria-fornix transection. Proc Natl Acad Sci USA 83: 9231-9236.

AMYOTROPHIC LATERAL SCLEROSIS AS A TARGET DISEASE FOR GENE THERAPY USING THE ENCAPSULATION TECHNOLOGY

P. Aebischer[1], S.A. Tan[1], N.A. Pochon[1], E.E. Baetge[2], A.C. Kato [3]

[1]Division of Surgical Research, Centre Hospitalier Universitaire Vaudois, Lausanne University Medical School, 1011 Lausanne, SWITZERLAND. [2]Department of Neuroscience, Cytotherapeutics, Providence, RI 02900, U.S.A. [3]Department of Pharmacology and Division of Clinical Neurophysiology, Centre Medical Universitaire, Geneva University Medical School, 1211 Genève, SWITZERLAND

INTRODUCTION

Amyotrophic lateral sclerosis (ALS) is a progressive and fatal disease of the voluntary motor system. Its cause is unknown and no effective therapy has yet been devised. The natural history and range of expression of the disease is well known. It is a fatal disease, within 2 to 3 years of diagnosis. Although there is individual variation in the progression of the disease, quantitative measurement of the isometric force generated by a set of muscles allows one to follow the progression of the disease. This test, named the Tufts Quantitative Neuromuscular Exam (TQNE, Munsat et al., 1988), allows one to predict the individual patient's course of disease on the basis of 3 successive tests, each one a month apart. The results of a novel therapy may therefore be assessed by looking for change in disease course versus the prediction.

The cloning of several new neurotrophic factor genes offer potential treatments for this disease. Exposure to ciliary neurotrophic factor (CNTF) or insulin-like growth factor I (IGF-I) or to neurotrophins, brain-derived neurotrophic factor (BDNF), neurotrophin-3 (NT-3) or neurotrophin-4/5 (NT4/5) has been shown to prevent the death of rat and human embryonic motoneurons in vitro (Mangal et al.,1991; Henderson et al., 1993; Hughes et al., 1993). Kato and Lindsay (submitted) have now reported a synergistic effect of these factors on human embryonic ventral spinal cord neurons in culture. The systemic administration of some of these factors has also been reported to prevent the loss of motoneurons following

Neural Cell Specification: Molecular Mechanisms and Neurotherapeutic Implications
Edited by Juurlink *et al.*, Plenum Press, New York, 1995

275

early postnatal axotomy of peripheral nerves in rodent models (Sendtner et al., 1990, 1992a; Yan et al., 1992; Hughes et al., 1993). More recently the administration of CNTF, through intraperitoneal implantation of cells genetically engineered to release CNTF, has been reported to slow down the progressive loss of muscle function in the pmn/pmn mouse, a mutant mouse with progressive motor neuropathy resembling ALS (Sendtner et al., 1992b). These mice, however, had to be sacrificed because of tumor formation.

Such information led to a number of experimental human protocols to treat ALS. These include the systemic delivery of CNTF and more recently, BDNF and IGF-I through subcutaneous injections. The BDNF and IGF-I trials have just begun and no data are available at this time. Early results with systemic administration of CNTF (3 subcutaneous injections per week) (Brooks et al., 1993) have been encouraging; however, recent reports suggest systemic toxicity potential (Barinaga, 1994).

CNTF has a half-life of 2.9 minutes in vivo (Dittrich et al., 1994). Such a short half-life precludes the easy approach to a direct administration in the cerebral spinal fluid (CSF). It is hypothesized that the intrathecal delivery will be more efficacious than systemic administration since CNTF in the CSF can act directly on the cell body receptors of both lower motoneurons (spinal motoneurons) as well as on the upper motoneurons (cortical motoneurons: Betz cells). The latter are, of course, inaccessible by systemic delivery since the blood brain barrier inhibits the diffusion of CNTF directly into the central nervous system, although retrograde transport may allow a small fraction of systemically administered drug to reach the lower motoneurons. However, experiments in sheep have shown that CNTF is toxic when given in bolus intrathecally (personal communication from Dr. M. Sendtner). Furthermore, it is a very unstable molecule which makes its continuous delivery by pumps almost impossible. Specifically in pumps more than 90% of its activity is lost in 12 hours or less (personal communication from Dr. M. Sendtner).

The delivery of CNTF through gene therapy circumvents the problem of stability of the neurotrophic factor allowing its continuous systemic and/or intrathecal delivery. A gene therapy approach seems therefore a rational way to proceed. We are therefore investigating the administration of CNTF through the transplantation of cells transfected with the human CNTF gene. Such cells are loaded in a device that protects them from immune rejection, controls dosing and allows for the recovery of the cells, if needed.

GENE TRANSFER SYSTEM TO BE USED

Carrier

The cells we have chosen for the ex vivo gene transfer are baby hamster kidney (BHK) cells. Cell lines such as BHK offer several advantages including unlimited availability, the ease of screening in vitro for the presence of pathogens, and suitability for gene transfer using recombinant DNA techniques. Transplantation of cell lines may, however, risk the formation of tumors. This complication may be prevented by the transplantation of polymer encapsulated cells (Fig. 1), i.e. surrounding cells with a permselective membrane which sequesters the transplanted cells but allows diffusion of small molecules such as nutrients and trophic factors into and out of the polymer envelope (Aebischer et al., 1991a). Transplantation of a cell line within polymer capsules offers the advantage that the surrounding membrane prevents rejection of the transplanted cells by the host immune system if the capsule wall is endowed with an appropriate

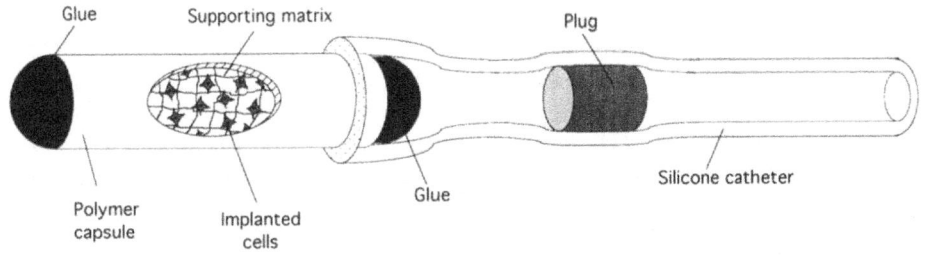

Figure 1. Schematic drawing of the polymer capsule used for intrathecal implantation of bovine chromaffin cells in humans.

molecular weight cut-off. Taking advantage of this immuno-isolation, the use of a cell line of xenogeneic origin provides additional advantages since the transplanted cells should be rejected by the host immune system in the case of a breakage of the capsule's wall. Other advantages of the transplantation of encapsulated cellular grafts are that it permits monitored and stabilized output of the desired cell product and retrieval of the transplanted cells in the event of side effects of CNTF administration.

The encapsulation system has been extensively studied both in rats and in primates for the transplantation of a dopaminergic cell line for the potential treatment of Parkinson's disease (Aebischer et al., 1991b, 1994; Winn et al., 1991; Tresco et al., 1992). Recently, several groups have studied the transplantation of encapsulated cells genetically engineered to release nerve growth factor (NGF) in a rat model of Alzheimer's disease (Hoffman et al., 1993; Winn et al., 1994). We have now applied this technology to humans by transplanting bovine chromaffin cells in terminal cancer patients suffering from intractable pain (Aebischer et al., submitted). The patients were implanted with a subarachnoid lumbar device where the tip of the implant is constituted by the permselective cylindrical membrane. The capsule was then mounted on a silicone catheter. Only the capsule part and a small segment of the silicone tubing was placed intrathecally. The remainder of the silicone catheter was fixed subcutaneously. The skin was closed to prevent any entry port for infection.

To date nine patients have been implanted with such devices (one patient implanted twice). Postoperative recovery was uneventful except in 3 patients who experienced postpunctural headaches. Explanted devices showed no host reaction either macro- or microscopically. All but one of the retrieved devices contained viable chromaffin cells throughout the capsule length and where analyzable released significant amounts of catecholamines. Some apparent analgesic effect was observed in 7 out of 9 analyzable patients (Aebischer et al., 1994).

A similar device would be used in ALS patients with the major difference revolving around the introduction of genetically engineered cells releasing CNTF.

Vector

The human CNTF gene was inserted into a dihydrofolate reductase (DHFR)-based expression vector designated pNUT (Baetge et al., 1986), which contains the entire pUC 18 sequence including the polylinker. The transcription of the cDNA encoding the mutant form of DHFR is driven by the SV40 promoter. The 3' end is fused with the 3' end of the hepatitis B virus gene (HBV 3') to ensure efficient polyadenylation and maturation signals.

The hCNTF gene was obtained by PCR amplification of the human DNA. The primers used contained EcoRI site at the position of the natural hCNTF initiation codon. The hCNTF gene was fused at its 5' extremity to 150 bp sequence coming from the mouse immunoglobulin (Ig) gene. The EcoRI site was used in such a way that the amino terminal part of the hCNTF protein corresponds to the first 18 amino acids of the Ig gene. A 325 bp hGH AvaI fragment containing the polyadenylation sequence and other sequences important for maturation of the mRNA was cloned at the 3' extremity of the hCNTF gene. Briefly, this fragment was introduced in the SpeI site of the Bluescript polylinker creating a BamHI and Not I site at the 5' and 3' end, respectively. Then the BamHI site was ligated to the BglII site engineered at the 3' end hCNTF.

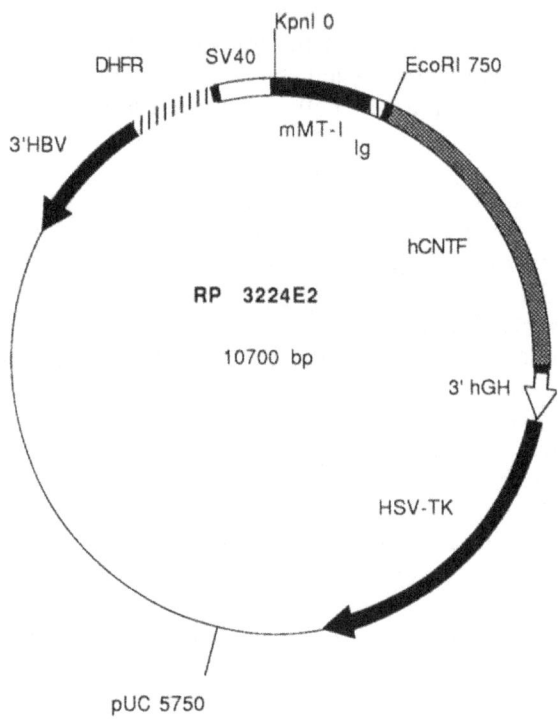

Figure 2. Map of the RP3224E2 vector.

This construction was inserted at the position +6 of the mouse MT-I promoter and the entire 3050 bp MT/Ig/hCNTF/hGH KpnI- NotI fragment was inserted in the KpnI-NotI site of the pNUT vector. Finally the HSV-TK gene was cloned in the NotI site of the vector, thus separating it from the DHFR gene by the entire pUC-18 plasmid. This final construction is named RP3224E2 (Fig. 2).

Encapsulation

The encapsulation procedure is based on the following technique. Hollow fibers are spun from a poly(acrylonitrile vinylchloride) solution by a spinning technique (Aebischer et al., 1991c). The resulting hollow fiber has an outside

278

diameter of less than 900 μm and a wall thickness of about 100 μm. The molecular weight cut-off is around 100,000 Daltons. The fibers are impregnated with glycerin for storage purposes. In order to make implantable capsules, lengths of fiber are first cut into 5 cm long segments and the distal extremity of each segment is adhesively sealed. Following sterilization with ethylene oxide and outgassing, the fiber segments are loaded with a suspension of 2×10^5 transfected cells/μl of collagen solution (Zyderm®) via a Hamilton syringe and a 25 gauge needle. The proximal end of the capsule is then sealed. A silicone tether (Speciality Silicone Fabrication) (OD: 1.25 mm) is placed over the proximal end of the fiber allowing handy manipulation and retrieval of the device.

EXPERIMENTAL RESULTS

Cell lines

Various cell lines releasing human CNTF have been created using the RP3224E2 plasmid. The genomic clone for the human CNTF has been separately subcloned into the dihydrofolate reductase (DHFR)-based expression vector system as described above and transfected by calcium/phosphate precipitation into BHK cells. The transfected cells were then exposed to methotrexate selection over 8 weeks to produce stable clones. Following this selection, the BHK engineered cells have been maintained in vitro without drug selection for several months. Clones have been selected and have shown no loss of growth factor expression as assessed by both bioassays and ELISA.

The constitutive release of the various cell lines created is in the order of $1ng/10^3$ cells/hour. The bioactivity of polymer encapsulated CNTF engineered cells was tested in vitro on rat E14 embryonic motoneurons and on E8 chick ciliary ganglion neuronal cultures. The motoneuron cultures showed an increase in choline acetyltransferase (ChAT) activity upon exposure to capsules containing CNTF engineered cells as compared to the capsules containing the BHK parent line (Fig. 3). An increase in the number of ciliary ganglion neurons extending neurites in cultures containing the encapsulated CNTF engineered cells but not the BHK parent line was also observed (Fig. 4).

Animal models

The in vivo bioactivity of the encapsulated cell system was tested in various models. In the first model, encapsulated CNTF producing cells prevented the degeneration of motoneurons after facial nerve axotomy in post-natal day 2 rats as compared to capsules containing the BHK parent line (Tan et al., submitted). The same observation was made with the transection of sciatic nerve in post-natal day 4 rats (Table 1).

In further experiments, the motoneurons were labeled with an in situ hybridization probe for choline acetyltransferase to demonstrate that the counted surviving cells were indeed motoneurons (Fig. 5).

In current experiments, we are testing the efficacy of implanting encapsulated neurotrophic factor releasing cells in the pmn/pmn mutant mouse model. These mice are characterized by an autosomal mutation expressed as a progressive motoneuronopathy resembling ALS. The homozygotes develop a weakness of the hind limbs beginning in the third week of life and rapidly progressing rostrally, leading to death during the seventh week of life (Sagot et al., submitted). Mice that received CNTF releasing capsules at the onset of the illness

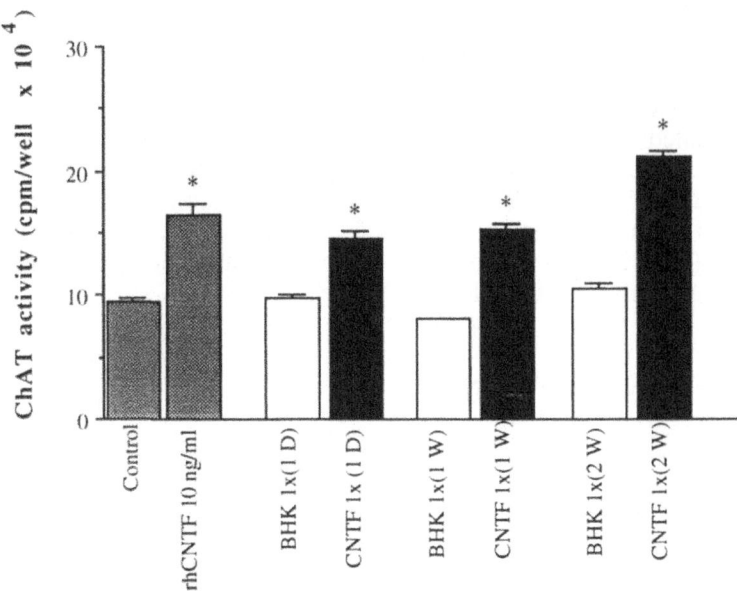

Figure 3. Representative graph of ChAT activity from embryonic spinal cord motoneuron cultures. ChAT activity is expressed as cpm of [^{3}H] acetylcholine per culture formed in 20 minutes. The culture exposed to capsules received one capsule per well. Capsules were kept in vitro for 1 day (1 D), 1 week (1 W) and 2 weeks (2 W) before the bioassay. The values show a mean activity obtained from two wells and error bars indicate SEM of cpm counts. Statistical analysis (paired Student's t test) was done against control well that did not receive any trophic factor (* $p<0.02$).

show longer survival times than mice receiving empty capsules or capsules loaded with the parent BHK line (Sagot et al., submitted). The CNTF mice also demonstrated improved motor behaviour compared to control mice. Preliminary results appear to demonstrate that a combination of cells releasing CNTF, BDNF and NT-3 produced even longer survival than any single factor (Sagot et al., submitted).

To insure that no adverse toxic effects will occur in ALS patients, a series of experiments have been performed. For example, six sheep have received one subarachnoid implant containing 6×10^6 CNTF transfected BHK cells over a 4 week period. No systemic toxicity, such as, fever, weight loss or neurological symptoms were observed in these animals. Fever and weight loss were seen in experimental animals and humans overdosed with soluble CNTF.

SAFETY ISSUES FOR A HUMAN CLINICAL TRIAL

Before initiating a human gene therapy trial for ALS using the encapsulation technology, the safety issue must be addressed. From a theoretical point of view, the safety of encapsulated and transfected cells is believed to be suitable for the following reasons: 1) the vector is stably integrated in the cell line, thus avoiding independent replication events or transmission of a vector; 2) the semipermeable membrane surrounding the transplanted cells prevents any significant genetic exchange between the transfected cells and the host human cells; 3) in case of capsule breakage, the cells should be rejected by the host immune system; 4) furthermore, should the capsule break, the insertion of the Herpes simplex virus thymidine kinase (HSV-TK) gene, a suicide gene, in the RP3224E2 plasmid provides an additional mechanism to eliminate any surviving cells through the systemic administration of ganciclovir.

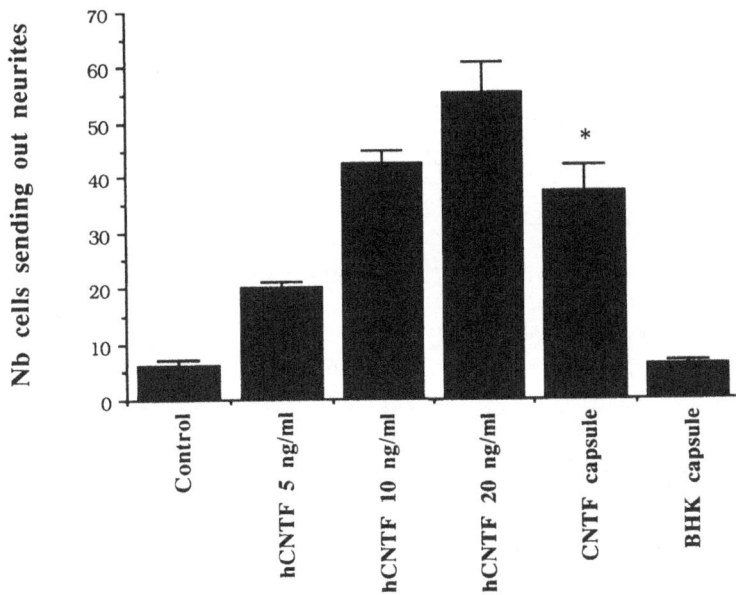

Figure 4. Representative graph of a ciliary ganglion bioassay. The number of cells sending out neurites longer than one cell diameter per visual field of 0.25 mm^2 were counted 24 hours post-plating. Control wells did not receive any trophic factor. Recombinant human CNTF (5, 10 and 20 ng/ml) was used as positive control. Average number from ten visual fields was taken with error bars representing SEM. Statistical analysis (paired Student's *t* test) was done against control well that did not receive any trophic factor (* $p < 0.001$).

To insure that the unencapsulated BHK cells are indeed rejected, the following experiments were performed. Unencapsulated BHK cells were implanted in both the striatum and ventricle of rats (2×10^5 cells per injection site). The animals were sacrificed after 1 or 3 months. Neurohistopathological examination of the animals was performed using vimentin and glial fibrillary acidic protein (GFAP). BHK cells are vimentin positive and GFAP negative. Astrocytes are vimentin negative and GFAP positive. Double labelling analysis at 1 month revealed few or no remaining BHK cells at the injection site. No BHK cells could be identified at 3 months.

CONCLUSION

This paper provides the rational basis for a gene therapy approach to ALS using the encapsulation technology. Both our data and theoretical considerations suggest that encapsulated CNTF cells could be efficacious. Preliminary experiments in the wobbler mouse model (Mitsumoto et al., 1993) suggest that it might be advantageous to release a cocktail of trophic factors (Sagot et al., submitted). It seems however prudent to start the trials with the release of a single trophic factor. Other trials might not be limited to trophic factors alone. For example, abnormal enzyme activity may also be involved in ALS. A mutation in the

Table 1. Number of motoneurons counted from the ventral horn of rats whose sciatic nerve was unilaterally transected at postnatal day 4.

Treatment	Number of Neurons		
	Lesioned	Non-lesioned	% survival†
Cresyl-violet staining: 1 week survival:			
BHK treatment (n=6)	1791±195	3183±207	56.2±4.7
CNTF-BHK treatment (n=7)	2547±600	3333±651	75.9±1.8*
Cresyl-violet staining: 2 weeks survival:			
BHK treatment (n=9)	1121±163	2083±330	55.0±4.2
CNTF-BHK treatment (n=6)	1217±146	1739±195	69.5±1.9**
ChAT in situ hybridization staining: 1 week survival:			
BHK treatment (n=5)	1025±173	1883±255	53.8±2.4
CNTF-BHK treatment	1439±143	2255±214	63.7±0.9#

† The percent survival represents the number of motoneurons on the lesioned side compared to the number on the contralateral side. Statistical analysis was done using a paired Student's t test (* $p<0.002$, ** $p<0.02$, # $p<0.005$).

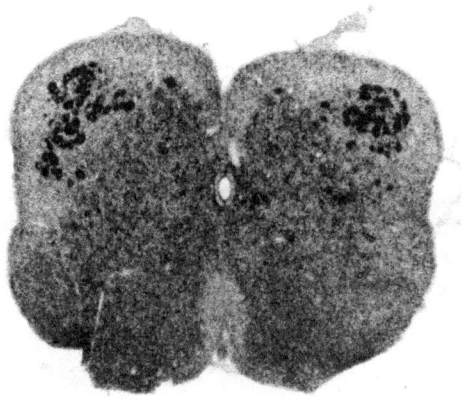

Figure 5. Micrograph of a rat spinal cord cross-section implanted with two capsules containing BHK parent cells stained using a digoxigenin in situ hybridization technique for ChAT. The motoneurons are visible in the ventral area of the spinal cord. The non-lesioned side is marked by the knife mark on the dorsal side of the spinal cord and shows a significantly higher number of stained motoneurons compared to the lesioned side. Scale bar = 250 μm.

superoxide dismutase gene has been recently identified in individuals suffering from its familial form (Rosen et al., 1993). We look forward to the initial trial results that step by step could provide a significant new approach to this deadly disease.

REFERENCES

Aebischer P Winn SR Tresco PA Greene LA and Jaeger CB (1991a): Transplantation of polymer encapsulated neurotransmitter secreting cells: effect of the encapsulation technique. J Biomech Eng 113: 178-183.

Aebischer P Winn SR Tresco PA Greene LA and Jaeger CB (1991b): Long term cross-species brain transplantation of a polymer encapsulated dopamine-secreting cell line. Exp Neurol. 3: 269-275.

Aebischer P Wahlberg L Tresco PA and Winn SR (1991c): Macroencapsulation of dopamine-secreting cells by coextrusion with an organic polymer solution. Biomaterials 12: 50-56.

Aebischer P Goddard M Signore A and Timpson R (1994): Functional recovery in MPTP lesioned primates transplanted with polymer encapsulated PC12 cells. Exp Neurol 126: 1-8.

Aebischer P Buchser E Joseph JM Heyd B Padrun V Zurn A Favre J de Tribolet N and Goddard M (1994): Xenotransplantation of encapsulated bovine chromaffin cells for the treatment of intractable pain: a phase I clinical study. Soc Neurosci Abst 696.10.

Baetge EE Suh YH and Joh TH (1986): Complete nucleotide and deduced amino acid sequence of bovine phenylethanolamine N-methyltransferase: partial amino acid homology with rat tyrosine hydroxylase. Proc Natl Acad Sci USA 83: 5454-5458.

Barinaga M (1994): Neurotrophic factors enter the clinic. Science 264: 772-774.

Brooks BA Sanjak M Mitsumoto H Szirony K Neville H Ringel S Brinkmann J Pestronk A Florence J Cedarbaum JM Charatan M Stampier N Wittes J and Brittain E (1993): Recombinant human ciliary neurotrophic factor (rhCNTF) in amyotrophic lateral sclerosis (ALS) patients: Dose selection strategy in phase I-II safety, tolerability and pharmacokinetic studies. Can J Neurol Sci 20: 5-83.

Dittrich F Thoenen H and Sendtner M (1994): Ciliary neurotrophic factor: pharmacokinetics and acute-phase response in rat. Ann Neurol 35: 151-163.

Henderson CE Camu W Mettling C Gouin A Poulsen K Karihaloo M Rullamas J Evans T McMahon SB Armanini MP Berkemeier L Phillips HS and Rosenthal A (1993): Neurotrophins promote motor neuron survival and are present in embryonic limb bud. Nature 363: 266-270.

Hoffman D Breakfield XO Short P Greene LA and Aebischer P (1993): Transplantation of polymer encapsulated cells genetically engineered to release NGF. Exp Neurol 122: 100-106.

Hughes RA Sendtner M and Thoenen H (1993): Members of several gene families influence survival of rat motoneurons *in vitro* and *in vivo*. J Neurosci Res 36: 663-671.

Kato AC and Lindsay RM: Overlapping and additive effects of neurotrophins and CNTF on cultured human spinal cord neurons. Submitted.

Magal E Brunham P and Varon S (1991): Effects of ciliary neurotrophic factor on rat spinal cord neuron *in vitro*: survival and expression of choline acetyltransferase and low-affinity nerve growth factor receptors. Devel Brain Res 63: 141-150.

Mitsumoto H Ikeda K Wong V Cedarbaum JM and Lindsay RM (1993): Ciliary neurotrophic factor (CNTF) and Brain-derived neurotrophic factor (BDNF) co-administration arrests loss of motor function in wobbler mice. Soc Neurosci Abstr 19: 83.11.

Munsat TL Andres PL Finison L Conlon T and Tibodeau L (1988): The natural history of motoneuron loss in amyotrophic lateral sclerosis. Neurol 38: 409-413.

Rosen DR Siddique T Patterson D Figlewicz DA Sapp P Hentati A Donaldson D Goto J et al. (1993): Mutation in Cu/Zn superoxide dismutase gene are associated with familial amyotrophic lateral sclerosis. Nature 362: 59-62.

Sagot Y Tan SA Baetge EE Schmalbruch H Kato AC and Aebischer P: Polymer encapsulated cell lines secreting neurotrophic factors can rescue motoneurons in an animal model of motoneuron disease. Submitted.

Sendtner M Kreutzberg GW and Thoenen H (1990): Ciliary neurotrophic factor prevents the degeneration of motor neurons after axotomy. Nature 345: 440-441.

Sendtner M Holtmann B Kolbeck R Thoenen H and Barde YA (1992a): Brain-derived neurotrophic factor prevents the death of motoneurons in newborn rats after nerve section. Nature 360: 757-759.

Sendtner M Schmalbruch H Stöckli KA Carroll P Kreutzberg GW and Thoenen H (1992b): Ciliary neurotrophic factor prevents degeneration of motor neurons in mouse mutant progressive motor neuronopathy. Nature 358: 502-504.

Tan SA Zurn AD Padrun P Baetge EE Hammang J Kato AC and Aebischer P: Rescue of motoneurons from axotomy induced cell death from polymer encapsulated cells genetically engineered to release ciliary neurotrophic factor. Submitted.

Tresco PA Winn SR Tan S Jaeger CB Greene LA and Aebischer P (1992): Polymer-encapsulated PC12 cells: long-term survival and associated reduction in lesion-induced rotational behavior. Cell Transpl 1: 255-264.

Winn SR Tresco PA Zielinski B Greene LA Jaeger CB and Aebischer P (1991): Behavioral recovery following intrastriatal implantation of microencapsulated PC12 cells. Exp. Neurol 113: 322-329.

Winn SR Hammang JP Emerich DF Lee A Palmiter RD and Baetge EE (1994): Polymer-encapsulated cells genetically modified to secrete human nerve growth factor promote the survival of axotomized septal cholinergic neurons. Proc Natl Acad Sci 91: 2324-2328.

Yan Q Elliott J and Snider WD (1992): Brain-derived neurotrophic factor rescues spinal motor neurons from axotomy-induced cell death. Nature 360: 753-755.

MODULATION OF CELL DEATH IN NEURAL CELL LINES AND TRANSPLANTS

Dale E. Bredesen

Program in Aging, La Jolla Cancer Research
Foundation, 10901 N. Torrey Pines, La Jolla, CA 92037,
U.S.A.

INTRODUCTION

The majority of transplanted neural cells fail to survive for more than a few days or weeks, and in many transplants, 90 to 98% of transplanted cells undergo cell death within this time period. Therefore, the study of the mechanisms controlling neural cell death is of relevance to the optimization of neural transplants. The predominant mode of neural cell death following trophic factor withdrawal is apoptosis. Gene products controlling neural apoptosis in both a positive and negative way have been discovered and characterized. Some of these gene products also control non-apoptotic cell death; therefore, an investigation of the role(s) of pro-apoptotic genes in limiting transplant survival is warranted. In addition, an evaluation of the effects of expressing anti-apoptotic genes in limiting transplant survival is also warranted. An evaluation of the effects of expressing anti-apoptotic genes in neural transplants is, thus, appropriate, and preliminary results in this area have been obtained.

Neural transplantation holds great promise for treating Parkinson's disease and other neurological disorders. However, with typical rates of neural cell death following transplantation being in the range of 90 to 95%, it has been estimated that approximately 20 fetuses would be required to replace the normal number of nigrostriatal neurons lost in each Parkinson's disease patient (Björklund, 1992). Such neural cell death often occurs in the absence of signs of transplant rejection, suggesting that it is not immunologically mediated. The enhanced survival of grafted neural cells reported in some studies in the presence of secreted growth factors suggests that the cell death in neural transplants may be apoptotic, in at least some cases. Therefore, it is appropriate to consider the emerging picture of the genetic control of neural cell death.

Neural Cell Specification: Molecular Mechanisms and Neurotherapeutic Implications
Edited by Juurlink *et al.*, Plenum Press, New York, 1995

285

APOPTOSIS

In 1951, Glücksmann pioneered morphological studies of cell death. In 1964, Lockshin, studying insect morphogenesis, proposed the term *programmed cell death* for the reproducibly timed cell deaths occurring during insect development (Lockshin, 1964). In 1966, this process was shown to require protein synthesis (Tata, 1966). In 1972, Kerr, Wyllie and Currie coined the term apoptosis, from the Greek *apo = away from* and *ptosis = falling*. Kerr and colleagues noted this phenomenon in both physiological and pathological states, and suggested that the underlying process was similar in the two situations. They also proposed that the process may act in conjunction with mitosis to regulate cell numbers and therefore maintain organ structure.

Apoptosis is marked by the early formation of plasma membrane blebs, which may be related to alterations in cytoskeletal proteins (Mirabelli et al., 1988). Chromatin condensation occurs, as well as margination of chromatin with subsequent fragmentation of the nucleus. The cell displays movements that are reminiscent of boiling, with subsequent budding of apoptotic bodies, which may or may not contain fragments of nucleus. The membrane remains intact until the completion of apoptosis, at which time, if the cell is in cell culture, secondary necrosis with cellular lysis may appear. In vivo, however, prior to the completion of apoptosis, phagocytosis of the apoptotic cell occurs, typically by nearby cells that are often of the same cell type rather than being of the macrophage lineage. Signaling for phagocytosis occurs through a change in the asymmetric placement of phosphatidyl serine in the plasma membrane (Fakok et al., 1992). Organelles remain intact, at least morphologically, in contrast to the picture in necrosis.

At the biochemical level, internucleosomal fragmentation of the DNA occurs, leading to the appearance of a characteristic "apoptotic ladder" of DNA fragments in multiples of approximately 180 base pairs (Wyllie, 1980). Specific genes are up-regulated or down-regulated, depending to some extent on tissue type, but with a degree of conservation. For example, sulfated glycoprotein-2 (SGP-2) is up-regulated in prostate cells undergoing apoptosis as well as neural cells undergoing apoptosis. Expression of SGP-2 has also been shown to be increased in Alzehimer's disease (Duguid et al., 1989). The function of SGP-2 is unknown. Tissue transglutaminase is also up-regulated, and it has been suggested that this is responsible for cross-linking of proteins, decreasing cellular fragility and therefore decreasing the likelihood of lysis of cells undergoing apoptosis (Kerr and Harmon, 1991). RP-2 and RP-8 have also been shown to be up-regulated (Owens et al., 1991). The functions of these genes are unknown.

Genes controlling the death of cells within the mammalian nervous system are being defined. Both pro-apoptotic and anti-apoptotic genes have been described, and it is envisioned that these genes have opposite effects on a cellular "apostat" that determines the propensity of each cell to undergo apoptosis. The low affinity receptor for nerve growth factor (p75^NGFR) has been shown to play a role in apoptosis in at least some neural systems (Rabizadeh et al., 1993; Barrett and Bartlett, 1994). p75^NGFR is a member of the tumor necrosis factor receptor superfamily, which contains at least 12 members (Smith et al., 1994). Interestingly, whereas at least two members of this family - TNFR-I and FAS/Apo-I - induce apoptosis following ligand binding (Trauth et al., 1989; Itoh et al., 1991; Wong et al., 1991), p75^NGFR and a related superfamily member, CD40, induce apoptosis constitutively, with apoptosis being inhibited by binding of ligand or agonistic monoclonal antibody (Rabizadeh *et al.,* 1993; Ruan et al., submitted for publication). Expression of p75^NGFR in neural cell lines produced a dichotomous effect: the absence of nerve growth factor resulted in enhanced apoptosis, whereas the presence of nerve growth factor resulted in increased survival over the cells not expressing p75^{NGFR} (Rabizadeh et al., 1993). A similar dichotomous response was noted in primary cultures of dorsal root ganglion neurons: between embryonic day 19 (E19) and postnatal day 4 (P4), antisense inhibition of p75^{NGFR} expression resulted in resistance to apoptosis following the withdrawal of nerve growth factor support; however, at E12 to E13, antisense inhibition of p75^{NGFR} expression resulted in a lack of response to nerve growth factor (Barrett and Bartlett, 1994). Additional evidence for the role of p75^{NGFR} in apoptosis is provided by the derivation of PC12 mutants; those retaining expression of p75 retained apoptosis but those that lost p75 expression lost the potential to undergo apoptosis following serum and growth factor withdrawal (Rabizadeh et al., 1993). The role of p75^{NGFR} and other apoptotic genes in survival of neural transplants has not been explored, but results with the anti-apoptotic gene *bcl*-2 suggest that the effects of such genes should be explored further (see below). The recent discovery of a second pro-apoptotic neuronal surface determinant, this one expressed by cortical neurons, suggests that additional such determinants may exist (Zhong et al., in preparation).

A number of anti-apoptotic genes have been identified, some of which are expressed in the mammalian nervous system. *bcl*-2 was originally identified at the translocation break point of follicular B-cell lymphomas (Tsujimoto et al., 1984). Subsequently, the expression of *bcl*-2 was shown to support hematopoietic cell survival (Vaux et al., 1988), and to inhibit hematopoietic cell apoptosis following the withdrawal of interleukin-3 (Hockenbery et al., 1990). *bcl*-2 expression has been shown to inhibit apoptosis of neural cells following a wide range of insults, from serum and growth factor withdrawal to calcium ionophore addition, glucose withdrawal, addition of β-amyloid peptide, addition of membrane peroxidizing agents, and addition of free radical-inducing agents (Garcia et al., 1992; Mah et al., 1993; Zhong et al., 1993; Bredesen, in press). The expression of *bcl*-2 has been shown to decrease the net cellular generation of reactive oxygen species by neural cells (Kane et al., 1993), and in hematopoietic cells to inhibit damage from reactive oxygen species (Hockenbery et al., 1993). Recent results with neural cell lines demonstrate that the expression of *bcl*-2 enhances the efficacy of neural cells transplanted in an animal model of hemiparkinsonism (Anton et al., submitted for publication).

In summary, apoptosis is a common mode of cell death occurring during developmental, physiological cell death and pathological cell death in a wide range of paradigms. The apoptotic program is largely conserved across phyla and cell type. Both pro-and anti-apoptotic genes have been identified, and some progress made in understanding their mode of action. The importance of apoptosis in the loss of transplanted neural cells is unknown, but the enhancement of efficacy occurring with transplanted neural cells expressing *bcl*-2 suggests that further studies in this area are warranted.

REFERENCES

Barrett G and Bartlett P (1994): The p75 nerve growth factor receptor mediates survival or death depending on the stage of sensory neuron development. Proc Natl Acad Sci USA 91: 6501-6505.

Björklund A (1992): Brain transplant: Prospects for neuronal replacement and reconstruction of axonal pathways of the lesioned CNS. Decade of the Brain Plenary Session. American Academy of Neurology, San Diego.

Duguid JR, Bohmont CW, Liu N and Tourtellotte WW (1989): Changes in brain gene expression shared by scrapie and Alzheimer disease. Proc Natl Acad Sci USA 86: 7260-7264.

Fakok VA, Voelker DR, Campbell PA, Cohen JJ, Bratton DL and Henson PM (1992): Exposure of phosphatidylserine on the surface of apoptotic lymphocytes triggers specific recognition and removal by macrophages. J Immun 148: 2207-2216.

Garcia I, Martinou I, Tsujimoto Y and Martinou JC (1992): Prevention of programmed cell death of sympathetic neurons by the bcl-2 proto-oncogene. Science 258: 302-304.

Glücksmann A (1951): Cell deaths in normal vertebrate ontogeny. Biol Rev 26: 59-85.

Hennet T, Bertoni G, Richter C and Peterhans E (1993): Expression of bcl-2 protein enhances the survival of mouse fibrosarcoid cells in tumor necrosis factor-mediated cytotoxicity. Cancer Res 53: 1456-1460.

Hockenbery D, Oltvai ZN, Yin X, Milliman CL and Korsmeyer SJ (1993): bcl-2 functions in an antioxidant pathway to prevent apoptosis. Cell 75: 241-251.

Itoh N, Yonehara S, Ishii A, Yonehara M, Mizushima S, Sameshima M, Hase A, Seto Y and Nagata S (1991): The polypeptide encoded by the cDNA for human cell surface antigen Fas can mediate apoptosis. Cell 66: 233-243.

Kane DJ, Sarafian TA, Anton R, Hahn H, Butler-Gralla E, Selverstone-Valentine J, Ord T and Bredesen DE (1993): bcl-2 inhibition of neural death: Decreased generation of reactive oxygen species. Science 262: 1274-1277.

Kerr JFR and Harmon BV (1991): Definition and incidence of apoptosis: An historical perspective. In Tomei LDE and Cope FO (eds): "Apoptosis: The molecular basis of cell death." Plainview, New York: Cold Spring Harbor Laboratory Press, pp 5-29.

Lockshin RA and Williams CM (1964): Programmed cell death. II. Endocrine potentiation of the breakdown of the intersegmental muscles of silkmoths. J Insect Physiol 10: 643.

Mah SP, Zhong LT, Liu Y, Roghani A, Edwards RH and Bredesen DE (1993): The protooncogene bcl-2 inhibits apoptosis in PC12 cells. J Neurochem 60: 1183-1186.

Mirabelli F, Salis A, Marinoni V, Finardi G, Bellomo G, Thor H and Orrenius S (1988): Menadione-induced bleb formation in hepatocytes is associated with the oxidation of thiol groups in actin. Arch Biochem and Biophysics 264: 261-269.

Owens GP, Hahn WE and Cohen JJ (1991): Identification of mRNAs associated with programmed cell death in immature thymocytes. Mol Cell Biol 11: 4177-4188.

Rabizadeh S, Oh J, Zhong L, Yang J, Bitler C, Butcher L and Bredesen D (1993): Induction of apoptosis by the low-affinity NGF receptor. Science 261: 345-348.

Ruan Y, Camerini D and Bredesen D: Expression of CD40 induces neural apoptosis. (submitted for publication)

Smith C, Farrah T and Goodwin R (1994): The TNF receptor superfamily of cellular and viral proteins: Activation, costimulation, and death. Cell 76: 959-962.

Tata JR (1966): Requirement for RNA and protein synthesis for induced regression of tadpole tail in organ culture. Dev Biol 13: 77.

Trauth BC, Klas C, Peters AM, Matzku S, Moller P, Falk W, Debatin KM and Krammer PH (1989): Monoclonal antibody-mediated tumor regression by induction of apoptosis. Science 245: 301-305.

Tsujimoto Y, Finger L, Yunis J, Nowell P and Croce C (1984): Cloning of the chromosome breakpoint of neoplastic B cells with the t(14;18) chromosome translocation. Science 226: 1097-1099.

Vaux DL, Cory S and Adams JM (1988): bcl-2 gene promotes haemopoietic cell survival and co-operates with c-myc to immortalise pre-B cells. Nature 335: 440-442.

Wong G, McHugh T, Weber R and Goeddel D (1991): Tumor necrosis factor alpha selectively sensitizes human immunodeficiency virus-infected cells to heat and radiation. Proc Natl Acad Sci USA 88: 4372.

Wyllie AH (1980): Glucocorticoid-induced thymocyte apoptosis is associated with endogenous endonuclease activation. Nature 284: 555.

Zhong L, Kane D and Bredesen D (1993): bcl-2 blocks glutamate toxicity in neural cell lines. Molecular Brain Research 19: 353-355.

ABSTRACTS

LOSS OF DLX HOMEOBOX GENE EXPRESSION IN ZEBRAFISH EMBRYOS TREATED WITH RETINOIC ACID. M. Ekker. Loeb Institute and Departments of Medicine and Anatomy & Neurobiology, University of Ottawa, Ottawa, ON, K1Y 4E9, CANADA

Three zebrafish homeobox genes of the *dlx* family, *dlx2*, *dlx3*, and *dlx4*, participate in the development of the forebrain, the branchial arches, the ears and the fins. At the present time, nothing is known about the regulation of the *dlx* genes in vertebrates. In order to determine if retinoic acid (RA), a known teratogen, can be a factor that influences *dlx* gene expression, we treated zebrafish embryos for 2 hr with various doses of all-trans RA administered at three different developmental stages, 6-8 hr, 8-10 hr, and 10-12 hr. *Dlx* gene expression was determined by in situ hybridization on whole-mount embryos. At 10^{-6} M RA, expression of all three *dlx* genes is abolished everywhere except in the median fin fold, a region of the embryo whose development does not seem to be perturbed by RA treatment. The *dlx3* pattern of expression in gastrulating embryos is only lost in embryos that received 10^{-6} M RA between 6-8h. At 10^{-7} M RA, expression of the three genes is still abolished in the arches, *dlx2* is almost completely lost in migrating neural crest cells and we observe abnormal patterns of expression of *dlx3* and *dlx4* in the inner ear and of *dlx2* and *dlx4* in the migrating crest cells of embryos that received RA between 6 and 8 hr. When more than one *dlx* gene is expressed in a given structure, they seem to be equally affected by RA treatment. These observations suggest that treatment with RA leads, either directly or indirectly, to a tissue-specific loss of expression of three *dlx* genes and that the three genes respond similarly to the same factor or, alternatively, cross-regulate one another. Supported by the MRC, the NCI and NSERC.

EGF-RESPONSIVE FETAL PROGENITOR CELLS IN CULTURE AND IN GRAFT. Y.H. Chiang and F.C. Zhou. Program in Medical Neurobiology and the Department of Anatomy, Indiana University School of Medicine, Indianapolis, IN, 46202, U.S.A.

Multipotential fetal progenitor cells responsive to epidermal growth factor (EGF) were screened from embryonic mouse brains. The potential of these cells to differentiate into neurons has been tested in vitro (Reynolds et al., 1992), but not in vivo. The aim of this study was to examine the phenotypic expression of these progenitor cells after being transferred to laminin-coated non-EGF culture or after transplantation into adult mouse brain. Three types of fetal mouse brain tissues - brainstem, striatum and cortex - were isolated from E14-15 mouse brain. The cells were dissociated and plated ($50,000/cm^2$) in 25 cm^2, non-coated tissue culture flasks. DMEM/F12 medium supplemented with N2 and EGF (20 ng/ml) was used to isolate progenitor cells which were screened between 14 to 96 days. Floating neurospheres (~4,000 cells/sphere) were obtained for culture assay or for transplantation. Immunocytochemistry shows that glutamic acid decarboylase (GAD) positive GABAergic neurons, 5-HT 1a receptor positive neurons, 5-HT-positive neurons and GFAP-positive glial cells were present in these cultures. The cell proliferating marker, PCNA, indicated that a large number of cells are capable of dividing. In one month old brainstem grafts in striatum or nigra, a small number of 5-HT and TH immuno-reactive cells were found.

EXPRESSION OF *WNT8B* IN THE DEVELOPING ZEBRAFISH HINDBRAIN.
G.M. Kelly and R.T. Moon. Department of Pharmacology, SJ-30, University of Washington, Seattle, Washington 98195, U.S.A.

The specification of the vertebrate body plan is dependent on numerous signaling molecules including members of the *Wnt* family. Most *Wnt* molecules are expressed in discrete, yet overlapping regions in the developing central nervous system (CNS). To better understand the role of the *Wnt* molecules in patterning the zebrafish CNS we have used a polymerase chain reaction strategy to identify *Wnt* genes expressed during embryogenesis. We have identified one *Wnt* (*wnt8b*) which, based on its highly restricted spatial pattern of expression, is a likely candidate involved in patterning the rostral CNS. By late gastrulation, *wnt8b* is spatially restricted in two stripes in the prospective rostral neuroepithelium. Concurrent with the ongoing mediolateral shortening of the embryonic shield the *wnt8b* signals fused over the dorsal midline, in the region of the presumptive midbrain-hindbrain boundary (MHB). During the segmentation period of development, *wnt8b* signals were evident at the boundaries between the forebrain and midbrain, and the midbrain and hindbrain. *wnt8b* staining at the MHB overlapped with *wnt1*, the vertebrate orthologue of the *Drosophila* segment polarity gene *wingless*. However, despite the significant overlap there was not a direct correlation between the two *Wnts* since the *wnt8b* signal was more broad in the dorsal region and it did not extend ventrally as far as the *wnt1* signal. Three broad stripes of *wnt8b* expression were also present in the hindbrain. Interestingly, the zebrafish *krox20* gene, which encodes a zinc finger protein, co-localized with *wnt8b* in rhombomeres 3 and 5. In late stages, *wnt8b* was readily detected in the rostral-most region of the forebrain. The spatial expression pattern of *wnt8b* would suggest that this signaling molecule is involved not only in patterning specific regions in the forebrain and midbrain, but also in the rhombomere organization in the hindbrain.

REGIONAL DIFFERENCE IN THE CEREBRAL CORTEX OF RAT EMBRYOS IN THE POTENTIAL TO PRODUCE LATEXIN-EXPRESSING NEURONS. K. Takiguchi-Hayashi, M. Ishida and Y. Arimatsu. Mitsubishi Kasei Institute of Life Sciences, 11 Minamiooya, Machida-shi, Tokoyo 194, JAPAN.

A subset of neurons in the infragranular layers of the lateral, but not dorsal, cerebral cortex of adult rat expresses a 29 kD protein named latexin, which is recognized by a monoclonal antibody PC3.1. To clarify the timing of the cortical regional specification, we examined the potential for latexin expression of cortical anlage at various developmental stages using organotypic slice cultures. When lateral portions of E12 cortices were cultured a substantial number of neurons developed latexin expression. In contrast, many fewer neurons developed expression of latexin in cultures derived from dorsal cortices. These results indicate that some specification events have already occurred within the cortex by E12. To elucidate whether this regional difference is due to the difference of potential of neuronal precursor cells for latexin expression, cells from lateral or dorsal portions of E13 cortex, which had been labeled with BrdU, were dissociated and cultured with an excess of non-labeled cells from either lateral or dorsal cortex. Substantial and similar proportions of the labeled lateral neurons developed latexin expression regardless of the culture conditions. In contrast, a much lower proportion of labeled dorsal neurons expressed latexin when cultured with non-labeled cells from either cortex. These results suggest that neuronal precursor cells are regionally specified as to the potential for latexin expression at least by E13. Whether latexin-expressing neurons are clonally related or not is a very important question and studies on this point are now in progress.

CEK-2, A RECEPTOR FOR FIBROBLAST GROWTH FACTORS IS REGIONALLY EXPRESSED IN THE DEVELOPING CHICK BRAIN. J.M. Richman*, S. Gubbels and J. Schwartz. Department of Preventive Dental Science, University of Manitoba, Winnipeg, MN, R3E OW2, CANADA

*Present Address: Department of Clinical Dental Sciences, Faculty of Dentistry, University of British Columbia, 2199 Wesbrook Mall, Vancouver, BC, V6T 1Z3, CANADA

Fibroblast growth factors (FGF) are important in several early embryonic events such as induction of mesoderm and axis development. Four high affinity receptors for FGFs have been cloned. In this study we have mapped the expression of chicken embryo kinase-2 (Cek-2, related to FGFR-3). ^{35}S-riboprobes were made from a portion of the cDNA coding for the extracellular domain of Cek-2. Sections of chick embryos from stage 9 to 31 (Hamburger and Hamilton, 1951) were hybridized to cRNA probes and dipped in photographic emulsion. Adjacent control sections were hybridized to mRNA probes for type II collagen. Stage 9 (7 somites) and stage 10 (10 somites) embryos had high levels of Cek-2 transcripts in the neural epithelium of the mesencephalon. At stages 15, 20, and 24, expression became restricted to the junction of the diencephalon and mesencephalon, and within the myelencephalon. The posterior mesencephalon and metencephalon had lower transcript levels for Cek-2. The neural tube also expressed Cek-2 in the luminal side of the ventral floor plate. At stage 28 signal was concentrated in the most superficial layer of the neural epithelium across the isthmus between the diencephalon and mesencephalon. At stage 31, signal for Cek-2 RNA is greatly down regulated. There is significant overlap between the domain of expression of Cek-2 and int-2 (FGF-3, a potential ligand for Cek-2) in the embryonic rhombencephalon. Other potential ligands such as FGF-4 are not expressed in the presumptive brain. Our data is a first step towards a clearer picture of the spatial relationships between receptors and ligands for the FGF family in the early embryo. Funded by MRC grant MT-11618. JMR is a Dental Clinician Scientist.

NEUROBLASTOMA CELLS DIFFERENTIATION SEEMS TO MAKE THEM COMPETENT FOR THE APOPTOTIC PATHWAY. I.I. Kruman, M.A. Kostenko, R.Ya. Gordon, V.I. Popov. and S.R. Umansky. Institute of Theoretical and Experimental Biophysics, Russian Academy of Sciences, 142292, Moscow Region, RUSSIA.

Using fluorescent and electron microscopy, DNA electrophoresis and differentiation analysis we have shown that different nonspecific factors (γ-irradiation, dimethylsulphoxide [DMSO], serum deprivation, hypotonic shock) are able to induce the differentiation and subsequent death by apoptotis of neuroblastoma N1E115 cells. Duration of both differentiation and apoptosis is dependent on the nature of stimuli used. The protein synthesis inhibitor cycloheximid does not prevent differentiation and apoptosis of neuroblastoma cells induced by DMSO and even accelerates both processes. With all treatments used differentiation of the cells preceded apoptosis. Regular replacement of the inducing medium by growth medium prevented death of the differentiated cells, at least for 30 days. Differentiation of neuroblastoma cells seems to make them competent for the apoptotic pathway but differentiation and apoptosis are not induced by the same signal. The data demonstrating that inhibition of protein synthesis does not prevent morphological differentiation and apoptosis indicate that N1E115 cells have all the protein factors necessary for expression of the differentiated phenotype and subsequent induction of apoptosis, but they need a signal, possibly first for differentiation and inhibition of proliferation and then for apoptosis of differentiated cells.

DEVELOPMENTALLY REGULATED EXPRESSION OF A PUTATIVE RECEPTOR TYROSINE KINASE GENE, *tyro*-4, INCLUDES EXPRESSION BY A SUBSET OF MOTOR NEURONS. T.J. Kilpatrick, C. Lai, M. Gassmann and G.E. Lemke. Molecular Neurobiology Laboratory, The Salk Institute, La Jolla, CA 92037, U.S.A.

The receptor tyrosine kinases (RTKs) represent a family of signal transduction molecules, many of which have been shown to play important roles in the regulation of developmental processes. By using PCR, our laboratory has previously identified 13 novel putative RTKs (1). One of these genes, *tyro*-4, encodes a kinase domain that exhibits significant homology to the equivalent region of members of the *eph* subfamily of RTKs, transmembrane proteins which contain cysteine-rich regions and fibronectin type III-like repeats in their extracellular domains. Northern blot analysis has shown that *tyro*-4 is predominantly expressed within the CNS, with a peak in expression at E14/17 in rat, a sharp drop at birth and a low steady-state level after P10. In situ hybridization reveals that at E12, tyro-4 is expressed in olfactory bulb, optic stalk, basal telencephalon and spinal cord, but not in the ventro-lateral region of the cord. In contrast, at both E15 and E17, *tyro*-4 is specifically expressed in ventro-lateral spinal cord. This reciprocity of expression suggests that *tyro*-4 may be initially expressed by proliferating cells but that it is subsequently expressed specifically by a subset of post-mitotic motor neurons suggesting that it may be implicated in motor neuron axonogenesis. At both E15 and E17, *tyro*-4 is also expressed in forebrain (i.e. in the striatum, thalamus and within the subeventricular zone and subplate of the cerebrum). The significance of this distribution remains to be established although, given the composition of the extracellular domain of homologous molecules, it is possible that the protein encoded by *tryo-4* subserves a cell adhesive function and that it may play a role in neuroblast migration.
1. Lai C and Lemke G (1991): Neuron 6:691-704.

POLYMER-CELL HYBRID DEVICE FOR NEURAL TISSUE REPLACEMENT. S. Woerly[1], G.W. Plant[2] and A.R. Harvey[2]. [1]Québec Biomaterials Institute, Hôpital Saint-Francois d' Assise, Québec, G1L 3L5 CANADA, and [2]Department of Anatomy and Human Biology, The University of Western Australia, Nedlands, WA 6009, AUSTRALIA.

Cells of the developing nervous tissue were immobilized within crosslinked N-(2-hydroxypropyl methacrylamide) hydrogel matrices by gel entrapment. Schwann cells and cells of dissociated cerebral hemispheres of rat embryos were suspended in various monomer concentrations which were dissolved in the culture medium that contained collagen. Cell-based polymer hydrogels were produced by radical block copolymerization (10^6 cells/g gel). Cell viability and differentiation were studied over 6 days in vitro by immunofluorescence using antibodies to the low affinity nerve growth factor receptor, S100 and laminin for Schwann cells, GFAP for astrocytes and neurofilaments (NF) for neurons. The immobilized cells showed intact morphology and were distributed throughout the gel matrices either individually or in small aggregates. The cells, including non-myelinating Schwann cells, possessed typical phenotypic characteristics. Lamin immunoreactivity indicated that embryonic astrocytes may have been able to synthesize new proteins. NF-positive neurons gave rise to dendritic processes that could form contacts with neighbouring cells. Such cell growth behavior was dependent upon the crosslinking density and the swelling capacity of the gels. Polymer matrices with optimal characteristics and incorporating differentiated functioning cells have the potential for replacing lost neural tissue and promoting axon regeneration.

REGULATION OF SENSORY NEURON DEVELOPMENT FROM NEURAL CREST CELLS; ROLES FOR FGF, LIF AND NGF.

M. Murphy, K. Reid and P.F. Bartlett. The Walter and Eliza Hall Institute of Medical Research, Post Office, Royal Melbourne Hospital, Victoria, 3050, AUSTRALIA.

We have been studying the regulation of proliferation of neural crest cells and their subsequent development into sensory neurons. We present data suggesting that these events are regulated by discrete growth factors. Fibroblast growth factor 2 (FGF-2) stimulates the proliferation of at least 94% of cells in cultures of mouse neural crest in the presence of serum. Subsequent treatment of the cultures with leukemia inhibitory factor (LIF) results in differentiation of 50% of the cells into neurons. The neurons that develop in these cultures morphologically resemble sensory neurons and are immunoreactive for the sensory markers calcitonin gene related peptide (CGRP) and substance P. We have also analysed sensory development after the proliferative phase in cultures of embryonic dorsal root ganglia (DRG). In very early DRG (embryonic day 12, E12), primary neuronal differentiation is promoted by LIF, and survival and further maturation of the newly differentiated neurons depends on nerve growth factor (NGF). At the time of target innervation, the majority of neurons survive in response to NGF, but not LIF. From E16, the number of neurons maintained by LIF increases to >90% by birth. Thus, development of sensory neurons from their precursors may proceed under the regulation of different growth factors; FGF for proliferation, LIF or a related cytokine for differentiation and NGF at the time of target innervation. Further development or survival of sensory neurons may rely on either LIF or NGF.

TURNING CELLS INTO NERVOUS SYSTEM: THE ROLE OF THE SINGLE-MINDED TRANSCRIPTIONAL ACTIVATOR.

S.T. Crews[1], R. Franks[2], Y. Kasai[2] and K. Wharton[2]. [1]Department of Biochemistry and Biophysics, University of North Carolina, Chapel Hill, NC. 27599; [2]Department of Biology, University of California, Los Angeles, CA 90024, U.S.A.

The Drosophila embryonic ventral nerve cord consists of two symmetrical hemiganglia separated by a discrete set of midline nerve cells and glia. These CNS midline cells have a distinct developmental origin, include functional motoneurons and interneurons, and are important developmentally for the formation of the ventral epidermis, musculature, and axon commissures. We have focused on questions concerning how embryonic cells commit to the CNS midline lineage. Genetic and ectopic expression experiments have shown that the single-minded (sim) basic-helix-loop-helix-PAS (bHLH-PAS) transcription factor acts as a genetic switch; its expression in neuroectodermal cells drives those cells into the CNS midline lineage. The *sim* gene is specifically activated in the midline precursor cells just before gastrulation occurs. Genetic, biochemical, and germline transformation experiments suggest that the *sim* gene is activated in the CNS midline precursors cells by gradients of transcriptional activators in combination with a transcriptional repressor. Once activated, *sim* regulates CNS midline transcription, and maintains its own synthesis via autoregulation. We have identified a CNS midline enhancer element (ACGTG) and putative binding site of sim in three genes: *sim*, *slit*, and *Toll*. Mutational studies of all three genes have shown that this sequence is required and sufficient for CNS midline expression. This element is related in sequence to the mammalian xenobiotic response element (GCGTC) recognized by the bHLH-PAS proteins that constitute the aromatic hydrocarbon (dioxin) receptor. Thus a similar molecular system is utilized for controlling both CNS development and carcinogen metabolism.

POSSIBLE INVOLVEMENT OF NEURON-SPECIFIC c-src GENE PRODUCT, pp60$^{c\text{-src}(+)}$ IN SYNAPTOGENESIS. S. Atsumi, X.-Y. Zhai and T. Kawate. Department of Anatomy, Yamanashi Medical University, Tamaho, Yamanashi, 409-38, JAPAN.

Neuron-specific c-src gene is known to be expressed mainly in neuronal cells. Its gene product, pp60$^{c\text{-src}(+)}$, has tyrosine kinase activity and interacts with the plasma membrane with the N-terminal myristic acid. Although morphologically pp60$^{c\text{-src}(+)}$ has been reported to be localized in neuronal cell bodies and growth cones of developing neurons, its function is still unknown. In order to elucidate the function of pp60$^{c\text{-src}(+)}$, we examined the localization of pp60$^{c\text{-src}(+)}$ in cultures of spinal neurons and spinal ganglion cells. First we raised the specific antibody against hexapeptide RKVDVR which was a unique amino acid sequence of the neuron-specific c-src gene product, and affinity-purified the antibody. Immunocytochemical detection of pp60$^{c\text{-src}(+)}$ was performed by indirect immunofluorescent staining or PAP technique. Neurons were dissected from 7-10 day-incubated embryos and cultured in MEM containing chick embryo extract and horse serum. In 1-day culture, neuroblasts extended neurites and formed growth cones at the tip of neurites. At this stage, immunoreactivity of pp60$^{c\text{-src}(+)}$ was weak or moderate in both cell bodies and growth cones. As the culture proceeded, neurites extended and made contact with other neurites or non-neuronal cells where several varicosities were observed to form. At this stage, the intense immunoreactivity of pp60$^{c\text{-src}(+)}$ was observed in neuronal cell bodies, extending neurites and especially in varicosities in contact area. These findings suggest that pp60$^{c\text{-src}(+)}$ plays an important role in synaptogenesis.

NGF-RESPONSE OF AN EGF-DEPENDENT PROGENITOR CELL FROM THE HUMAN NEURAL CREST. V. Silani, S. Bernasconi, A. Sampietro, C. Ghezzi, C. Mariani and G. Scarlato. Institute of Neurology, University of Milan Medical School, 20122 Milan, ITALY.

A central feature of development is the production of differentiated progeny from multipotent progenitor cells. Little is known about the cellular mechanisms that control these processes in the human neural crest derivatives. Extrinsic signalling molecules are thought to play a significant role in determining the fate of the progenitor cells. The sympathetic chain was demonstrated in 7th to 13th week post-conception fetuses by staining for low affinity nerve growth factor (p^{75}-NGFR) receptor (Silani, et al., 1993). Following anatomical isolation and dissection, neurospheres were obtained after culturing dissociated cells for 10 days in Sato's medium in the presence of human recombinant epidermal growth factor (EGF) and in absence of the substratum according to Reynolds and Weiss (1992). After the second passage neurospheres of EGF-dependent progenitor cells were subsequently mechanically dissociated and plated on poly-L-lysine. Single cells after treatment with the specific neurotrophin for the neural crest derivatives, i.e. nerve growth factor (NGF), were demonstrated to significantly increase neurite extension and long term survival. These effects were protracted and required 7 days in vitro to become evident. Cells were immunostained for tyrosine hydroxylase (TH) and immunopositivity observed after plating on the poly-L-lysine. EGF treatment did not produce significant effects. This study demonstrates the possibility to obtain an EGF-dependent progenitor cell from the human neural crest that, after proliferation in the presence of EGF, has the characteristics both of expressing TH enzyme and responding to NGF with neurite extension and increased survival.

298

TRANSGENIC EXPRESSION OF A DOMINANT-NEGATIVE FORM OF SCIP (SUPPRESSED cAMP INDUCED *POU*) RESULTS IN THE DISRUPTION OF SCHWANN CELL-AXON INTERACTIONS. D.E. Weinstein and Greg E. Lemke. Molecular Neurobiology Laboratory, The Salk Institute for Biological Studies, La Jolla, CA 92037, U.S.A.

The myelin structural genes encoding myelin basic protein, proteolipid protein and protein zero (P0), are expressed exclusively in the myelinating cells of the central and peripheral nervous systems under tight developmental control. SCIP, a *trans*-acting transcriptional regulator expressed predominantly in oligodendrocyte and Schwann cell progenitors has been identified and cloned in our laboratory. Co-transfections of SCIP and the P0 promotor driving a reporter CAT gene strongly suggests that SCIP represses P0 activated CAT expression in cultured Schwann cells. During development, SCIP expression precedes the expression of the myelin genes. In sciatic nerve transection experiments, SCIP is rapidly and transiently reinduced and the myelin genes are transiently turned off, suggesting a reciprocal pattern of expression consistent with the expression in dividing progenitors. To study the role of SCIP in the development of the peripheral nervous system we have made transgenic mice that express a dominant-negative form of the gene under the transcriptional regulation of the P0 promoter. This form of the SCIP protein binds cognate DNA but fails to act as a transcriptional regulator. We have generated two independent lines of mice which harbor the transgene. In heterozygotes the animals show a mild phenotype indicative of a peripheral neuropathy, and consistent with this, demonstrate histological and ultrastructural abnormalities in their sciatic nerves. Interestingly, it is the unmyelinated fibers that are normal, resulting in poor ensheathment and the articulation of numerous axons with no Schwann cell membrane interposed. Homozygotes are more severely affected, exhibiting developmental delay and growth retardation.

EXPRESSION OF RAB-GDI AND RAB3A IN BRAIN AND CNS DURING MOUSE EMBRYO DEVELOPMENT. M. Gulisano[1,3], F. Tamanini[2], E. Boncinelli[1] and D. Toniolo[2]. 1DIBIT Scientific Institue H.S. Raffaele, Milan; 2IGBE, CNR, Pavia; 3Institute of General Biology, University of Catania, Catania, ITALY.

We have cloned human and mouse Rab-GDI (Rab-GDP Dissociation Inhibitor) (1). This protein seems to play a major role in the regulation of the mechanism of vesicle formation and exocytosis. *In vitro*, Rab proteins are removed from membranes by RabGDI and this leads to formation of a cytosolic complex of Rab with the inhibitor protein specifically involved in synaptic vesicle exocytosis (3). We cloned, by RT-PCR, the mouse Rab3a cDNA to compare the expression of the two genes (Rab-GDI and Rab3a) by in situ hybridization. Both are already expressed at E9. During embryo development both show very similar patterns of expression. Rab3a is exclusively confined to post mitotic neurons both in brain and spinal cord. RabGDI, instead, shows a two-intensity expression pattern. It is expressed at high levels in the same territories of Rab3a, but shows a faint hybridization signal almost everywhere in the nervous system. Then they are both expressed at very high levels in post mitotic cells since their appearance, thus supporting the idea that they are involved in mechanisms of synaptic communication. It is tempting to speculate that high expression of RabGDI is needed early in post mitotic cells in order to sequestrate Rab3a protein to down regulate synaptic vesicle exocytotic degranulation in a period in which neurons are not yet involved in synapses formation.

(1) Bione et al. 1993: PNAS 90:10977-10981 (2) Ullrich et al. 1993: JBC 268:18143-18150 (3) Oberhauser et al. 1992: Nature 360:270-273.

ELECTRON PARAMAGNETIC RESONANCE STUDY OF THE NEUROPROTECTIVE EFFECTS OF U-78517F AND HYPOTHERMIA ON FREE RADICALS IN THE ISCHEMIC-REPERFUSED GERBIL HIPPOCAMPUS.

W. Zhao[1], J.S. Richardson[1,2], M.J. Mombourquette[3], J.A. Weil[3], S. Ijaz[4], and A. Shuaib[4]. Departments of [1]Pharmacology, [2]Psychiatry, [3]Chemistry and [4]Medicine, University of Saskatchewan, Saskatoon, SK, S7N 0W0, CANADA.

Oxygenic free radicals are considered to be the major contributors to the pathological processes induced by ischemia/reperfusion insult. The present study used electron paramagnetic resonance (EPR) spectroscopy at a temperature of 100 K to determine the formation of free radicals in brain tissue during reperfusion following forebrain ischemia in gerbils, and the effects of hypothermia or the antioxidant compound U-78517F on these free radicals. Thirty-two male gerbils were randomly divided into four groups and forebrain ischemia was induced by bilateral carotid artery occlusion (BCO). The Ischemia Group received 15 minutes of ischemia followed by 15 minutes of reperfusion. The U-78517F Group received 15 minutes of ischemia followed by a dose of 15 mg/kg body weight of U-78517F $i.p.$ at the beginning of the 15 minutes reperfusion period. The Hypothermia Group was maintained at scalp temperature of $35.5 \pm 0.5°$ C and rectal temperature of $34.5 \pm 0.5°$ C during the ischemia and the reperfusion periods. The Control Group was sham-operated without BCO. Compared to samples from the Control Group, samples of hippocampus taken from gerbils of Ischemia Group, frozen in liquid nitrogen less than 20 seconds after sacrifice and scanned by low temperature (100 K) EPR, show a significant increase in oxygen-based free radicals ($p<0.05$) and a decrease in carbon-based ubiquinone-like free radicals ($p<0.01$). Both the intraperitoneal injection of U-78517F at the start of reperfusion, and the hypothermia treatment during ischemia/reperfusion, prevent the increase in oxygenic free radicals seen in the Ischemia Group but do not alter the reduction in the ubiquinone-like free radicals. These results suggest that the neuroprotective actions of U-78517F and hypothermia treatment include a direct inhibitory effect on oxygenic free radicals.

CONTRIBUTORS

P. Aebischer
Division of Surgical Research
Centre Hospitalier Universitaire
Vaudois, Lausanne University
Medical School, 1011 Lausanne,
SWITZERLAND

S. Ahmed
Neuroscience Research Group,
Departments of Anatomy and
Pharmacology & Therapeutics,
University of Calgary Faculty of
Medicine, Calgary, AB, T2N 4N1
CANADA

B. Andersen
Eukaryotic Regulatory Biology
University of California, San Diego
9500 Gilman Drive
La Jolla, CA 92093-0648, U.S.A.

D.J. Anderson
Howard Hughes Medical Institute,
California Institute of Biology,
Pasadena, CA 91125, U.S.A

D.K. Anderson
Departments of Neuroscience and
Neurosurgery, University of Florida
College of Medicine
Gainesville, FL 32610, U.S.A.

C.P. Austin
Howard Hughes Medical Institute
and Department of Genetics
Harvard Medical School
200 Longwood Avenue
Boston, MA 02115, U.S.A.

E.E. Baetge
Department of Neuroscience,
Cytotherapeutics, Providence, RI
02900, U.S.A.

S.J. Birren
Department of Biology
Center for Complex Systems
Brandeis University
Waltham, MA 02254, U.S.A.

A. Björklund,
Department of Medical Cell Research
Section of Neurobiology, University
of Lund, Biskopsgatan 5, S-223 62
Lund, SWEDEN

B.C. Blanchard
Department of Neuroscience, The
Chicago Medical School, Chicago, IL
60064, U.S.A.

E. Boncinelli
DIBIT, Istituto Scientifico H.S.
Raffaele, Via Olgettina 60
20132 Milano, ITALY

D.E. Bredesen
Department of Neurology
Brain Research Institute and
Molecular Biology Institute
University of California, Los Angeles
710 Westwood Plaza
Los Angeles, CA 90024-1769, U.S.A.

V. Broccoli
DIBIT, Istituto Scientifico H.S.
Raffaele, Via Olgettina 60
20132 Milano, ITALY

M.A. Burke
Department of Anatomy and Cell
Biology, University of Illinois School
of Medicine, Chicago, IL 60612,
U.S.A.

C.L. Cepko
Howard Hughes Medical Institute
and Department of Genetics, Harvard
Medical School, 200 Longwood
Avenue, Boston, MA 02115, U.S.A.

P. Chambon
Laboratoire de Génétique Moléculaire
des Eucaryotes du CNRS, Unité 184
de Biologie Moléculaire et de Génie
Génétique de l'INSERM, Institut de
Chimie Biologique, Faculté de
Médecine, 11 rue Humann
67085 Strasbourg Cedex, FRANCE

G. Charron
Centre for Research in Neuroscience,
The Montreal General Hospital
Research Institute, McGill University
Montreal, PQ H3G 1A4, CANADA

M.C. Colbert
The Children's Hospital Medical
Center, Division of Molecular
Cardiovascular Biology, The
Children's Hospital Research
Foundation, 3333 Burnett Ave,
Cincinnati, OH 45229, U.S.A

J.-F. Collard
Centre for Research in Neuroscience,
The Montreal General Hospital
Research Institute, McGill University
Montreal, PQ H3G 1A4, CANADA

D. Collazo
LMB-NINDS, Bldg 36, Rm 3D02, NIH,
Bethesda, MD 20892, U.S.A.

T.J. Collier
Departments of Pharmacology and
Psychiatry, Yale University School of
Medicine, New Haven, CT 06510,
U.S.A

F. Côté
Centre for Research in Neuroscience,
McGill University The Montreal
General Hospital Research Institute,
Montreal, PQ H3G 1A4, CANADA

S. Dietrich
Department of Molecular Cell Biology
Max Planck Institute for Biophysical
Chemistry, Am Fassberg
37077 Göttingen, GERMANY

P. Dollé
Laboratoire de Génétique Moléculaire
des Eucaryotes du CNRS, Unité 184
de Biologie Moléculaire et de Génie
Génétique de l'INSERM, Institut de
Chimie Biologique, Faculté de
Médecine, 11 rue Humann, 67085
Strasbourg Cedex, FRANCE

T. Edlund
Department of Microbiology, Umeå
University, Umeå, SWEDEN

J.D. Elsworth
Department of Neurological Sciences,
Rush College of Medicine, Chicago,
IL 60612, U.S.A.

L. Erkman
Eukaryotic Regulatory Biology
University of California, San Diego,
9500 Gilman Drive
La Jolla, CA 92093-0648, U.S.A.

G.J. Fishell
The Skirball Institute
550 1st Avenue
New York, NY 10016, U.S.A.

G. Garriga
Department of Molecular and Cell
Biology, 401 Barker Hall, University
of California,Berkeley, CA 94720-
3204, U.S.A.

P. Gorry
Laboratoire de Génétique Moléculaire
des Eucaryotes du CNRS, Unité 184
de Biologie Moléculaire et de Génie
Génétique de l'INSERM, Institut de
Chimie Biologique, Faculté de
Médecine, 11 rue Humann, 67085
Strasbourg Cedex, FRANCE

P. Gruss
Department of Molecular Cell Biology
Max Planck Institute for Biophysical
Chemistry, Am Fassberg
37077 Göttingen, GERMANY

A.K. Groves
Division of Biology, 216-76
California Institute of Technology
Pasadena, CA 91125, U.S.A.

M. Gulisano
DIBIT, Istituto Scientifico H.S.
Raffaele, Via Olgettina 60
20132 Milano, ITALY

D.R. Howland
Departments of Neuroscience and
Neurosurgery, University of Florida,
College of Medicine, Gainesville, FL,
32610, U.S.A.

Y.N. Jan
Howard Hughes Medical Institute
and Departments of Physiology &
Biochemistry, University of
California, San Francisco, San
Francisco, CA 94143-0724, U.S.A.

A.P. Jarman
Institute of Cell and Molecular
Biology, University of Edinburgh
Darwin Building, King's Buildings
Edinburgh EH9 3JR, U.K.

T.M. Jessel
Howard Hughes Medical Institute,
Department of Biochemistry and
Molecular Biophysics, Center for
Neurobiology and Behavior,
Columbia University, New York, NY
10032, U.S.A

J.-P. Julien
Centre for Research in Neuroscience
The Montreal General Hospital
Research Institute, McGill University
Montreal, PQ H3G 1A4, CANADA

A.C. Kato
Department of Pharmacology and
Division of Clinical Neurophysiology,
Centre Medical Universitaire, Geneva
University Medical School, 1211
Genève, SWITZERLAND

J.H. Kordower
Department of Neurological Sciences
and Rush Alzheimer's Disease Center
Rush Presbyterian
St. Lukes Medical Center
Chicago, IL 60612, U.S.A.

R. Krumlauf
Laboratory of Developmental
Neurobiology, National Institute for
Medical Research
The Ridgeway, Mill Hill,
London NW7 1AA, U.K.

A. Kuroiwa
Department of Molecular Biology,
School of Science, Nagoya
University,Chihusa-Ku, Nagoya,
JAPAN

A.-S. LaMantia
Departments of Neurobiology
Duke University Medical Center
Durham, NC 27712, U.S.A.

P. Li
Biomedical Sciences Graduate
Program, University of California San
Diego, 9500 Gilman Drive
La Jolla, CA 92093-0648, U.S.A.

C.R. Lin
Eukaryotic Regulatory Biology
Program, University of California San
Diego, 9500 Gilman Drive, La Jolla,
CA 92093-0648, U.S.A.

S.-C. Lin
Eukaryotic Regulatory Biology
Program, University of California San
Diego, 9500 Gilman Drive, La Jolla,
CA 92093-0648, U.S.A.

E. Linney
Departments of Microbiology
Duke University Medical Center
Durham, NC 27712, U.S.A.

T. Lufkin
Laboratoire de Génétique Moléculaire
des Eucaryotes du CNRS, Unité 184
de Biologie Moléculaire et de Génie
Génétique de l'INSERM, Institut de
Chimie Biologique, Faculté de
Médecine, 11 rue Humann, 67085
Strasbourg Cedex, FRANCE

J.D. Macklis
Department of Neurology and
Program in Neuroscience
Harvard Medical School,
Neuroscience Division
Mental Retardation Research Center
Children's Hospital
Boston, MA 02115, U.S.A.

H. Marshall
Laboratory of Developmental
Neurobiology, National Institute for
Medical Research
The Ridgeway, Mill Hill,
London NW7 1AA, U.K.

M. Mark
Laboratoire de Génétique Moléculaire
des Eucaryotes du CNRS, Unité 184
de Biologie Moléculaire et de Génie
Génétique de l'INSERM, Institut de
Chimie Biologique, Faculté de
Médecine, 11 rue Humann, 67085
Strasbourg Cedex, FRANCE

R. McEvilly
Biology Graduate Program,
Department of Medicine
University of California, San Diego,
9500 Gilman Drive, La Jolla
CA 92093-0648, U.S.A.

R. McKay
LMB-NINDS
Bldg 36, Rm 3D02, NIH
Bethesda, MD 20892, U.S.A.

E.J. Mufson
Department of Neurological Sciences
and Rush Alzheimer's Disease Center,
Rush Presbyterian/ St. Lukes Medical
Center, Chicago, IL 60612, U.S.A.

M. Noble
Ludwig Institute for Cancer Research,
91 Riding House Street, London W1P
8BT, U.K.

S.L. Pfaff
Howard Hughes Medical Institute
Department of Biochemistry and
Molecular Biophysics, Center for
Neurobiology and Behavior
Columbia University
New York, NY 10032, U.S.A.

N.A. Pochon
Division of Surgical Research, Centre
Hospitalier Universitaire Vaudois,
Lausanne University Medical School,
1011 Lausanne, SWITZERLAND

H. Pöpperl
Laboratory of Developmental
Neurobiology, National Institute for
Medical Research
The Ridgeway, Mill Hill,
London NW7 1AA, U.K.

G. Raisman
Norman and Sadie Lee Research
Centre, Laboratory of Neurobiology
National Institute for Medical
Research, Mill Hill
London NW7 1AA, U.K.

D.E. Redmond, Jr
Department of Neurological Sciences,
Rush College of Medicine, Chicago,
IL 60612, U.S.A.

P.J. Reier
Departments of Neuroscience and
Neurosurgery, College of Medicine,
University of Florida, Gainesville, FL
32610, U.S.A.

B.A. Reynolds
NeuroSpheres Ltd, Calgary, AB
CANADA.

F.M. Rijli
Laboratoire de Génétique Moléculaire
des Eucaryotes du CNRS, Unité 184
de Biologie Moléculaire et de Génie
Génétique de l'INSERM, Institut de
Chimie Biologique, Faculté de
Médecine, 11 rue Humann
67085 Strasbourg Cedex, FRANCE

M.G. Rosenfeld
Eukaryotic Regulatory Biology
Program, Howard Hughes Medical
Institute, University of California, San
Diego, 9500 Gilman Drive
La Jolla, CA 92093-0648, U.S.A.

R.H. Roth
Department of Neurological Sciences,
Rush College of Medicine, Chicago,
IL 60612, U.S.A.

M. Schonemann
Biology Graduate Program,
Department of Medicine
University of California, San Diego,
9500 Gilman Drive
La Jolla, CA 92093-0648, U.S.A.

J. Sladek, Jr.
Department of Neuroscience
The Chicago Medical School
Chicago, IL 60064, U.S.A.

F. Spada
DIBIT, Istituto Scientifico H.S.
Raffaele, Via Olgettina 60
20132 Milano, ITALY

A. Stoykova
Department of Molecular Cell Biology
Max Planck Institute for Biophysical
Chemistry, Am Fassberg
37077 Göttingen, GERMANY

E.T. Stuart
Department of Molecular Cell Biology
Max Planck Institute for Biophysical
Chemistry, Am Fassberg
37077 Göttingen, GERMANY

M. Studer
Laboratory of Developmental
Neurobiology, National Institute for
Medical Research
The Ridgeway, Mill Hill,
London NW7 1AA, U.K.

S.A. Tan
Division of Surgical Research, Centre
Hospitalier Universitaire Vaudois,
Lausanne University Medical School,
1011 Lausanne, SWITZERLAND

J.R. Taylor
Department of Neurological Sciences,
Rush College of Medicine, Chicago,
IL 60612, U.S.A.

P. Tremblay
Department of Molecular Cell Biology
Max Planck Institute for Biophysical
Chemistry, Am Fassberg
37077 Göttingen, GERMANY

E. Turner
Eukaryotic Regulatory Biology
Program, Howard Hughes Medical
Institute, University of California, San
Diego, 9500 Gilman Drive
La Jolla, CA 92093-0648, U.S.A.

J.M. Verdi
Howard Hughes Medical Institute,
California Institute of Biology,
Pasadena, CA 91125, U.S.A

A. Vescovi
Instituto Neurologica "C Besta", Via
Celoria, Milan, ITALY

S. Weiss
Neuroscience Research Group
Departments of Anatomy and
Pharmacology & Therapeutics
Faculty of Medicine
University of Calgary
Calgary, AB T2N 4N1, CANADA

H. Widner
Restorative Neurology Unit
Department of Neurology
University Hospital, S-221 85 Lund
SWEDEN and The Parkinson's
Institute, 1170 Morse Avenue
Sunnyvale, CA 94089-1605, U.S.A.

T. Yamada
Centre for Molecular Biology and
Biotechnology, University of
Queensland, Brisbane, AUSTRALIA

INDEX

The manufacturer's authorised representative in the EU is Springer
Nature Customer Service Centre GmbH, Europaplatz 3, 69115 Heidelberg,
Germany. If you have any concerns regarding our products, please
contact ProductSafety@springernature.com

Printed and bound by CPI Group (UK) Ltd, Croydon, CR0 4YY
23/04/2026
02095607-0016